Introduction to Orbital Mechanics

Introduction to Orbital Mechanics

A Concise and Practical Approach

James C. Hermanson
William E. Boeing
Department of Aeronautics & Astronautics,
University of Washington,
Seattle, WA, USA

WILEY

Published by John Wiley & Sons, Inc., Hoboken, New Jersey.

The manufacturer's authorized representative according to the EU General Product Safety Regulation is Wiley-VCH GmbH, Boschstr. 12, 69469 Weinheim, Germany, e-mail: Product_Safety@wiley.com.

For general information on our other products and services or for technical support, please contact our Customer Care Department within the United States at (800) 762-2974, outside the United States at (317) 572-3993 or fax (317) 572-4002.

Wiley also publishes its books in a variety of electronic formats. Some content that appears in print may not be available in electronic formats. For more information about Wiley products, visit our web site at www.wiley.com.

Library of Congress Cataloging-in-Publication Data applied for

Hardback: 9781394190157
Epub: 9781394190164
Epdf: 9781394190171

Cover Design: Wiley
Cover Image: © Stocktrek Images/Getty Images

Set in 9.5/12.5pt STIXTwoText by Straive, Pondicherry, India

For Andrea, Rolf, and Lisa
Soli Deo Gloria

Contents

About the Author

James C. (Jim) Hermanson is Professor Emeritus and former Chair of the William E. Boeing Department of Aeronautics & Astronautics at the University of Washington in Seattle Washington, USA. Hermanson received a Bachelor of Science in Aeronautics & Astronautics from the University of Washington in 1977, and Master's and Doctoral degrees from the California Institute of Technology in 1980 and 1985, respectfully. Before joining the University of Washington in 2002, he was a professor of Mechanical Engineering at Worcester Polytechnic Institute in Worcester, MA. In addition, Professor Hermanson has nine years of industrial experience at Boeing Aerospace Company and United Technologies Research Center. He has also worked abroad as a visiting professor/researcher at the Universität Göttingen, in Göttingen, Germany, ZARM (Center for Applied Space Technology and Microgravity) in Bremen, Germany, and at the Royal Institute of Technology in Stockholm, Sweden.

Professor Hermanson and his research groups are known for fundamental work in fluid mechanics and propulsion. This has included fuel injection in compressible flow, multi-phase flow, heat transfer, and shock wave interactions. His research has also involved turbulent combustion and flame structure, flame stability, exhaust emissions, alternative fuels, and cryogenic propellant vaporization and combustion. The combustion and multi-phase flow research has included testing on reduced-gravity platforms including drop towers and parabolic-trajectory aircraft. Professor Hermanson is a Fellow of the American Society of Mechanical Engineers and an Associate Fellow of the American Institute of Aeronautics and Astronautics. He is also a Blue Angels Key Influencer for the United States Navy. Hermanson is the author or co-author of more than 150 technical publications and conference papers and has won numerous awards for teaching and research. He is also a passionate amateur astronomer, having logged more than 400 observation sessions in Washington State and elsewhere.

Preface

This text is intended to provide a systematic, progressive presentation of the fundamentals of orbital mechanics. This includes the basic underlying physics and the classic description of planetary motion, more modern considerations of the motion of satellites and spacecraft in Earth orbit, and the orbital mechanics of interplanetary probes. Practical examples are provided throughout. The foundational principles offered by this work should be suitable for a first course in orbital mechanics on either a quarter or semester basis, and will leave students well prepared for more advanced, more comprehensive presentations on orbital mechanics and spacecraft dynamics.[1] An overview of the structure of this book and introduction of key topics follows.

Chapter 1, the introductory chapter, presents a history of the study of the motion of the planets, which led over the centuries to the development of the correct models for planetary motion. This history begins with the Earth-centered scheme postulated by Ptolemy. The correct statement of a Sun-centered Solar System by Copernicus, verified by the breakthrough observations of Galileo, combined with the detailed astronomical observations of Brahe, led Kepler to develop mathematically correct descriptions of the motion of the planets. The reasoning on gravitation provided by Newton then provided physical reasoning for the orbits correctly modeled by Kepler. The extension of orbital mechanics from planetary motion to the motion of spacecraft in Earth orbit and interplanetary probes is then discussed. Finally, some examples of ancient, non-Western astronomy are presented (specifically Babylonian, Chinese, and Mayan astronomy), all of which made striking advances with discoveries in some cases considerably in advance of their European counterparts.

Chapter 2 describes the basic physics that govern the motion of planets and spacecraft and the resulting orbits and trajectories. This is based on the classic two body problem, which assumes that the distance between the two bodies (e.g., the Sun and a planet) is very large compared with the sizes of the objects themselves. Based on the differential equation of motion dictated by Newtonian gravitation, principles for the conservation of angular momentum and energy are developed, leading in turn to the derivation of the Orbit Equation that describes the motion of objects on all trajectories, be they circular or elliptical orbits, parabolae, or hyperbolae. The essential geometric properties of all orbits and trajectories are then presented.

Chapter 3 extends the descriptions of orbital motion developed in Chapter 2 to address the question of where a spacecraft or planet is in its trajectory at a given time. The discussion begins with the trivial example of a circular orbit, where the orbit speed and radial distance are constant. Determining the times for elliptical orbits requires more complicated analysis. This is done both from a

1 Excellent examples of comprehensive texts on orbital mechanics include Curtiss, H. *Orbital Mechanics for Engineering Students (Aerospace Engineering)*, 4th edition, Butterworth-Heinemann, 2019 and Vallado, D.A. *Fundamental of Astrodynamics and Applications*, 4th edition, Microcosm Press/Springer, 2007.

geometric point of view (based on Kepler's laws of planetary motion presented in Chapter 2) as well as from a strictly mathematical viewpoint. Two cases are considered, the direct situation, where the positions on an orbit are known and the time between those positions is to be determined, and the more complicated inverse problem, where a time interval is given and the change in orbital position needs to be determined. For the case of parabolic trajectories, a simple mathematical formulation describes the position versus time, applicable to both the direct and inverse situations. The analysis of time on hyperbolic trajectories is comparable to the complexity of the case of elliptical orbits, and it is also addressed using both a geometric and analytical approach.

Chapter 4 extends the planar orbits and trajectories addressed in Chapters 2 and 3 to the general case of motion in three dimensions. Consideration of motion in three dimensions introduces additional orbital elements beyond those needed for planar orbits and trajectories. This requires different frames of reference, depending on whether the motion of the object is considered relative to the Sun (the heliocentric frame of reference), the center of the Earth (geocentric frame), the Earth's surface (topocentric frame), or the plane of the object's orbit itself (perifocal reference frame). The necessity of converting from one reference frame to another, for example the motion of an object seen at the Earth's surface (topocentric) to that referenced to the center of the Earth (geocentric), which describes the motion in the object's planar orbit, is explained. Another example is the motion of a planet relative to the sun, as described by Orbit Equation, but observed from the Earth (topocentric + geocentric reference frames). The basic principle of transformation between reference frames is simply illustrated in two dimensions, then extended to three dimensions. The specific examples of transformations between perifocal and geocentric coordinates, and between the topocentric and geocentric reference frames, are discussed.

Chapter 5 describes how orbits and trajectories can be modified by spacecraft maneuvers, an essential topic for both satellite positioning and interplanetary probes. The first maneuver, the Hohmann Transfer, represents the simplest change from one orbit to another. The limiting cases for transfers between two orbits are then considered, ranging from the most efficient Hohmann Transfer (that is, needing the least total velocity change) to highly-energetic hyperbolic trajectories, which require larger velocity changes and lead to departure from the Earth on interplanetary missions. Several specific orbit-change scenarios are discussed. These include the clever but long-duration Bi-Elliptic Hohmann Transfer and a Phasing Maneuver that allows for repositioning a spacecraft in the same orbit, for example to accomplish an orbital rendezvous. In the more general case, changing both the velocity and direction of a spacecraft's trajectory allows for a faster way to intercept an object in the same orbit, leading to a Chase Maneuver, as well as rotating the axis of an orbit in an Apse Line Rotation. All of these maneuvers are in a single orbital plane. The important case of a Plane Change Maneuver, needed for example to bring an object into an equatorial orbit from an orbit launched from a nonequatorial site, is then discussed. Finally, combined maneuvers, such as a plane change combined with raising an orbit, are considered.

Chapter 6 applies the spacecraft maneuvers discussed in Chapter 5 to interplanetary missions. The discussion first accounts for where a target planet needs to be at the moment of departure from Earth (planetary phasing). A practical approach to modeling interplanetary missions is the Patched Conic Approximation, which considers a space probe first in the vicinity of the Earth, then far from the Earth but not near the target planet, then in the vicinity of the target planet during planetary approach. Whether an interplanetary spacecraft is more influenced by the Earth, planet, or the sun is evaluated in terms of the Sphere of Influence concept. Departures from Earth orbit on interplanetary trajectories are discussed in both two and three dimensions. A Hohmann Transfer between planets illustrates the important consideration of reference point, that is, whether spacecraft motion is viewed by the Sun or the target planet, which will be different. The features of planetary approaches in general are then addressed, leading to the details of planetary encounters, such as planetary flybys and insertion into planetary orbits. It is seen that Planetary flybys can be in various orientations relative to the target planet, such as on the sunlit side, the dark side, or over a planetary pole. The change in spacecraft trajectory due to planetary passage can lead to a gravity assist, effectively boosting the spacecraft's speed relative to the Sun, is then described.

Chapter 7 addresses satellite movement in Earth orbit, is also described. This discussion centers on the ground track of a satellite over the Earth, and how that track is affected by the Earth's rotation. The special cases of geosynchronous satellites that have an orbital period that matches the rotational period of the Earth, and the practically important case of geostationary satellites, which remain over a fixed point on the Earth, are discussed. Perturbations to orbits are then introduced as an advanced topic. The perturbing effects of a nonspherical Earth are addressed, representing an important deviation from the two-body point-mass approach considered up to this point. Perturbations to orbits due to atmospheric drag, including satellite lifetime, are also discussed.

Chapter 8 presents methods used to determine the properties of an orbit based on information from observations. This could be considered another advanced topic. The simplest case of having simultaneous velocity and radius vectors known at a point in time readily allows for determining the elements of an orbit, as discussed in Chapters 2 and 4. The mathematical complexity of the solution increases considerably for the case where three position vectors are known (commonly termed Gauss' or Gibb's method) and for Lambert's problem, where two position vectors, and the time between them, are known. In both Gauss' and Lambert's problem, the position information is assumed to be known in geocentric coordinates. Often, information on an orbit is only available by observation from the Earth's surface, necessitating transformation between topocentric to geocentric, and possibly, then to heliocentric coordinates. Another situation is where range, azimuth, and elevation (and their rates) are available, for example from radar observations. Alternatively, the azimuths and elevations of an object might be known at multiple times – this is the method employed by astronomers to categorize new solar system objects and is quite complicated mathematically. All of these cases are presented and discussed.

Chapter 9 discusses two other advanced topics. The first is the relative motion between two spacecraft in Earth orbit, a critical consideration for spacecraft rendezvous. This development dictates a mathematical formulation more complex than simple use of the Orbit Equation. Several approximations are required to simplify the mathematical analysis and provide solutions reasonable for performing an on-orbit rendezvous maneuver. The second topic concerns spacecraft motion within the Earth-Moon system. This three body problem (spacecraft, Earth, and Moon) considers spacecraft continuously under the influence of both the Earth and the Moon's gravity. This leads to trajectories very different from the two-body orbits discussed previously. Consideration of the three-body system also establishes the existence of Lagrange points, where a spacecraft can remain at a fixed position within the Earth-Moon system.

Chapter 10 addresses the "how" of orbital maneuvers described in previous chapters by considering the performance of rockets. The discussion begins with the physical principles of how rocket thrust is produced, and how rocket performance is characterized. Given that spacecraft mass is critical to any space mission, the use of staging to decrease overall rocket mass, while producing a required velocity change, is developed and discussed. A special case of rocketry is that of a sounding rocket, which is typically a suborbital mission that includes consideration of the rocket boost phase, coasting to apogee, and descent. The above discussion centers on the case of high-thrust chemical rocket systems. Electric spacecraft propulsion is then introduced. These systems can provide for long-duration, very efficient, low-mass propulsion, but with limited thrust levels. Missions using electric propulsion are considered.

Finally, the Appendix presents key properties of planets and planetary orbits for reference.

It is hoped that this text will provide a useful and enjoyable introduction to the fascinating field of Orbital Mechanics and prepare engineers and other interested professionals to address exciting problems related to Earth-orbiting spacecraft and interplanetary missions. The fundamentals presented here should also serve as a solid foundation for more advanced studies in orbital mechanics and spacecraft dynamics.

Acknowledgments

The contribution of Dr. Dieter Zube of Aerojet Rocketdyne of the electric propulsion section of Chapter 10 is gratefully acknowledged. My thanks also to Professor Nicholas Apazidis of the Royal Institute of Technology in Stockholm, Sweden, for his guidance in initiating this project. The extensive work of Andrea Hermanson in preparing the graphics was essential to this work and is lovingly acknowledged. Finally, I thank the hundreds of students that I have had the privilege to teach over the years in my Orbital Mechanics class (AA 310) at the University of Washington. Their interest and enthusiasm for the topic were the primary inspiration for this textbook.

1 Introduction

The topic of Orbital Mechanics is both very old, and at time same time, quite contemporary. The antiquity stems from the universal human experience of observing the nighttime sky, and the realization that some objects appear to move relative to the apparent fixed background of the stars. Tracking these moving objects, the planets, led over the centuries to the development of the fundamentals of orbital mechanics that govern planetary motion. The historical development of orbital mechanics might be considered a case study in the advancement of science, with observers, theorists, and mathematicians building on the works of the scientists preceding them until a correct, fundamental understanding of the science was attained. More recently, the emergence of the *Space Age* led to many essential modifications and extensions of basic orbital mechanics, allowing for the application of these analytical and computational tools to spacecraft of all kinds, including Earth-orbiting satellites, missions to the Moon, and spacecraft that have explored all of the known major planets as well as comets and asteroids.

The outline of this chapter is as follows:

- Astronomy and the development of orbital mechanics
  - Astronomy and the orbital mechanics of planetary motion
  - The development of the orbital mechanics of spacecraft
- A first estimate of planetary distances
- Historical examples of non-Western astronomy
  - Babylonian
  - Chinese
  - Mayan

1.1 Astronomy and the Development of Orbital Mechanics[1]

A notable early western astronomer was Claudius Ptolemaeus, commonly known as "Ptolemy" (A.D. 90–168), who proposed one of the first models of the solar system, based solely on direct observation of the motion of the stars and planets without any optical aids. As was commonly assumed at the time, Ptolemy's model took the Earth to be fixed in space, with the sky rotating around it ... a

1 One good reference for the history of the development of orbital mechanics is Thurston, H. *Early Astronomy*, Springer Verlag, New Yok, 1994.

Introduction to Orbital Mechanics: A Concise and Practical Approach, First Edition. James C. Hermanson.

completely understandable assumption since standing on the Earth gives no sense of motion due to the Earth's rotation or revolution about the sun. The resulting model took the sky to a be a rotating dome of fixed stars, with the Sun, Moon, and planets revolving around the earth on a diurnal basis. A representation of the Ptolemaic system is shown in Fig. 1.1.

An objective, cursory examination of the Ptolemaic system points to an obvious problem. Although the model recognized that the relationship of Mercury and Venus to the Sun was different than with other planets, the model has those two planets appearing at times opposite the Sun as seen by the Earth, which does not occur. More subtle astronomical observations would reveal a second problem, that the planets appear occasionally to move backwards relative to the background of fixed stars, moving for a time to the east instead of the usual motion towards the west. This phenomenon, termed *retrograde motion*, is shown schematically in Fig. 1.2.

Despite these complications, Ptolemy's model remained for centuries as the leading model of the solar system. Early astronomers attempted to account for the retrograde motion by the introduction of *epicycles*, local circles on which the planets would move as they revolved in their orbits about the sun. An example epicycle is shown in Fig. 1.3. The concept of epicycles is quite ancient, going as far back as Eudoxus (approx. 408–368 B.C.), and built upon by the philosopher Aristotle (384–322 B.C.) and the

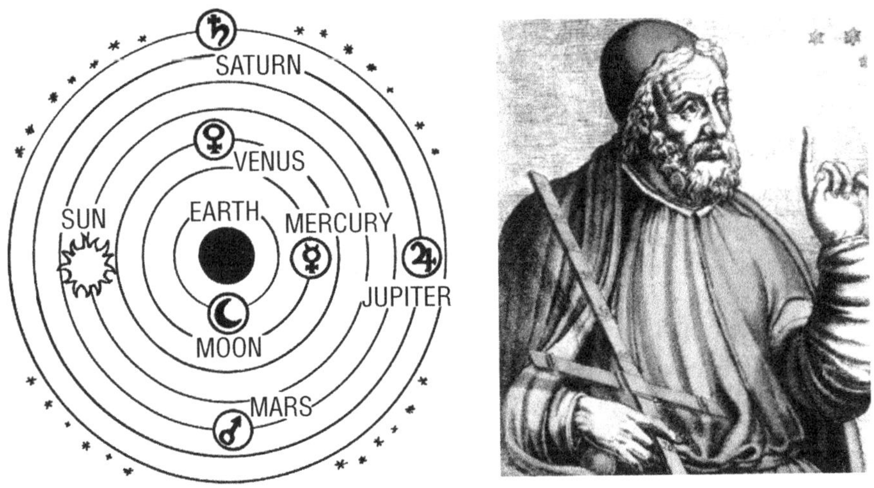

Figure 1.1 Ptolemaic system (and Ptolemy). *Source:* Unknown author / Wikimedia Commons / Public Domain / https://commons.wikimedia.org/wiki/File:PSM_V78_D326_Ptolemy.png / last accessed on October 09, 2025.

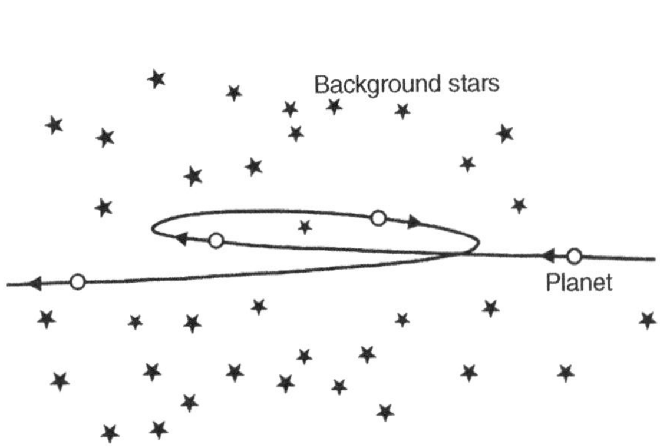

Figure 1.2 Perceived retrograde motion illustrated.

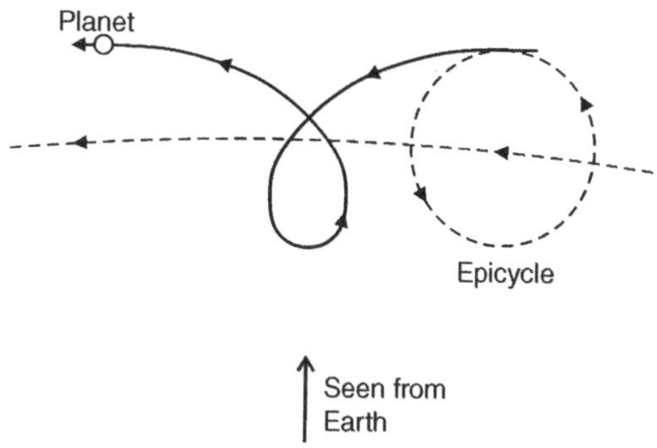

Figure 1.3 An epicycle illustrated.

Greek astronomer and mathematician Callippus (370–310 BC). More detailed calculations were performed by Hipparchus c. 190–120 BC.

Although the introduction of epicycles partially accounted for the phenomenon of retrograde motion, calculations using single epicycles were unable to accurately match the observed motion of the planets. This led to further attempts incorporating multiple epicycles, or epicycles on top of epicycles, eventually resulting in more than 50 individual cycles distributed throughout the solar system. Still, such a complicated model proved only adequate to model planetary motion for at most a very few years.

A breakthrough occurred finally in the 16th century, when Mikołaj Kopernik (1473–1543) commonly known as "Copernicus" postulated that the sun, not the Earth, was at the center of the system, with all planets, including Earth, revolving around it. This system, combined with the (still controversial at the time) concept of the Earth's rotation, correctly explained the observed motion of the Sun, Moon, all planets, and the fixed stars. A representation of the Copernican system, along with its creator, is shown in Fig. 1.4.

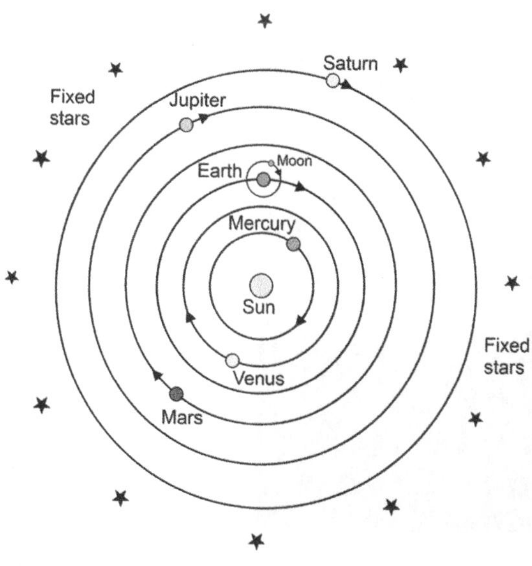

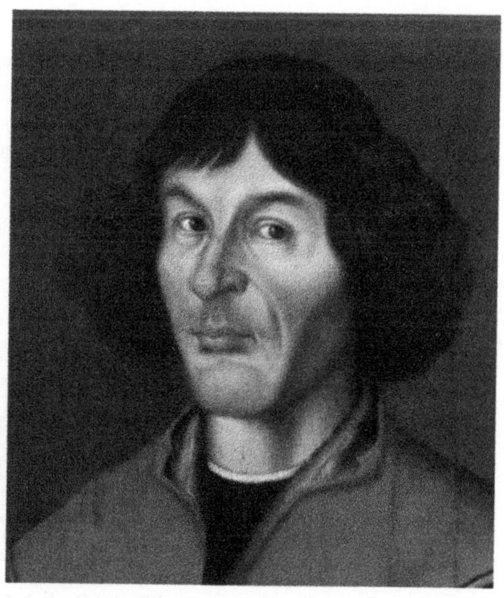

Figure 1.4 Copernican system (and Copernicus). *Source:* Nicolaus Copernicus / Wikipedia / Public Domain.

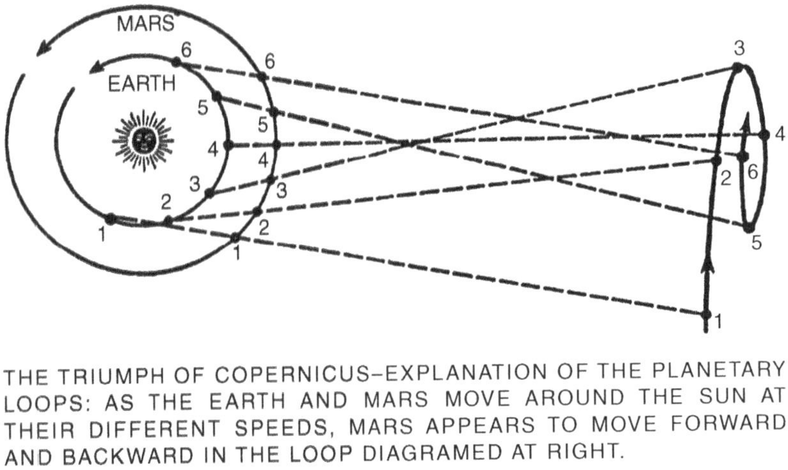

Figure 1.5 Retrograde motion of Mars in the Copernican system. *Source:* Wikipedia stock image.

An additional plus of the Copernican system is that it correctly accounted for the observed retrograde motion of the planets, without the need for epicycles, as shown schematically in Fig. 1.5.

The Danish astronomer Tyge Ottesen Brahe "Tycho" (1546–1601) used quadrants, sextants, and armillary spheres (the last invented in China) to measure planetary positions from the Royal Observatory in Copenhagen. The detailed angle information, combined with improvements in timekeeping, allowed Tycho to compile a comprehensive database of planetary positions which proved key to advancing the understanding of planetary motion. Based on his observations, Tycho proposed a different model for the solar system than Copernicus by (curiously) again assuming that the Earth was fixed in space with the heavenly objects rotating about it.

Tycho's system is shown in Fig. 1.6. The system was an improvement over the Ptolemaic system, in that it correctly accounted for the fact that Mercury and Venus are always observed to be comparatively close to the Sun, never opposite it. One problem with this approach is the substantial

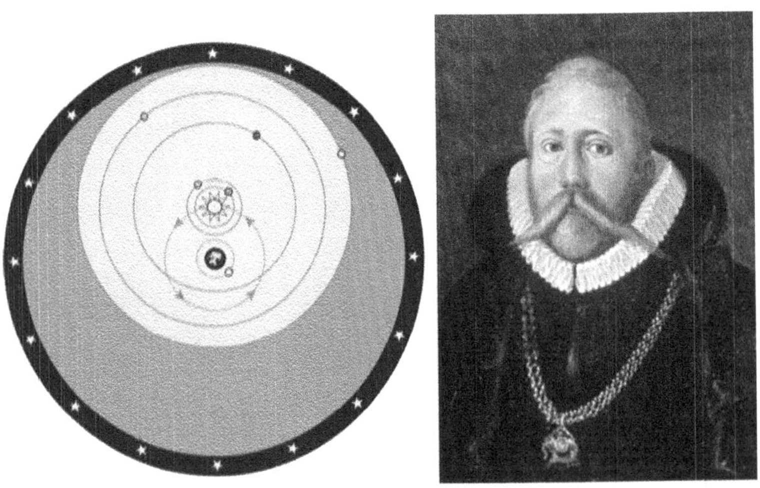

Figure 1.6 Tychonian system (and Tycho Brahe). *Source:* Eduard Ender / Wikimedia Commons / Public domain.

change in distances from the Earth to the outer planets as those planets move about the Sun. This would lead to profound changes in the brightness of the planets beyond those actually observed, though techniques for the precise measurements of the brightness of astronomical objects were not available at that time.

A student of Tycho's, Johannes Kepler (1571–1630), developed a mathematical model of the planets' motion that would explain the detailed observations of his mentor. Insisting on achieving rigorous agreement with *all* of the observations of Tycho led Kepler to correctly formulate the motion of the planets as ellipses, with the Sun at one focus, as shown in Fig. 1.7. The essential features of an elliptical orbit include the *periapsis*, or distance of closest approach, the *apoapsis*, the farthest separation, the *major axis* (the sum of the two distances), and the locations of the two foci. This description of planetary motion was a significant update to previous modeling attempts that assumed that all orbits were circular (with the possible addition of circular epicycles, as discussed above).

Kepler's analysis also led to his three laws of planetary motion: 1. All orbits are ellipses with the Sun at one focus, 2. the square of the orbital period is proportional to the cube of the semi-major axis of the ellipse, and 3. a planet "sweeps out" equal areas in its orbit over equal times. These principles, and the underlying physics, are developed formally in Chapters 2 and 3.

At about the same time, the famous scientist Galileo Galilei (1564–1642) made a major breakthrough in astronomy when he assembled optics to produce an astronomical telescope.[2] The device was very small by today's standards, with an objective lens diameter of only 37 mm (about ½ inches).[3] Nonetheless Galileo was able to use his telescope to observe for the first time the four largest moons of Jupiter, the phases of Venus, many lunar features, and sunspots (the latter causing him to lose most of his eyesight!). The discovery of the large Jovian moons Io, Europa, Ganymede,

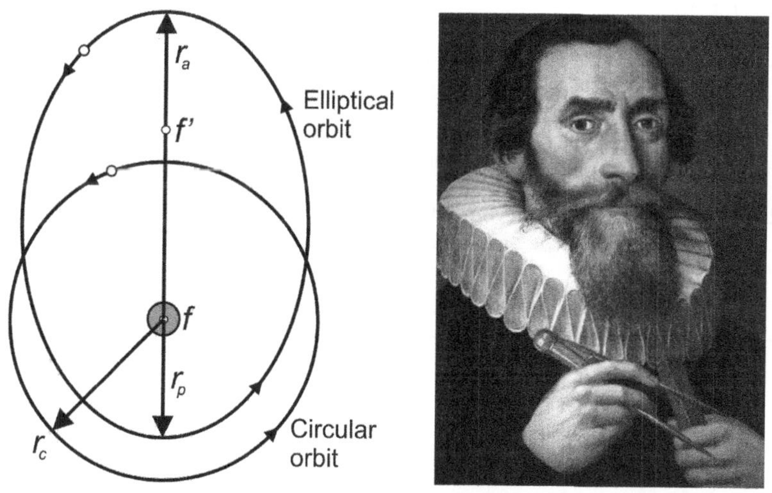

Figure 1.7 Elliptical vs. circular orbits (and Johannes Kepler). r_p and r_a are respectively the periapsis and apoapsis radii of the elliptical orbit; r_c is a circular orbit radius. The primary body is located at the focus *f*; the other focus of the ellipse is indicated as *f′*. *Source:* Wikipedia stock image.

2 The first telescope was made in the Netherlands in around 1608 by Jacob Metius and Hans Lipperhey.

3 The aperture of modern telescopes for amateur astronomy is typically roughly in the range of 100–400 mm; observatories have telescopes as large a 40 *m* in diameter.

and Callisto (termed the Galilean moons of Jupiter), and observing their motion around Jupiter, struck Galileo as being strikingly similar to Copernicus' model of the solar system. Even more compelling were his observations of the phases of Venus, which were fully consistent with the Copernican model, but not that of Ptolemy. Taken together, the model of Copernicus, the calculations of Kepler, and the observations of Galileo, cemented the understanding of our Sun-centered solar system. Going back to Tycho, observations involving the angles of objects above the Earth's horizon at various times are still used today to determine the orbital elements of comets, asteroids, and other solar system objects (see Chapter 8).

With the configuration of the solar system established, a natural scientific question was "why?" ... what are the physical principles that account for the planetary motions? The physicist and mathematician Sir Isaac Newton (1643–1727) provided the fundamental understanding of the behavior of gravity, that the gravitational force depended on the masses of the bodies involved and the inverse square of the distance of their separation. Newton also proposed his three laws of motion: 1. An object at rest remains at rest, and an object in motion moves with a constant velocity, unless acted upon by an external force, 2. The acceleration of an object depends on the mass of the object and the external force, and 3. for each reaction force there is an equal and opposite reaction. The second principle will be used to explain planetary motion in Chapter 2; the third underlies the basic operational principle of rockets (see Chapter 10).

Beyond the physical principle of gravitation, a question remained as to what accounted for the observed distances of each planet from the Sun? Whether by random chance or by divine design, John Elert Bode in 1772 and others attempted to develop a mathematical model that would account for those distances. The formulation of Bode, commonly termed *Bode's Law*, took the radius of each planet's orbit in terms of *Astronomical Units* (*AU*), where 1 AU is the radius of the Earth's orbit, to follow the series relation:

$$AU = ((0, 3, 6, 12, ...) + 4)/10 \quad (1.1)$$

where the index 0 corresponds to Mercury, 1 for Venus, and so on. The distances calculated by Bode's Law are compared with the actual mean orbital radii of the planets in the table below.

Table 1.1 Bode's Law planetary distances compared with actual values.

Planet	Bode's Law orbital radius (AU)	Actual mean orbital radius (AU)
Mercury	0.4	0.39
Venus 0.7 0.72	0.7	0.72
Earth	1	1
Mars	1.6	1.52
Main asteroid belt (average)	2.8	2.65
Jupiter	5.2	5.20
Saturn	10.0	9.52
Uranus	19.6	19.28
Neptune	38.8	30.17

It can be seen from the Table that Bode's Law provided very good agreement with the mean orbital radii of planets known at the time (out to Saturn) and even predicted with reasonable accuracy the orbital radius of Uranus, a planet not known at the time. Bode's Law, however, fails for Neptune, as well as for the minor planets farther out (Bode's Law does however provide a reasonable value for the main asteroid belt).

1.2 The Development of the Orbital Mechanics of Spacecraft

The development of space exploration in the 20th century led to many new considerations in orbital mechanics that extended the fundamental description of planetary orbital motion described above. Although the official beginning of the Space Age is commonly attributed to the launch of the Soviet satellite *Sputnik 1* on October 4, 1957 (first artificial object to orbit the Earth), some theoretical extensions to basic orbital mechanics were developed earlier in the century. The launch of *Sputnik 1* also signified the beginning of the "space race" between the United States (US) and the Soviet Union (USSR). Since *Sputnik 1* the number of Earth-orbiting satellites has increased steadily, with as of this writing number approximately 10,000, with many of these the *Starlink* constellation of satellites launched by SpaceX. The next major milestone in space exploration was the *Vostok 1* mission on April 12, 1961, in which Soviet cosmonaut Yuri Gagarin became the first human in space. The United States was able to achieve a crewed orbital flight on February 20, 1962, with astronaut John Glenn in the *Friendship 7* capsule as part of Project Mercury.

The accurate tracking of objects in Earth orbit required new considerations in orbital mechanics due to the perturbations resulting from the non-spherical shape of the Earth (discussed in Chapter 7). These perturbations can result in significant shifts in the plane of an object in Earth orbit, which must be accounted for in any application where the precise position of the object over ground needs to be known for communications, weather forecasting, Earth observation and surveillance, as well as for re-entry and landing targeting.[4]

Another innovation in orbital mechanics involves using on-board propulsion to change the plane of an orbit (see Chapter 5), such as inserting an object into an equatorial orbit or an orbit that passes over both poles. The first polar orbit was achieved on April 1, 1960, by the *TIROS 1* satellite, used for weather forecasting by imaging cloud formations. Another variation on basic orbital mechanics was the exploitation of a *geostationary* orbit, where the satellite essentially remains over a fixed point on the Earth' surface by matching its orbital speed with the rotation rate of the Earth (see Chapter 7). The first successful geostationary satellite was *Syncom 3*, launched August 19, 1964. Presently there are more than 500 satellites in geostationary orbit, most for communications and weather forecasting.

Given its comparatively close distance to the Earth, the Moon quickly became a target for both uncrewed probes and crewed spacecraft. Missions to the Moon required another extension of orbital mechanics, to take into account the simultaneous gravitational pull of both the Earth and the Moon, which leads to trajectories very different from elliptical orbits. This "three body" problem is discussed in Chapter 9. The first spacecraft to fly past the Moon was the Soviet *Luna 1* on January 4, 1959. That probe passed within roughly 6000 km of the lunar surface and after the flyby became the first object to leave the Earth-Moon system and enter an orbit around the

4 Though landings and splashdowns on Earth normally occur within a few kilometers of the targeted point, the *Friendship 7* project Mercury mission on February 20, 1962, overshot the planned landing point by nearly 400 km, for a variety of operational reasons.

Sun (a heliocentric orbit). The first object to hit the Moon (termed a "hard" landing) was *Luna 2* on September 13, 1959. The Americans did not succeed in achieving a hard landing on the Moon until April 26, 1962. Subsequently the US *Ranger 7* probe reached the moon on July 31, 1964, transmitting thousands of close-up images of the near-side lunar surface. The first probe to send images of the far side of the moon was the Soviet probe *Luna 3* on October 7, 1959.

As spaceflight technology advanced, the next step in lunar exploration ... a "soft" landing on the surface, was first achieved by *Luna 9* on February 3, 1966. The corresponding American landing occurred with *Surveyor 1* on June 2 of the same year. Both space probes provided panoramic images from the surface on the near side of the moon. The culmination of lunar exploration in the 20th century was the landing of a two-person crew on the moon by *Apollo 11* on July 20, 1969. Several other successful crewed lunar landings occurred during the Apollo program, the last being the landing of *Apollo 17* on December 11, 1972. All the Apollo landings resulted in significant amounts of lunar material being returned to the Earth. A smaller amount of material was returned from the lunar surface by the un-crewed Soviet *Luna 16* mission on September 24, 1970.

The spacecraft performing soft landings on the moon were typically first inserted into lunar orbit (see Chapter 5) by changing the inbound trajectory from the Earth. Another critical consideration for the Apollo program was the need to perform rendezvous in between two different spacecraft. This advance in orbital mechanics produced the perhaps surprising result that to increase the velocity of a spacecraft in orbit, the spacecraft must perform a braking maneuver (see Chapter 9). The first successful rendezvous in orbit was performed by the American *Gemini 6* and *Gemini 7* crewed spacecraft on December 15, 1965. The Soviet crewed spacecraft *Vostok 3* and *Vostok 4* reached identical orbits on August 12, 1962, but did not perform a rendezvous. Currently rendezvous in Earth orbit are routinely performed by spacecraft reaching the *International Space Station* (ISS), the Chinese *Tiangong* Station, or previously, the Soviet/Russian *Mir* Station.

Missions to nearby planets were conducted beginning in the early 1960's.[5] The general progression of those uncrewed missions was similar to the exploration of the Moon, first consisting of planetary flybys, leading to insertion into orbit around a planet, and/or landing on the surface (see Chapter 6). The earliest planetary probe was the Soviet *Venera 1*, which flew past Venus on May 19–20, 1961, at a distance of roughly 100,000 km, but did not transmit useful data. The first successful flyby of Venus was the American *Mariner 2* spacecraft, which flew to withing 35,000 km of the surface of Venus on December 14, 1962. *Mariner 4* flew by Mars on July 15, 1965, capturing the first close-up images of the Martian surface. *Mariner 9* was successfully inserted into Mars orbit on November14, 1971 and provided images of 85% of the surface.

The first successful soft landings on Mars occurred in the 1970's with the Soviet *Mars 3* probe, consisting of an orbiter and a lander, reaching Mars on December 2, 1971. The soft landing on Mars was successful but data from the lander were only transmitted for a few seconds; the orbiter however transmitted images and other data for several months. The American *Viking 1* probe entered Martian orbit on June 19, 1976, and the lander touched down on the surface the next day and operated for four years on the Martian surface. *Viking 2* entered Mars' orbit on August 7, 1976, with the lander performing a soft landing on September 3, 1976. More recent missions to the surface of Mars include the American *Pathfinder* probe (landed July 4, 1997) which deployed the *Sojourner* rover. The *Spirit* and *Opportunity* rovers landed on Mars on January 4 and January 25, 2004, respectively, with both spacecraft transmitting data for many years. The *Curiosity* rover landed on Mars on August 6, 2012.

---

5 A timeline of planetary exploration can be found here: https://en.wikipedia.org/wiki/Timeline_of_Solar_System_exploration

Soft landings on the planet Venus proved to be more problematic due to the severe environment of the Venutian atmosphere. The only landings on Venus to date were by the Soviet probes *Venera 7* and *Venera 8*. *Venera 7* reached on the surface of Venus on December 15, 1970, but with an unintended hard landing. *Venera 8* successfully landed on Venus on July 22, 1972, producing for about an hour the only surface images in that hostile environment. The first Venus orbiter was *Venera 9*, which reached orbit on October 22, 1975.

Another consideration in interplanetary orbital mechanic is the "gravity assist" made possible by planetary flybys (see Chapter 6). This maneuver was first performed by *Mariner 10* which in 1974 used a flyby of Venus to enable multiple flybys of Mercury. The first flyby of Jupiter was performed by the American *Pioneer 10* spacecraft, making its closest approach (132,250 km) to that gas giant planet on December 4, 1973. *Pioneer 10* subsequently reached the orbit of Saturn (though not the planet) in February 1976; the orbit of Neptune in June 1983, more than a decade after the launch on March 2, 1972. The times involved on orbits and trajectories are addressed in Chapters 3, 5, and 6. Subsequently *Pioneer 11* performed flybys of both Jupiter and Saturn, the former on December 3, 1974; the latter, on September 1, 1979, making use of the gravity-assist maneuver. The famous American *Voyager* probes were launched towards the outer planets in the late 1970's. *Voyager 1* performed a flyby of Jupiter on March 5, 1979, and of Saturn on November 12, 1980. *Voyager 2*, launched at about the same time as *Voyager 1*, used multiple gravity assists to take advantage of an unusual alignment of the outer planets to perform a flyby of each of them (known at the time as a "grand tour" of those planets). The flybys were as follows: Jupiter, July 1979, Saturn, August 1981, Uranus January 1986, and Neptune, August 1989. *Voyager* 2 has thus far been the only space probe to visit Uranus and Neptune. *Voyagers* 1 and 2 have since become the most distant human-made objects from the Earth, having entered interstellar space in August 2012 and November 2018, respectively.

More recently the US *Galileo* mission, launched in October 1989, reached Jupiter in December of 1995, after multiple flybys involving the Earth and Venus, taking advantage of the gravity assist associated with each flyby. The *Galileo* mission included an entry probe that was launched into the atmosphere of Jupiter.[6] A diagram of the *Galileo* trajectory to Jupiter is shown in Fig. 1.8.

Another notable interplanetary mission was the US-European *Cassini-Huygens* mission that reached Saturn on June 30, 2004 after a flyby of Jupiter. That mission included the soft landing of the *Huygens* probe on the surface of Saturn's largest moon, Titan (the only moon known to have an atmosphere). There have also been spectacular missions to comets and asteroids. These include the European *Giotto* flyby of Halley's comet on March 13–14, 1996, the US *Near Shoemaker* probe which flew by the asteroids 253 Mathilde and made a soft landing on 422 Eros on February 12, 2001, and the joint Japanese/German/French *Hayabusa 2* which performed a landing on the asteroid 162,173 Ryugu, returning material samples on December 6, 2020.

There have been many other successful planetary missions, too many to be listed here. Other notable examples include the US *Mercury Messenger*, which reached orbit around Mercury on March 11, 2011, the US *New Horizons* spacecraft, which accomplish a flyby of Pluto in July 14, 2015, and many other successful interplanetary probes launched by Japan, India, China, the United Arab Emirates, as well as by the US and Russia. All of these missions depended on a detailed understanding of orbital mechanics.

6 The author, working at Boeing Aerospace Company, helped to develop the Inertial Upper Stage (IUS) that launched the Galileo probe from the payload bay of the Space Shuttle Atlantis in October of 1989.

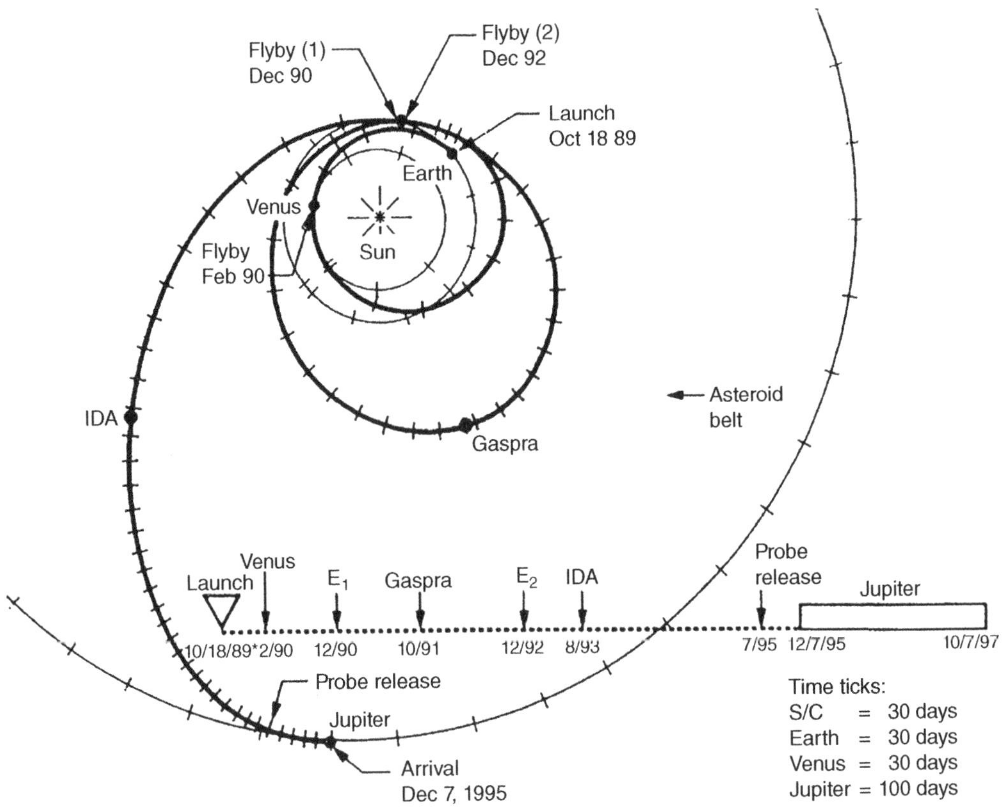

Figure 1.8 Galileo mission trajectory, showing multiple planetary flybys. *Source:* Wikipedia.

1.3 A First Estimate of Planetary Periods

As a prelude to the orbital mechanics developed in this text, one might consider what can be concluded about orbital mechanics without resorting to detailed calculations or derivations. One way to first address practical problems in physics and engineering to use *Dimensional Analysis*. The fundamental principle that underlies Dimensional Analysis is that the result of a calculation cannot be dependent on the units of the included variables. For example, the calculation of the orbital period of a planet (in years or other measure of time) must yield the same result whether SI or Imperial or other units are employed for the included variables. Many detailed discussions of Dimensional Analysis can be found in the literature.[7]

As a simple example of this approach, consider the problem of estimating the orbital periods of the planets. The period might be expressed in a simple functional form as

$$T = f(r, M, m, ?) \tag{1.2}$$

where T is the orbital period, r the radius of the orbit (assumed for simplicity to be circular), M the mass of the Sun, and m the mass of the planet. The question mark leaves open the possibility that

7 A good reference for dimensional analysis is Gibbings, J.C. *Dimensional Analysis*, Springer-Verlag, London, 2011.

other variables may come into play, and the actual functional form f is to be determined. The physical units associated with the variables in Eq. 1.2 are listed below, where the brackets indicate the appropriate units. A second assumption is that the mass of the planet, m, being vastly less than that of the Sun, does not play a substantial role in the calculation.

Period $[\![T]\!] = \mathrm{s}$
Orbital radius $[\![r]\!] = \mathrm{m}$
Solar mass $[\![M]\!] = \mathrm{kg}$

Taking only the orbital radius and solar mass into account, would suggest $T = f(r, M)$. This clearly fails, since the units of r and M do not contain time, as required for the orbital period. A second attempt could be to incorporate the acceleration,

$$[\![a]\!] = \mathrm{m}/s^2,$$

which also fails because it is not possible to eliminate the solar mass using only r, M, and a.

Consider next the attractive force due to Newtonian gravitation, expressed as $F = \frac{GMm}{r^2}$ with the force having units

$$[\![F]\!] = kg - \mathrm{m}/s^2,$$

where G is the gravitational constant. It was previously argued that the planet mass m cannot figure into the final result, so consider the gravitational constant by itself, which has the units

$$[\![G]\!] = \frac{\mathrm{m}^3}{\mathrm{s}^2 kg}.$$

This then suggests the functional form

$$T = f(r, M, G). \tag{1.3}$$

The necessary powers to which each variable needs to be raised to result in units of time in Eq. 1.3, can be solved algebraically. In this simple example, it can be noted that

$$[\![MG]\!] = \frac{\mathrm{m}^3}{\mathrm{s}^2},$$

which eliminates the unit kg. Next, including the cube of the radius r resulting in the unit m^3 gives

$$\left[\!\!\left[\frac{MG}{r^3}\right]\!\!\right] = \frac{1}{\mathrm{s}^2},$$

which now produces the needed unit of time only. Completing the dimensional analysis by incorporating T^2 then gives the final result,

$$\left[\!\!\left[\frac{MG}{r^3}T^2\right]\!\!\right] = 1, \tag{1.4}$$

which is non-dimensional (unitless), so formulation is independent of the units used, as required. Separating the orbital period from the other variables then suggests the proportionality $T^2 \sim \frac{r^3}{MG}$, which simplifies to

$$T \sim \sqrt{\frac{r^3}{MG}}. \tag{1.5}$$

The exact solution for the period of a circular orbit (derived in Chapter 4) is

$$T = 2\pi\sqrt{\frac{r^3}{MG}}. \tag{1.6}$$

This necessarily has the same functional form as the result of the dimensional analysis, but with an added multiplicative constant. An important utility of dimensional analysis is the ability to scale a formulation to different conditions. For example, if all variables are known for one case, for example for the Earth's orbit, the result for any other orbit (using the same assumptions) the scaling result in Eq. 1.5 can be used to determine the relationship between the orbital period and the orbital radius for any other planet.

1.4 Examples of Non-Western Astronomy

Considering the universal human experience of observing heavenly objects, it comes as no surprise that, in addition to the European astronomers discussed above, many ancient civilizations also made astronomical observations. Three examples of non-western astronomy are presented below, one each from modern-day Middle East, China, and Mexico/Central America.[8] Detailed observations recorded contained much information on the motions of the sun, moon, stars, and planets, and led to major discoveries in all cases, such as correct explanations of eclipses, accurate determination of the length of the solar year, in addition to comprehensive astronomical almanacs. All of those observations were made visually, since the telescope was unknown, from observatories, in some cases developing and using early astronomical instruments.

1.4.1 Babylonian

One of the earliest known systematic observations of the heavens were performed by the Babylonians,[9] with historical records going back as far as roughly 1800 BC. Limited to visual observation, the Babylonians made detailed records on stone tablets of a number of astronomical events including some pertaining to the moon and planets. By around 1100 BC, Babylonian astronomers had developed a catalog of the rising times of selected stars near the north celestial pole, the celestial equator and to the south. A listing of stars appearing opposite in the sky, along with the lengths of daylight and nighttime throughout the year helped to develop a sense of timing for astronomical events. Their calendar was lunar based, defining months on the cycle of the Moon, rather than the annual cycle of the Sun. The Babylonians also recorded the time when stars reach their highest points in the sky (culmination). Another important contribution divided the sky into 360° of arc (roughly corresponding to the length of the solar year, 365 ¼ years). An example Babylonian stone tablet listing information about the stars, including their locations withing each constellation, is shown in Fig. 1.9.

8 One detailed presentation of the history of non-Western astronomy is Thurston, H. *Early Astronomy*, Springer-Verlag, New York, 1994.
9 Geographically, Babylon roughly corresponded to modern day Iraq and parts of other nearby Middle Eastern nations.

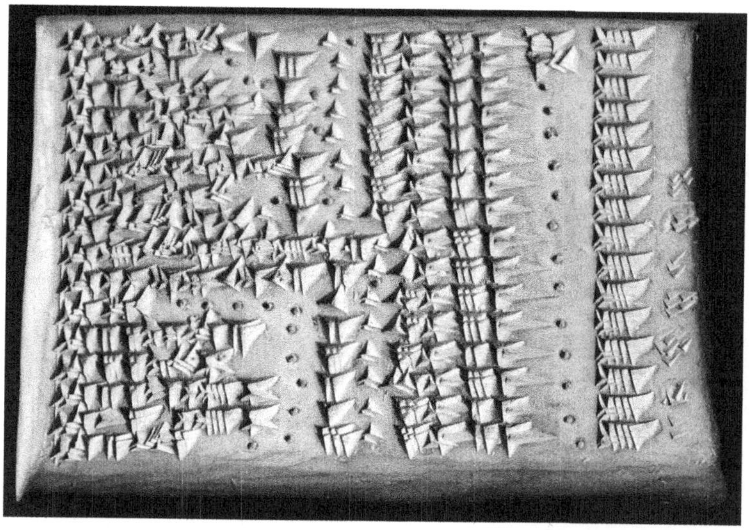

Figure 1.9 Babylonian star list from around 320–150 BC. *Source:* Anagoria / Wikimedia Commons / Public domain.

Starting around 700 BC, the Babylonians recorded observations of the path of the Moon through the stars, leading to the definition of the familiar 12 constellations of the Zodiac. They also understood the phenomena of eclipses and recorded some information pertaining to those. The Babylonians also noted that the motion of the moon was in a different plane than the sun, using the sun to define the ecliptic plane (the earth's orbit about the sun). The Babylonians noted that the motions of the observable planets were different from that of the stars, and they developed detailed almanacs of the time that included the observed positions of the planets, including their positions in the sky and their rising and setting times.

1.4.2 Chinese

The Chinese made many impressive discoveries of astronomical phenomena, based on visual observations combined with early astronomical instruments. The earliest records of Chinese astronomy go as far back as roughly 1500 BC. Chinese astronomers kept comprehensive records of the motions of the planets, leading to many detailed astronomical almanacs. They developed a calendar based on both the positions of the moon and of the sun. These early Chinese astronomers also correctly determined the causes of ecliptics (around 20 BC) and even the existence of sunspots, by viewing the sun through thin slices of jade around 3 BC.[10] The Chinese also developed one of the early concepts of cosmology, taking the earth and the heavens to be on separate spheres.

Chinese astronomy continued for centuries, perhaps reaching its peak during the Yuan dynasty in the 13th century AD. By then the Chinese had precisely determined the time of the summer solstice and the time throughout the year and were able to provide an accurate estimate of the length of the solar year, 365.24 days, which compares extremely well with the exact average value of 365.24219 days. This was centuries ahead of similar accuracy being achieved in Europe, where European astronomers, well aware of the existence of China, consulted the detailed Chinese almanacs. A Chinese observatory from the 13th century is shown in Fig. 1.10.

---

10 As a practical note one should *never* look at the sun without proper eye protection.

Figure 1.10 Chinese observatory in Goacheng. *Source:* takwing.kwong / Wikimedia Commons / CC BY SA 2.0.

The observatory shown included an observation platform at the top. On the ground was a low wall with markers to allow determination of the length of the Sun's shadow. Observations in this period made use of astronomical instruments. One such instrument was the *armillary sphere*, a hollow sphere with rings that allowed for the measurement of the latitudes of celestial bodies, including establishing the ecliptic plane. The Chinese also employed sextants and quadrants for measuring angles, well before similar instruments were used by Tycho Brahe. An image of a 17th century Chinese armillary sphere is shown in Fig. 1.11. Chinese astronomy began to decline with the increasing interactions with westerners beginning in the 17th century.

Figure 1.11 Reproduction of Ming Dynasty armillary sphere. *Source:* Matyas Rehak/Adobe Stock Photos.

1.4.3 Mayan

The Mayan empire existed in what is now Mexico and Central America from roughly 250 AD to 900 AD. Based entirely on visual observations, the Maya developed detailed astronomical calendars on cloth produced from tree bark and in stone. Some of the stone calendars are still in existence today.

The Maya oriented many buildings with astronomical events, such as the positions of the Sun and Moon. Their observatories typically included sight lines for equinoxes, solstices and eclipses, and also vertical wells for establishing the times of the overhead passage of the sun, moon, selected stars, and planets. Such overhead passages of the Sun occurred twice a year for the Maya who were situated in the tropics. Together the observations also led to the conclusion that the solar year was of 365 ¼ days duration. An example of a Mayan observatory, the Caracol observatory from approximately 900 AD, is shown in Fig. 1.12.

Astronomical observations made during the Mayan period included in detail the motion of the Moon and of the planets Mercury, Venus, and Mars. The lunar observations provided some information on eclipses. Mayan astronomers also detected and recorded the retrograde motion of the planets, and planetary conjunctions (the time at which a planet is visually closest to another) were recorded in detail. The Mayans identified a zodiac of 13 constellations, which is technically correct (in addition to the customary 12 constellations, the Sun also transits the constellation Ophiuchus). Mayan astronomers also recorded *synodic periods* (the times between the earth and planet being in a given relative position). Those observations produced a synodic period of 584 days for Venus and 780 days for Mars, in close agreement with the currently accepted values (synodic periods are discussed in Chapter 6). Cosmologically, the Maya believed the Earth to be the center of the universe, a common model at the time in both western and non-western astronomy.

Figure 1.12 El Caracol observatory (ca. AD 906). *Source:* Daniel Schwen/ Wikimedia Commons / Public domain / CC BY-SA 4.0.

1.5 Problem

1.1 The average, actual distances from the sun for Mercury, Venus, Mars, Jupiter, Saturn, and Uranus were given in Table 1.1 above. Using this information only, calculate the expected orbital periods for each planet (years). State all assumptions and do your own analysis, without making use of Eq. 1.6. Compare your calculated orbital periods with tabulated values. Comment specifically on the results for Mercury and Mars (noting that the orbit of Mars is rather elliptical; for Mercury, the orbit is *really* elliptical).

2

Orbits and Trajectories

The gravitational force between two bodies discussed in the introduction is the central physical principle in classical orbital mechanics. Taking that force and applying suitable geometric considerations allows for developing the analytical descriptions of the motions of objects both on closed orbits (such as the planets) and open trajectories (needed for interplanetary missions). This fundamental analysis also leads to the determination of the velocity and direction of an object at any point on any trajectory. The chapter outline is as follows:

- Gravitational force and idealized 2-body motion
- Derivation of the governing equation
- Conservation of angular momentum and energy
- The orbit equation
- Geometries of orbits and trajectories, and velocities

2.1 Gravitational Forces, the 2-Body Model and Newton's Laws of Motion

The dominant force in orbital mechanics is the gravitational attraction between bodies. This can be termed "Newtonian" gravitation, having the basic characteristics as recognized by Newton.[1] Important implications for this gravitational force for orbital mechanics include:

- It is a long-range force that acts between the Sun, planets, moons, and other objects (including spacecraft) over scales ranging from orbits around planets to the entire solar system.
- The force is proportional to the product of the masses of the bodies involved.
- The force is inversely proportional to the square of the distance between the bodies.
- Gravity is a central force that acts only in the direction of the line between two bodies.

The above statements can be examined in a bit more detail. Gravitational attraction is practically unlimited in range and impacts the motions of stars and galaxies. It is reasonable to expect that the acceleration of a body will depend directly on the mass of the body attracting it. Defining force as the mass times the acceleration of a body then suggests that the gravitational force will depend on

1 Sir Isaac Newton, 1643–1727, is widely considered the "father" of classic physics. In addition to formulating the Laws of Motion, Newton also made major contributions to optics and mathematics (he was the inventor of differential calculus).

Introduction to Orbital Mechanics: A Concise and Practical Approach, First Edition. James C. Hermanson.

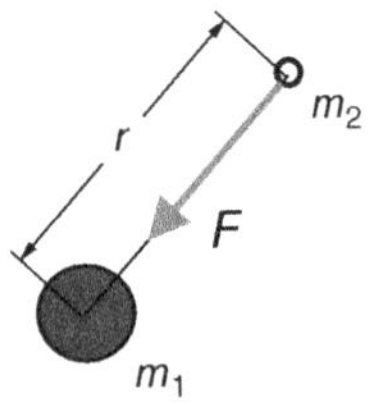

Figure 2.1 Central-force gravitational attraction force on object m_2 due to the presence of m_1.

the product of the masses of the object and the attracting body. The inverse square relation applies to many physical situations. One example is the intensity of uniform radiation from a central body. Though the total energy flow is constant through any sphere centered on the body, the energy flux, or intensity, decreases inversely with the square of the radius. Finally, the central force concept is strictly valid if the gravitational attraction occurs between two points. Deviations from this will be important when considering the orbital perturbations due to nonspherical objects in Chapter 7.

The central-force nature of gravitational attraction is shown in Fig. 2.1, where F is the force and r the distance between the two bodies. The direction of the force is exactly between the two masses, as shown. Using the principles of Newtonian gravitation described earlier, the magnitude of the force vector can be given by:

$$F = \frac{Gm_1m_2}{r^2}, \tag{2.1}$$

where m_1 and m_2 are the masses of the two bodies involved. The mass m_1 can be taken to be the mass of the primary body (the Sun, or a planet) and m_2 that of the secondary body (such as a planet orbiting the Sun, or a satellite orbiting a planet). To ensure the correct units of force (kg/m-s^2) requires insertion of the gravitational constant, G. The constant has the value $G = 6.67430 \times 10^{-11}$ m^3/kg-s^2. This value appears to be an intrinsic property of our universe, such as, the speed of light in a vacuum, Planck's constant, and others. The force vector acting on the secondary body m_2 is properly expressed as follows:

$$\vec{F}_{21} = -\frac{Gm_1m_2}{r^2}\frac{\boldsymbol{r}}{r}, \tag{2.2}$$

where $\boldsymbol{r}/r$, the vector distance divided by its magnitude, is simply the unit vector pointing in the direction of m_2 from m_1. The subscript 21 denotes the force applied on body 2 due to the gravitational attraction from body 1. The corresponding force exerted by body 2 on body 1 is then simply

$$\vec{F}_{12} = \frac{Gm_1m_2}{r^2}\frac{\boldsymbol{r}}{r}. \tag{2.3}$$

Since both bodies are subject to the same Newtonian gravitational attraction, the forces are exactly antisymmetric as seen in Eqs. 2.2 and 2.3, with the sign change merely resulting from the use of the same vector $\boldsymbol{r}$ in both cases here.[2] Though the *forces* must have the same magnitude, the *accelerations* of the two bodies can clearly be different. For the case of a smaller body (such as a satellite or planet) orbiting a larger, much more massive body (a planet, or the Sun), the acceleration of the more massive body can usually be neglected.

As an aside, one might consider what would be the implications of simply setting $G = 1$ (unitless). In this case, the magnitude of the gravitational force would be $F = m_1m_2/r^2$, with the (seemingly strange!) units of kg^2/m^2. This contrasts with the conventional expression for force $F = ma$, where a is the acceleration. Can the gravitationally based force just defined be considered valid? Yes, if it is recognized that force is a *defined quantity* (as is acceleration and very many other physical quantities in engineering). If the defined "gravitational force" is then related to the acceleration, the force would need to be expressed as $F = Cma$, where a constant C would be required to

2 In this text, all vector quantities will be shown in **bold** text; scalars are unbolded.

20/2/2026 8:09:1

produce the unusual units of kg^2/m^2. Thus keeping to the universally employed definition of force as $F = ma$ dictates the use of the gravitational constant G to get the consistent definitions given in Eqs. 2.1–2.3.

The gravitational forces are consistent with Newton's "laws" of motion.[3] These can be summarized as:

1) In the absence of an external force, an object at rest will remain at rest, or will be in uniform motion (constant velocity and direction);
2) The rate of momentum change of a body is equal to the net external force acting upon it;
3) For every action there is an equal, opposite reaction.

The first is the familiar statement pertaining to the inertia of an object. The second is simply the connection between the definitions of force and momentum. Defining momentum of an object of fixed mass m and with velocity V to be mV, the time rate of momentum change is $\frac{d}{dt}(mV) = m\frac{d}{dt}V = ma = F$, exactly the definition of force provided earlier. The third of Newton's laws is relevant to rocket propulsion (see Chapter 10), where it will be seen that the thrust force on a rocket is mostly due to the high flow rate of momentum in the rocket exhaust.

2.2 Further Justification, and Limitations, of the 2-Body Model

A key principle in the 2-body model is that the direction of gravitational attractive force is always exactly toward the center of the other body. Strictly speaking, this is exactly true only in two situations: (i) the body has all of its mass concentrated at a single point, or (ii) the body is a perfect sphere with a perfectly spherical distribution of mass (this does not require the mass be uniformly distributed, just that said distribution is spherically symmetric – such as a crust of uniform material and thickness overlying a core of uniform material). Scenario (i) represents a mathematical idealization, which is in fact often quite practical. The practicality of both approaches can be appreciated by considering, for example, the relative sizes and distances of objects in the solar system.

A sketch showing the relative sizes of the planets and the Sun, is shown in Fig. 2.2.[4] The actual distances between the planets are in fact much greater than those shown in that figure. For example, Mercury, the planet nearest to the Sun has a mean orbital distance from the Sun of 57.9×10^6 km. This amounts to 83 times the solar equatorial radius of 696,000 km. The Earth with its average orbital radius of 149.6×10^6 km is nearly 215 solar radii from the Sun. The separation between the Sun and Mercury is shown to scale (based on the size of the Sun) in Fig. 2.3. At such large distances, the mass of the Sun can effectively be modeled as being concentrated at a point from the standpoint of planetary orbital mechanics, and the 2-body model can be expected to work well.

---

3 The quotation marks around the term "law" are intended to serve a reminder that such relations are simply connections between consistently defined physical properties.

4 The author's view on the "what constitutes a planet" controversy is to go back to the Greek, where the term "planetes" simply means "wanderer." Simply categorizing the objects in the solar system as terrestrial planets, gas-giant planets, minor planets (asteroids), Kuiper Belt objects, and comets nicely takes care of this issue.

Figure 2.2 Relative sizes of the terrestrial (left), gas giant (center), and minor planets (right). The Earth-moon system is third from the left; the Sun is shown to scale in the background.

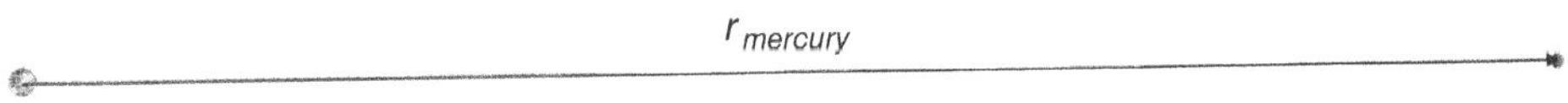

Figure 2.3 Actual mean orbital distance of Mercury, shown to scale with the diameter of the Sun (the diameter of Mercury is greatly exaggerated here to make it visible!).

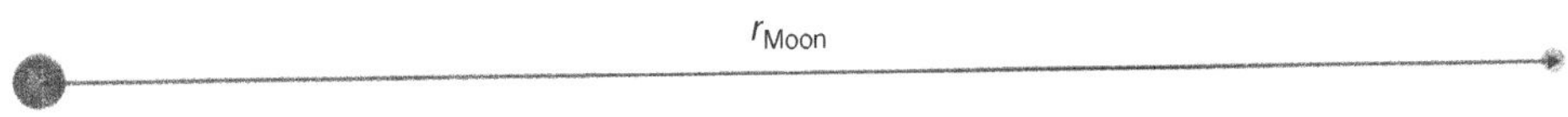

Figure 2.4 Actual mean orbital separation of the Moon and Earth, with both bodies shown to scale.

The Moon is on average at a distance of little more than 60 Earth radii, as shown to scale in Fig. 2.4. In this case, the Moon is actually close enough to experience perturbations to its orbit due to the nonspherical shape of the Earth (the Sun is in fact much closer to being perfectly spherical than the Earth). The perturbations to the orbits of the moon, and Earth-orbiting satellites, are discussed in Chapter 7.

As stated, the 2-body model described earlier is physically reasonable and useful in a great many situations in orbital mechanics. However, in some cases additional effects must be taken into account. These complications can include:

- Atmospheric drag on a spacecraft, for example, during planetary atmospheric entry for descent or aerobraking, is also responsible for the decay of low-earth orbits over time. Aerodynamic effects are discussed in Chapter 7.
- Perturbations to a trajectory. These may result from the gravitational forces exerted by other planets (e.g. Jupiter) on an interplanetary spacecraft, or from noncentral gravitational forces acting on a satellite (such as those from the nonspherical Earth discussed in Chapter 7).
- Thrust on a spacecraft. Interplanetary missions may employ long-duration thrust devices, such as electric propulsion or solar sails. A problem relating to solar sails is included at the end of this chapter.

Any of the above effects can result in significant deviations in the orbits or trajectories calculated using the 2-body model.

2.3 The Governing Equation for 2-Body Orbital Motion

A brief remark about the importance of *derivations* is offered here. Following and understanding the details of derivations of engineering processes are important for several reasons. First, understanding every step in a derivation leads to the certainty that the result can be used with confidence and does not suffer from any errors in analysis or presentation (these *do* occur, at any level). Second, any derivation requires developing a clear understanding of the assumptions, approximations, and limitations that underlie the result. The appropriateness of the 2-body approximation has already been discussed, as one example. Finally, understanding the details of a derivation allows extension to new, unfamiliar situations. This might include modifying the equation of orbital motion to take into account atmospheric drag, perturbations, or unusual propulsion methods. In many cases, such modifications can result in complex formulations that cannot be solved analytically and require numerical solutions.

The positions and velocities of bodies on their trajectories[5] in the 2-body model are governed by a single vector equation of motion, which is developed in this section. Consider the motion of the two bodies with masses m_1 and m_2, as seen from an inertial (nonmoving) reference frame, as shown in Fig. 2.5. The only forces acting in this system are those due to the gravitational attraction between the two bodies. The radii $\boldsymbol{R}_1$ and $\boldsymbol{R}_2$ are the vectors from the origin O of this Cartesian-coordinate reference frame; $\boldsymbol{R}_c$ is from the origin to the center of mass of the system (between the two bodies).

The forces experienced by either body as expressed in Eqs. 2.2 and 2.3 can be rewritten as follows:

$$\boldsymbol{F}_{12} = \frac{Gm_1m_2}{r^2}\frac{\boldsymbol{r}_{12}}{r} = m_1\ddot{\boldsymbol{R}}_1 \tag{2.4}$$

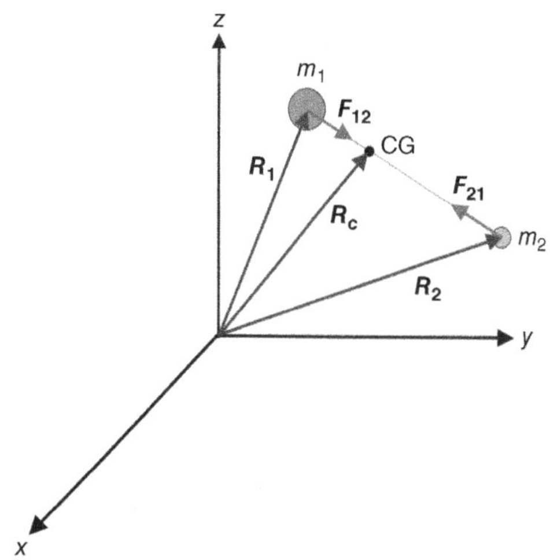

Figure 2.5 Vectors and forces in the 2-body problem.

5 The term "trajectory" will be used in this work generally; "orbit" will be used only to describe closed trajectories.

and

$$\boldsymbol{F}_{21} = \frac{Gm_1m_2}{r^2}\frac{\boldsymbol{r}_{21}}{r} = m_2\ddot{\boldsymbol{R}}_2, \tag{2.5}$$

where the vector radius from body 2 to body 1, $\boldsymbol{r}_{21} = -\boldsymbol{r}_{12}$, is simply the reverse of the vector from body 1 to body 2, with both vectors having the magnitude r. The expression $\boldsymbol{F} = m\boldsymbol{a} = \frac{d^2}{dt^2}\boldsymbol{R} = m\ddot{\boldsymbol{R}}$ was also invoked in both expressions.

The location of the center of mass of this 2-body system is determined by performing the mass weighting on the two bodies:

$$\boldsymbol{R}_c = \frac{m_1\boldsymbol{R}_1 + m_2\boldsymbol{R}_2}{m_1 + m_2}. \tag{2.6}$$

The corresponding acceleration of the center of mass is then given by:

$$\frac{d^2}{dt^2}\boldsymbol{R}_c = \ddot{\boldsymbol{R}}_c = \frac{m_1\ddot{\boldsymbol{R}}_1 + m_2\ddot{\boldsymbol{R}}_2}{m_1 + m_2} = 0. \tag{2.7}$$

Since there are no external forces acting on this 2-body system (the only forces being the gravitational attraction between them), the acceleration of the center of mass in this coordinate system must be zero – implying either no motion or motion with a constant velocity and direction. The latter case is unimportant and can be eliminated simply by allowing the coordinate system to move with the center of mass of the system. Thus the coordinate system can be taken to be fixed in space.

For the next step in the derivation, multiplying Eq. 2.5 by m_1 and Eq. 2.4 by m_2 and subtracting gives:

$$m_1m_2\left(\ddot{\boldsymbol{R}}_2 - \ddot{\boldsymbol{R}}_1\right) = \frac{Gm_1^2m_2}{r^2}\frac{\boldsymbol{r}_{21}}{r} - \frac{Gm_1m_2^2}{r^2}\frac{\boldsymbol{r}_{12}}{r}. \tag{2.8}$$

The quantity $\ddot{\boldsymbol{R}}_2 - \ddot{\boldsymbol{R}}_1 = \ddot{\boldsymbol{r}}_{12}$ is simply the relative acceleration of body 2 as seen by body 1. Dividing both sides of Eq. 2.8 by m_1m_2 then gives:

$$\ddot{\boldsymbol{r}}_{12} = \frac{Gm_1}{r^2}\frac{\boldsymbol{r}_{21}}{r} - \frac{Gm_2}{r^2}\frac{\boldsymbol{r}_{12}}{r}. \tag{2.9}$$

Since $\boldsymbol{r}_{21} = -\boldsymbol{r}_{12}$, this can be simplified further to:

$$\ddot{\boldsymbol{r}}_{12} = -\frac{G(m_1 + m_2)}{r^3}\boldsymbol{r}_{12}. \tag{2.10}$$

Although this expression is valid for the relative motion between any two bodies using the 2-body formulation defined here (for example, a binary system consisting of two stars of similar mass), the case of greatest interest is where the mass of the primary body, m_1, is much greater than that of the secondary body, m_2. This includes the practical situations of a planet orbiting the Sun, or a spacecraft trajectory near a planet. In this case, the only position vector of interest is $\boldsymbol{r}_{12}$, which for convenience will just be denoted as the vector. Combining the masses and gravitational constant into the single parameter $\mu = G(m_1 + m_2)$ then gives the fundamental differential vector equation of motion for the 2-body problem as follows:

$$\boxed{\ddot{\boldsymbol{r}} = -\frac{\mu}{r^3}\boldsymbol{r}.} \tag{2.11}$$

Equation 2.11 is the single, fundamental equation of motion for the 2-body problem, from which all characteristics of a trajectory, including its geometry, the velocity of the object at any point, and the associated times, can be determined.

In most practical situations, the mass of the orbiting body is much less than that of the primary body (i.e. $m_2 << m_1$) so that the gravitational parameter can be taken to be simply $\mu = Gm_1$. For example, the mass of the Earth compared to that of the Sun is $m_{\oplus}/m_{\odot} = 3.0 \times 10^{-6}$. By contrast, the mass ratio of Jupiter to the Sun is $m_{♃}/m_{\odot} = 0.001$. Although this value is small, it does imply a certain amount of "wobble" of the Sun due to the orbit of Jupiter, as the center of mass of the system is not exactly at the center of the Sun. This is in fact one way that exoplanets can be detected, by observing small shifts in the spectrum of a star due to the wobble caused by the motion of planets in that star's system. For the Earth/Moon system, the corresponding mass ratio is $m_{☽}/m_{\oplus} = 0.012$.

2.4 Conservation Laws of Angular Momentum and Energy

As is the case with many physical systems, conservation relations can be defined for orbital mechanics relating the defined concepts of energy and momentum in a self-consistent fashion. In particular, the definitions and conservation relations of angular momentum and energy are very useful in determining the parameters of orbits and trajectories and will be used to derive the geometric equation for all trajectories in the next section.

Given that velocity and momentum are vector quantities, developing the conservation relations and the equation for orbits and trajectories will require some vector operations. The familiar scalar, or "dot" product between the two vectors, $\boldsymbol{x} \cdot \boldsymbol{y}$, is related to their magnitudes and the angle between them, θ, by $\boldsymbol{x} \cdot \boldsymbol{y} = |\boldsymbol{x}||\boldsymbol{y}| \cos\theta$. The corresponding vector, or "cross" product, is $\boldsymbol{x} \times \boldsymbol{y} = |\boldsymbol{x}||\boldsymbol{y}| \sin\theta\, \boldsymbol{n}$, where $\boldsymbol{n}$ is the unit vector normal to the plane defined by $\boldsymbol{x}$ and $\boldsymbol{y}$, following the familiar right-hand rule. These basic definitions can then be used in developing the following list of vector identities, which will be used throughout this text.

Table of Some Useful Vector Identities

1) $\boldsymbol{x} \cdot \boldsymbol{y} = \boldsymbol{y} \cdot \boldsymbol{x}$
2) $\boldsymbol{x} \times \boldsymbol{y} = -\boldsymbol{y} \times \boldsymbol{x}$
3) $\boldsymbol{x} \cdot (\boldsymbol{y} + \boldsymbol{z}) = \boldsymbol{x} \cdot \boldsymbol{y} + \boldsymbol{x} \cdot \boldsymbol{z}$
4) $\boldsymbol{x} \times (\boldsymbol{y} + \boldsymbol{z}) = \boldsymbol{x} \times \boldsymbol{y} + \boldsymbol{x} \times \boldsymbol{z}$
5) $(\boldsymbol{x} \times \boldsymbol{y}) \cdot \boldsymbol{z} = (\boldsymbol{y} \times \boldsymbol{z}) \cdot \boldsymbol{x} = (\boldsymbol{z} \times \boldsymbol{x}) \cdot \boldsymbol{y}$
6) $(\boldsymbol{x} \times \boldsymbol{y}) \times \boldsymbol{z} = (\boldsymbol{x} \cdot \boldsymbol{z})\boldsymbol{y} - (\boldsymbol{y} \cdot \boldsymbol{z})\boldsymbol{x}$
7) $\boldsymbol{x} \times (\boldsymbol{y} \times \boldsymbol{z}) = (\boldsymbol{x} \cdot \boldsymbol{z})\boldsymbol{y} - (\boldsymbol{x} \cdot \boldsymbol{y})\boldsymbol{z}$
8) $\frac{d}{dt}(\boldsymbol{x} + \boldsymbol{y}) = \frac{d\boldsymbol{x}}{dt} + \frac{d\boldsymbol{y}}{dt}$
9) $\frac{d}{dt}(\boldsymbol{x} \cdot \boldsymbol{y}) = \frac{d\boldsymbol{x}}{dt} \cdot \boldsymbol{y} + \boldsymbol{x} \cdot \frac{d\boldsymbol{y}}{dt}$
10) $\frac{d}{dt}(\boldsymbol{x} \times \boldsymbol{y}) = \frac{d\boldsymbol{x}}{dt} \times \boldsymbol{y} + \boldsymbol{x} \times \frac{d\boldsymbol{y}}{dt}$

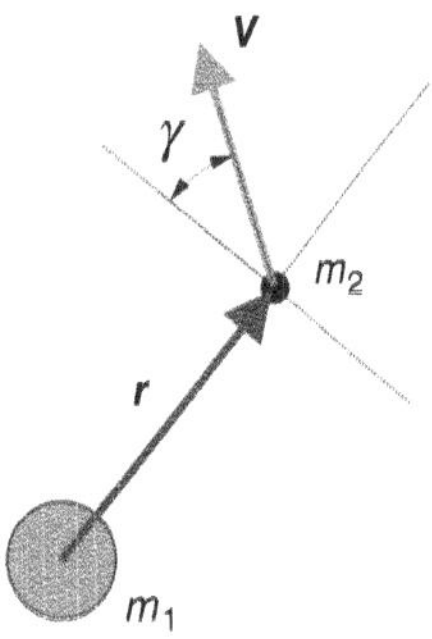

Figure 2.6 Geometry at a point showing the flight path angle.

2.4.1 Conservation of Specific Angular Momentum

The next step is to develop the conservation principle for angular momentum in orbital motion. Previously it was noted that the *linear* momentum of an object is, by definition, directly related to the applied force, i.e. $\frac{d}{dt}(m\boldsymbol{V}) = \boldsymbol{F}$. Since there is always a gravitational attractive force present, it is clear that the linear momentum of an object changes continually and cannot be conserved. Consider instead the *angular* momentum, defined to relate the object's moment of inertia and its angular velocity as $\boldsymbol{H} = \boldsymbol{r} \times m\boldsymbol{V}$, as shown in Fig. 2.6. This is actually the angular momentum of the entire 2-body system, since the primary body is taken to be effectively a single point in space and therefore has no angular momentum of its own. Also, since the angular momentum is the vector product of the radius and velocity vectors, its direction is exactly perpendicular to the plane of the orbit.

Since the mass of the orbiting object, m, is constant, it is convenient to use the specific angular momentum $\boldsymbol{h} = \boldsymbol{H}/m$.[6] This can be written in scalar form as $h = |\boldsymbol{h}| = rV\cos\gamma$, where γ is the flight-path angle, which is defined relative to the normal of the radius vector, as in Fig. 2.6. The change in the angular momentum with time can be evaluated by use of Vector Identity #10 listed earlier and noting that $\boldsymbol{V} = \dot{\boldsymbol{r}}$ to give:

$$\frac{d}{dt}(\boldsymbol{r} \times \boldsymbol{V}) = \frac{d}{dt}(\boldsymbol{r} \times \boldsymbol{r}) = \dot{\boldsymbol{r}} \times \dot{\boldsymbol{r}} + \boldsymbol{r} \times \ddot{\boldsymbol{r}}. \tag{2.12}$$

The first term on the right side is zero from the definition of the vector product. The second term can be evaluated by noting that the quantity $\ddot{\boldsymbol{r}}$ is related to $\boldsymbol{r}$ by the governing equation of motion (Eq. 2.11). Taking the vector product with $\boldsymbol{r}$ on both sides then gives:

$$\boldsymbol{r} \times \ddot{\boldsymbol{r}} = -\frac{\mu}{r^3}(\boldsymbol{r} \times \boldsymbol{r}), \tag{2.13}$$

which is also seen to be zero. Thus Eq. 2.12 gives the important result that:

$$\frac{d}{dt}(\boldsymbol{r} \times \boldsymbol{V}) = \frac{d}{dt}(\boldsymbol{h}) = 0. \tag{2.14}$$

This is the formal statement that, in the 2-body problem, the angular momentum of the system *remains constant*. This might also be expected from the definition of the torque on a system as $\tau = \frac{d\boldsymbol{H}}{dt}$, the time rate of change of the angular momentum. Since the forces in this system are all acting centrally, there is zero torque, so the angular momentum must remain unchanged.

That the angular momentum of the system must be constant has two key implications. First, the constant scalar magnitude of the angular momentum implies that $h = rV\cos\gamma = constant$. Second, since the *vector* angular momentum $\boldsymbol{h} = constant$ is perpendicular to the plane of the orbit, there can be no change in the plane of the orbit.

6 From this point onward, the term "angular momentum" will mostly be used; that this is the specific angular momentum is implied.

Example Problem 2a

A satellite in an elliptical orbit around the Earth as shown in Fig. 2a. The satellite has a velocity at *perigee* (the point in the orbit closest to Earth) of $V_p = 8.067$ km/s at a perigee radius of $r_p = 9800$ km. If the *apogee* (the point farthest from the earth) is at $r_a = 26{,}000$ km, what is the orbital velocity at that point, V_a?

Figure 2a An elliptical orbit around the Earth (shown to scale).

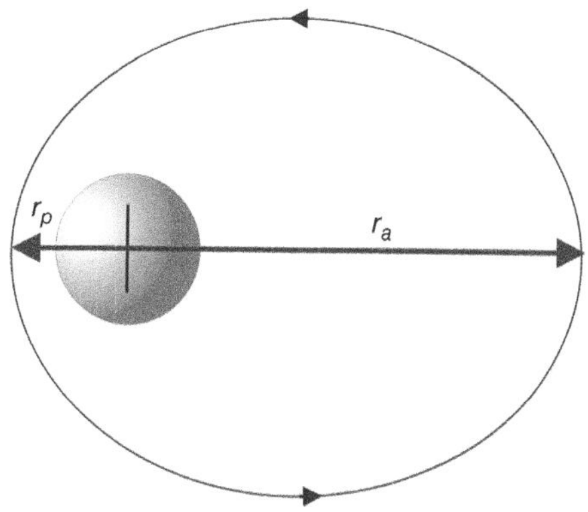

Solution: Since the angular momentum of the orbit is constant,

$$h = r_p V_p \cos\gamma_p = r_p V_p \cos\gamma_p.$$

The flight path angles at both perigee and apogee are exactly zero, so $r_p V_p = r_p V_p$ and the apogee velocity is then:

$$V_a = V_p \frac{r_p}{r_a} = (8.067\ \text{km/s})\,\frac{9800\ \text{km}}{39{,}200\ \text{km}} = 2.017\ \text{km/s}.$$

2.4.2 Conservation of Specific Mechanical Energy

In a similar fashion to the angular momentum, it could be reasonably expected that it is possible to define consistent forms of mechanical energy such that the overall energy of the system is also conserved. If the only energy forms to be accounted for are those due to the body's motion and gravity, and if there is no external energy input, the conservation principle can simply be stated as $KE + PE = constant$, where KE and PE are respectively the kinetic and potential energies. In ordinary laboratory dynamics, the kinetic energy is defined as $KE = \frac{1}{2}mV^2$ and the potential energy as $PE = mgz$, where g is the gravitational acceleration and z some reference height. The definition of kinetic energy is reasonable for a body in orbital motion, but that of the potential energy is not, given that the gravitational force and acceleration change significantly with distance, as seen in Eqs. 2.2–2.5. A different approach is thus required.

Returning to the governing equation of motion (Eq. 2.11) and taking the scalar product of that equation with the velocity vector $\boldsymbol{V} = \dot{\boldsymbol{r}}$ gives:

$$\dot{\boldsymbol{r}} \cdot \ddot{\boldsymbol{r}} = \boldsymbol{V} \cdot \dot{\boldsymbol{V}} = -\frac{\mu}{r^3}(\dot{\boldsymbol{r}} \cdot \boldsymbol{r}). \tag{2.15}$$

The scalar products can be evaluated by:

$$\frac{d}{dt}(\boldsymbol{r}\cdot\boldsymbol{r}) = \boldsymbol{r}\cdot\frac{d\boldsymbol{r}}{dt} + \frac{d\boldsymbol{r}}{dt}\cdot\boldsymbol{r} = 2\boldsymbol{r}\cdot\frac{d\boldsymbol{r}}{dt} = 2(\boldsymbol{r}\cdot\dot{\boldsymbol{r}}), \tag{2.16}$$

and also noting that $\boldsymbol{r}\cdot\boldsymbol{r} = r^2$ gives:

$$\frac{d}{dt}(\boldsymbol{r}\cdot\boldsymbol{r}) = \frac{d}{dt}(r^2) = 2r\frac{dr}{dt} = 2r\dot{r} \tag{2.17}$$

Thus from Eqs. 2.16 and 2.17, $\boldsymbol{r}\cdot\dot{\boldsymbol{r}} = r\dot{r}$. This manipulation applies for any vector, so $V\cdot\dot{\boldsymbol{V}} = V\dot{V}$, allowing Eq. 2.15 to be rewritten directly in scalar form as:

$$V\dot{V} = -\frac{\mu}{r^3}(r\dot{r}) = -\frac{\mu}{r^2}\dot{r}. \tag{2.18}$$

The above scalars can be rewritten as:

$$V\dot{V} = V\frac{dV}{dt} = \frac{d}{dt}\left(\frac{V^2}{2}\right) = -\frac{\mu}{r^2}\dot{r} = -\frac{\mu}{r^2}\frac{dr}{dt} = +\mu\frac{d}{dt}\left(\frac{1}{r}\right). \tag{2.19}$$

The third and last terms in Eq. 2.19 can then be recombined to give:

$$\frac{d}{dt}\left(\frac{V^2}{2} - \frac{\mu}{r}\right) = 0. \tag{2.20}$$

The time invariance of the expression in parentheses points then to this definition of specific mechanical energy for the 2-body system[7]:

$$E = \frac{V^2}{2} - \frac{\mu}{r} = constant. \tag{2.21}$$

This is the self-consistent definition of the specific mechanical energy (energy per unit mass), and the quantity is constant on any 2-body trajectory. The first term in the equation, $V^2/2$, is the familiar specific kinetic energy. The second term, $-\mu/r$, is the correct, consistent form of the potential energy for this case where the radius from the central body is variable.

Example Problem 2b

Verify that the apogee conditions determined in Example Problem 2a are consistent with the conservation of energy for this orbit.

Solution: Equation 2.21 dictates that $\frac{V_p^2}{2} - \frac{\mu}{r_p} = \frac{V_a^2}{2} - \frac{\mu}{r_a} = constant$. Evaluating each expression separately with $\mu = Gm = 398{,}600\ \text{km}^3/\text{s}^2$ for the Earth gives:

$$\frac{V_p^2}{2} - \frac{\mu}{r_p} = \frac{(8.067\ \text{km/s})^2}{2} - \frac{398{,}600\ \text{km}^3/\text{s}^2}{9800\ \text{km}} = -8.134\ \text{km}^2/\text{s}^2$$

$$\frac{V_a^2}{2} - \frac{\mu}{r_a} = \frac{(2.017\ \text{km/s})^2}{2} - \frac{398{,}600\ \text{km}^3/\text{s}^2}{39{,}200\ \text{km}} = -8.134\ \text{km}^2/\text{s}^2,$$

which are equal (to the number of significant figures considered here), so the conservation of energy is satisfied for the orbital parameters in Example Problem 2a.

7 As with the momentum, the simpler term "energy" will now be used for the specific mechanical energy.

2.5 The Orbit Equation

The conservation principles developed in the previous section provide important connections between properties on orbits and trajectories. These can allow, for example, determination of the velocity and flight path angle of an object if the radial distance and angular momentum and energy are known. The latter quantities, by themselves, do not however determine the shape or other geometric properties of the trajectory. To determine these requires development of the *Orbit Equation* as presented in this section.

The derivation begins by again using the governing equation of orbital motion, Eq. 2.11. Taking the vector product of that relation with the angular momentum vector[8] gives the following:

$$\ddot{\boldsymbol{r}} \times \boldsymbol{h} = -\frac{\mu}{r^3}\boldsymbol{r} \times \boldsymbol{h}. \tag{2.22}$$

The left side of this equation is simply equal to $\frac{d}{dt}(\dot{\boldsymbol{r}} \times \boldsymbol{h})$, since the angular momentum vector, $\boldsymbol{h}$, is constant as shown in the previous section. The right side is more complicated. Expanding the vector product and employing Vector Identity #4 gives:

$$\boldsymbol{r} \times \boldsymbol{h} = \boldsymbol{r} \times (\boldsymbol{r} \times \boldsymbol{V}) = \boldsymbol{r} \times (\boldsymbol{r} \times \dot{\boldsymbol{r}}) = (\boldsymbol{r} \cdot \dot{\boldsymbol{r}})\boldsymbol{r} - (\boldsymbol{r} \cdot \boldsymbol{r})\dot{\boldsymbol{r}}. \tag{2.23}$$

The quantity $\boldsymbol{r} \cdot \boldsymbol{r}$ on the right-hand side is equal to the scalar radius squared, r^2, and as seen in the previous section, $\boldsymbol{r} \cdot \dot{\boldsymbol{r}} = r\dot{r}$. Inserting these simplifications for the vector products in Eq. 2.23 then gives, using the chain rule of differentiation:

$$\frac{d}{dt}(\dot{\boldsymbol{r}} \times \boldsymbol{h}) = -\frac{\mu}{r^3}\left[(r\dot{r})\boldsymbol{r} - r^2\dot{\boldsymbol{r}}\right] = -\frac{\mu}{r^2}\dot{r}\boldsymbol{r} + \frac{\mu}{r}\dot{\boldsymbol{r}} = \mu\frac{d}{dt}\left(\frac{\boldsymbol{r}}{r}\right). \tag{2.24}$$

Integrating the far-left and far-right terms with time leads to:

$$\dot{\boldsymbol{r}} \times \boldsymbol{h} = \mu\frac{\boldsymbol{r}}{r} + \boldsymbol{B}, \tag{2.25}$$

where $\boldsymbol{B}$ is a vector constant. The final step is to derive the distance to the orbiting body as a function of its position along the trajectory. This can be done by taking the scalar product of both sides of Eq. 2.25 with the radius vector, resulting in:

$$(\dot{\boldsymbol{r}} \times \boldsymbol{h}) \cdot \boldsymbol{r} = \mu\frac{\boldsymbol{r} \cdot \boldsymbol{r}}{r} + \boldsymbol{B} \cdot \boldsymbol{r}. \tag{2.26}$$

The left side of this equation can be simplified, with the aid of Vector Identity #5, to give:

$$(\dot{\boldsymbol{r}} \times \boldsymbol{h}) \cdot \boldsymbol{r} = (\boldsymbol{r} \times \dot{\boldsymbol{r}}) \cdot \boldsymbol{h} = (\boldsymbol{r} \times \boldsymbol{V}) \cdot \boldsymbol{h} = \boldsymbol{h} \cdot \boldsymbol{h} = h^2, \tag{2.27}$$

where h^2 is simply the square of the scalar angular momentum. The first term on the right side of Eq. 2.26 simplifies immediately since $\boldsymbol{r} \cdot \boldsymbol{r}$ is the scalar quantity r^2. The second term can be written as $\cos\theta$, where θ is the angle between the radius vector **r** and the constant vector $\boldsymbol{B}$. Using these simplifications results in the scalar relation:

8 This may seem like yet another mathematical "trick" employed to get to a needed result. An applied mathematician once responded to objections from students about so many "tricks" being needed that such are only "tricks" the first time – after that, they become "techniques."

$$h^2 = r\,(\mu + B\cos\theta), \tag{2.28}$$

which can be rewritten as:

$$r = \frac{h^2/\mu}{1 + B/\mu\ \cos\theta}. \tag{2.29}$$

This equation then gives the radius of the object, r, as a function of its position on the trajectory, as given by the angle θ. Defining the constants $p = h^2/\mu$ and $e = B/\mu$ allows rewriting Eq. 2.29 to give:

$$\boxed{r = \frac{p}{1 + e\,\cos\theta}}. \tag{2.30}$$

This *Orbit Equation* applies to *all* 2-body orbits and trajectories. Furthermore, the Orbit Equation implies that all 2-body trajectories are *conic sections*. The angle θ is termed the *true anomaly* and indicates how far around the trajectory an object is located. It can be seen that since $-1 \le \cos\theta \le 1$, the smallest value of the radius r will occur for $\theta = 0$. This is the location of *periapsis* (the general term) or *perigee* (strictly speaking, this term is applicable only to orbits around the Earth). If the trajectory is an orbit, it follows that the farthest orbital distance, termed the *apoapsis* or *apogee*, will occur for $\theta = 180° = \pi$ radians. The other constants given in Eq. 2.30 are the *parameter*, p, which is indicative of the overall *size* of the trajectory, and the *eccentricity*, e, which indicates the *shape* of the trajectory, for example how close an orbit is to being perfectly circular. The geometric nature of conic sections and the implications of these parameters in the Orbit Equation are discussed in the next section.

The characteristics of 2-body orbits and trajectories discussed so far can be summarized as:

1) All 2-body orbits and trajectories are conic sections.
2) The focus of the orbits and trajectories is at the center of the primary body (this is for the practical case where $m_1 >> m_2$, otherwise the focus is at the barycenter of the system).
3) The specific mechanical energy of the motion, E, is constant.
4) The specific angular momentum (both the vector $\boldsymbol{h}$ and the scalar h) is constant.

2.6 The Geometry of 2-Body Orbits and Trajectories

All 2-body orbits and trajectories can be described by using the Orbit Equation as stated in Eq. 2.30. The classical conic sections are shown in Fig. 2.7. Conic sections are defined by taking sections ("cuts" or "slices") through a system consisting of two identical, coaxial cones joined at their apexes, as shown. Taking a section through either cone exactly normal to the system's axis results in a circle. If the section is taken somewhat obliquely to the axis, an ellipse results. Further inclining the section until it is exactly parallel to the sides of a cone then yields a parabola. Inclining the section beyond this results in the intersection of both cones, which results in a pair of curves, termed hyperbolae (the singular term is hyperbola).

The connection between the geometries of the various conic-section curves and the parameters of orbits and trajectories can be made starting with the formal, geometric definition of a parabola shown in Fig. 2.8. For an object on a parabolic trajectory, the distance from the focus (the location of the primary body, consistent with Kepler's first law) is exactly the same as that to a fixed line parallel to, and below, the horizontal axis, termed the *directrix*. Thus in this case, $r = r'$ at all points.

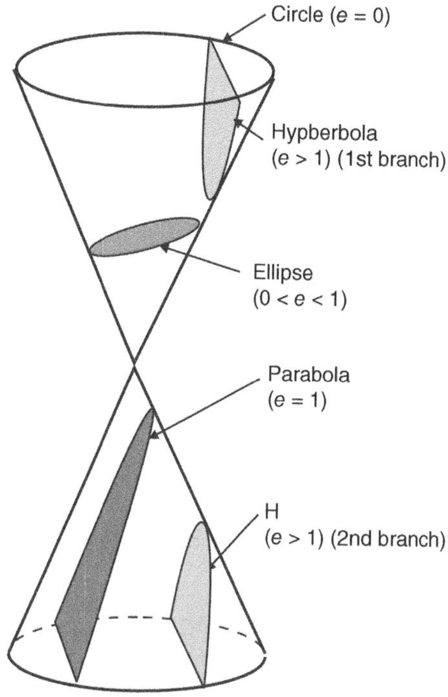

Figure 2.7 Classical conic sections.

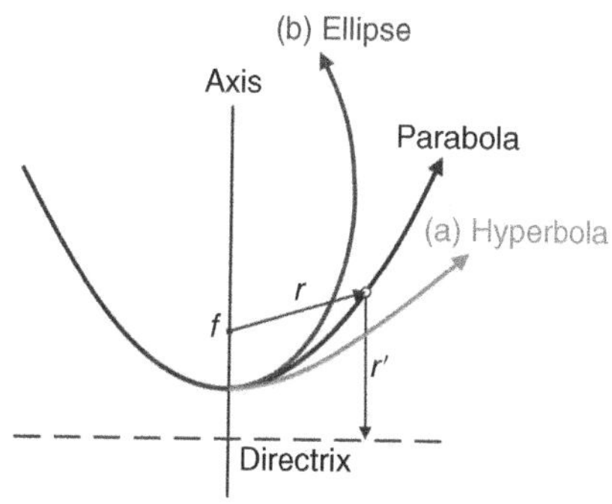

Figure 2.8 Geometric definitions of conic sections.

If the object is moving away on a trajectory at a faster rate than the parabola, as shown by the line designated a) in Fig. 2.8, $r > r'$ at all points the curve is a hyperbola. Correspondingly, for a slower rate of departure, as shown by lineb), $r < r'$ and the curve will be an ellipse. The *eccentricity* of the curves can be expressed as $e = r/r'$. For a parabola, the eccentricity is then unity ($e = 1$); for an ellipse, $e < 1$, and for a hyperbola, $e > 1$.

A comparison between trajectories for different values of eccentricity, but with the same focus and periapsis radius, is shown in Fig. 2.9. For $e < 1$, the trajectories are all closed, elliptical orbits, with the limiting case $e = 0$ being an exactly circular orbit. The parabola ($e = 1$) is the limiting case of an open, no-return trajectory. A further increase in eccentricity corresponds to hyperbolae, with the limiting case of the eccentricity becoming infinite being a straight line. The parameter, p, in this limiting case also becomes infinite, as can be inferred from the Orbit Equation (Eq. 2.30).

2.6.1 Elliptical Orbits

The elliptical orbit is of major importance given that it describes the motion of the planets, consistent with Kepler's laws of planetary motion, as well as satellites (both natural and artificial) in orbit about the Earth and other planets or moons. The characteristic features of an elliptical orbit are detailed in Fig. 2.10. One key parameter is the *semi-major axis*, a (b is the corresponding

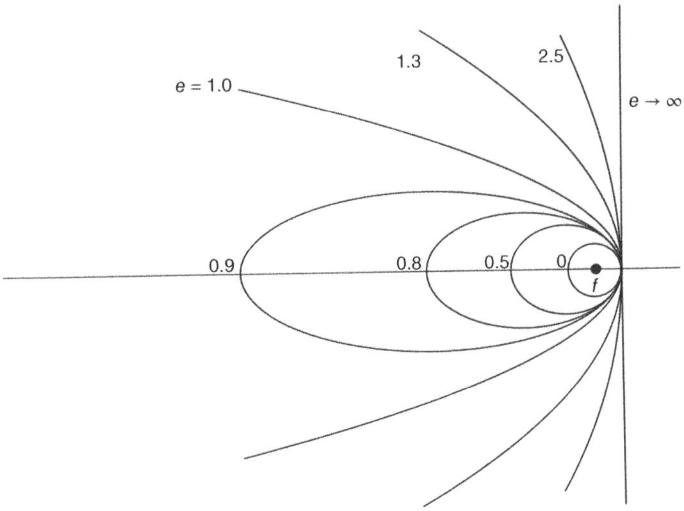

Figure 2.9 Trajectories with different eccentricities. *f* is the location of the primary body.

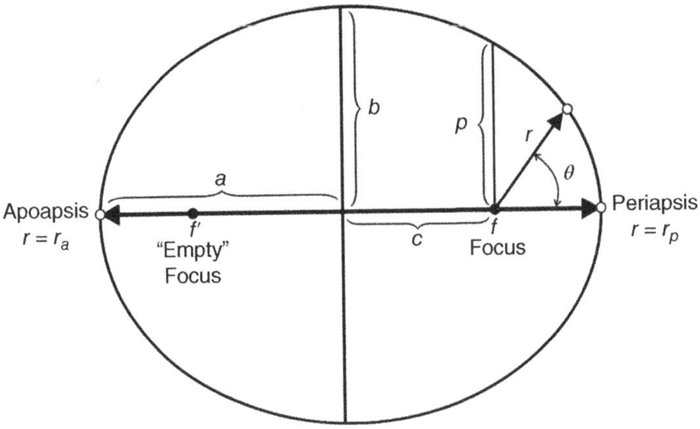

Figure 2.10 Characteristics of an elliptical orbit.

semi-minor axis). The orbital *parameter*, *p*, is the radial distance for an anomaly of $\theta = 90°$, as dictated by the Orbit Equation (Eq. 2.30). The semi-major axis can be thought of as being indicative of the "size" of the orbit – the orbital parameter of the orbit's "width." The sum of the apoapsis and periapsis radii is the entire major axis, so $r_p + r_a = 2a$.

For the case of an ellipse, the sum of the distances to the primary focus, *f* (the location of the primary body), and to f' (the "empty" focus), is always twice the semi-major axis, that is, $r + r^* = 2a$. This explains why an ellipse can be drawn by tracing a pencil around two thumbtacks using a loop of string of sufficient length.

The quantity *c* is the distance from the center, or origin, of the ellipse to either focus. This can be related to the distances to the periapsis and apoapsis, respectively, by $r_p = a - c$ and $r_a = a + c$. Since by definition $r'e = r$, the distance from the point of periapsis to the directrix is $r_{pd} = (a - c)/e$; the corresponding distance from apoapsis is then $r_{ad} = 2a + (a - c)/e$. This gives:

$$r_a = er_{ad} = e\left(2a + \frac{a - c}{e}\right) = a + c, \tag{2.31}$$

which simplifies to $e = c/a$. This allows writing the earlier relations for the periapsis and apoapsis radii directly in terms of the semi-major axis and the eccentricity to give these very useful relations:

$$\boxed{\begin{aligned} r_p &= a(1-e) \\ r_a &= a(1+e). \end{aligned}} \tag{2.32}$$

The parameter p can be determined from the Orbit Equation (2.30), or $p = r(1 + e\cos\theta)$. Evaluating this at periapsis, where $\theta = 0$, gives $p = r_p(1 + e) = a(1 - e^2)$ using Eq. 2.31. This expression is valid for all conic-sections; for the parabola $p = 2r_p$ since $e = 1$. This suggests another convenient form of the Orbit Equation:

$$r = \frac{a(1-e^2)}{1 + e\cos\theta}. \tag{2.33}$$

Also, Eqs. 2.32 can be combined to get this useful relation for the eccentricity of an elliptical orbit:

$$e = \frac{r_a - r_p}{r_a + r_p} \tag{2.34}$$

As an aside, consider the case of a circular orbit. In this case, $r =$ *constant*, so the Orbit Equation (Eq. 2.30) dictates that $e = 0$ and that $r = p$. Furthermore, Eqs. 2.32 indicate that $r_a = r_p = a$. Since in this case, the primary focus and empty focus occupy the same point, $c = 0$, and the distance from any point on the orbit to the directrix is $r_d = a/e = \infty$.

It may be apparent at this point that, for a 2-body orbit, knowing any two parameters allows determination of all the others. This can be seen, for example, from Eqs. 2.33 that knowing the semi-major axis and eccentricity, a and e, respectively, immediately gives both the periapsis and apoapsis radii, r_p and r_a, and also the parameter of the orbit, p. Similarly, knowing the periapsis and apoapsis radii then gives the eccentricity from Eq. 2.34, and the semi-major axis $a = r_a + r_p$, and so forth.

Example Problem 2c

An Earth-orbiting satellite is in an elliptical orbit with perigee and apogee radii of $r_p = 6500$ km and $r_a = 45{,}000$ km, respectively. At what value of the true anomaly, θ, does the satellite enter the Van Allen radiation belts, if that belt region starts at an altitude of $h = 640$ km?

Solution: First the eccentricity can be determined: $e = \dfrac{r_a - r_p}{r_a + r_p} = \dfrac{45{,}000\text{ km} - 6500\text{ km}}{45{,}000\text{ km} + 6500\text{ km}} = 0.748.$

The Orbit Equation can be written $r = \dfrac{r_p(1+e)}{1 + e\cos\theta}$ where the two parameters e and r_p are known. Using the Earth's radius and the given altitude, the orbital radius at belt entry is $r = r_\oplus + h = 6378\text{ km} + 640\text{ km} = 7018\text{ km}$. The true anomaly can then be extracted from the Orbit Equation to give: $\cos\theta = \left[\dfrac{r_p(1+e)}{r} - 1\right]\dfrac{1}{e} = \left[\dfrac{6500\text{ km}\,(1 + 0.748)}{7018\text{ km}} - 1\right]\dfrac{1}{0.748} = 0.619$. This gives for the true anomaly at radiation belt entry $\theta = 51.76°$. Noting that the cosine is a double-valued trigonometric function also gives as a solution $\theta = -51.76° = +308.24°$. Given the symmetry of this orbit and the radiation belt threshold, this location would be the point of belt *exit*, as sketched in Fig. 2c

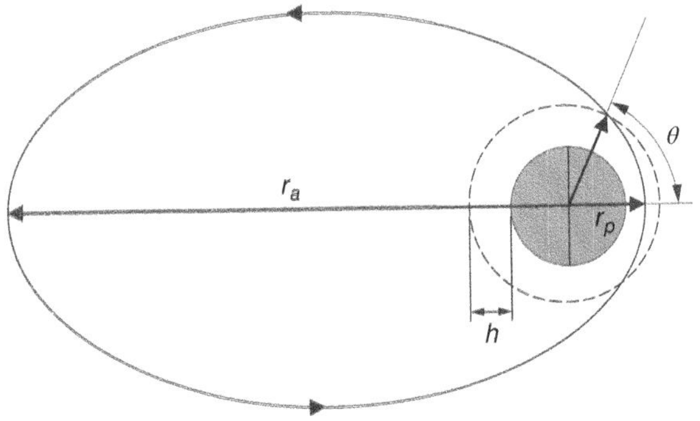

Figure 2c Orbit for Van Allen belt entry/exit (not to scale).

Note that this solution gives the *locations* of the key points in the orbit; it does not provide the spacecraft velocity or the relevant *times*, such as how long does the spacecraft take to fly from perigee to belt entry or how long the spacecraft remains within the radiation belts. Determining the velocities on a trajectory is discussed next; the times on an orbit are discussed in the next chapter (Chapter 3).

2.6.2 Parabolic Trajectories

Parabolic trajectories are also described by the Orbit Equation, as are all 2-body trajectories. In the case of a parabola $e = 1$ exactly, so the equation becomes:

$$r = \frac{p}{1 + \cos\theta} = \frac{2r_p}{1 + \cos\theta}. \tag{2.35}$$

The expression $p = a(1 - e^2)$ is not useful in this case, and simply implies that if the eccentricity is unity, the semi-major axis must be infinite to give the finite value $p = 2r_p$. In fact, parabolic trajectories are seldom encountered in engineering orbital mechanics – they are not useful as orbits and do not provide a sufficiently rapid departure from Earth for an interplanetary mission. Occasionally, a comet is observed to be on a parabolic trajectory relative to the Sun.

2.6.3 Relationships Between the Energy and Angular Momentum and the Eccentricity and Velocities of an Orbit

The energy can be expressed in terms of the conditions at periapsis, so Eq. 2.21 implies that:

$$E = \frac{V^2}{2} - \frac{\mu}{r} = \frac{V_p^2}{2} - \frac{\mu}{r_p}. \tag{2.36}$$

The scalar expression for the angular momentum, $h = rV\cos\gamma$, implies that at periapsis $h = r_pV_p$, so $V_p^2 = \frac{h^2}{r_p^2}$. Evaluating Eq. 2.30 at periapsis ($\theta = 0$) gives $h^2 = \mu p = \mu r_p(1 + e)$. Thus, the expression for the energy becomes:

$$E = \frac{\mu r_p(1 + e)}{2r_p^2} - \frac{\mu}{r_p} = \frac{\mu}{2r_p}[e - 1]. \tag{2.37}$$

Using $r_p = a(1 - e)$ simplifies this expression to:

$$\boxed{E = -\frac{\mu}{2a}.} \tag{2.38}$$

This expression is valid for all 2-body, conic-section orbits and trajectories and indicates that the overall size of an orbit is directly related to its energy. Using this definition, the energy is exactly zero for a parabolic trajectory ($a = \infty$). This can be thought of as the reference state against which the energies of other trajectories can be compared. For elliptical orbits, a is positive and finite, implying that the energy is *negative* but becomes more positive (closer to zero) as the size of the orbit is increased. The negative value simply implies that, for an elliptical orbit, the gravitational potential energy is greater than the kinetic energy.

The energy can be combined with the angular momentum to directly determine the eccentricity or shape of a trajectory or orbit. This can be seen as follows: nothing that $p = h^2/_{\mu} = a(1 + e^2)$ leads to this expression for the eccentricity:

$$e = \sqrt{1 - \frac{p}{a}} = \sqrt{1 - \frac{h^2}{\mu}\frac{1}{a}}. \tag{2.39}$$

Substituting $E = -\dfrac{\mu}{2a}$ then gives this very useful relation for the eccentricity:

$$\boxed{e = \sqrt{1 + \frac{2Eh^2}{\mu^2}}.} \tag{2.40}$$

For the case of a parabolic trajectory, $E = 0$, so the eccentricity $e = 1$, as expected. For elliptical orbits, $E < 0$ then gives $e < 1$. It can be seen from Eq. 2.39 that since $p \leq a$ on any orbit, $e \geq 0$. Thus on any closed orbit, elliptical or circular, $0 \leq e < 1$.

If the energy and angular momentum are known, for example if at a point the radius and velocity vectors $\boldsymbol{r}$ and $\boldsymbol{V}$, or the scalars r, V, and the flight path angle γ, are given, then both the semi-major axis (the size) and the eccentricity (the shape) can be determined.

The velocity of an object on a 2-body orbit can be simply determined from the conservation of energy. Taking $E = \dfrac{V^2}{2} - \dfrac{\mu}{r} = \dfrac{-\mu}{2a}$ and solving for the velocity gives:

$$V = \sqrt{\frac{2\mu}{r} - \frac{\mu}{a}}, \tag{2.41}$$

or, since for an elliptical orbit $a = (r_a + r_p)/_2$,

$$\boxed{V = \sqrt{2\mu\left(\frac{1}{r} - \frac{1}{r_a + r_p}\right)}.} \tag{2.42}$$

For a circular orbit, since $r_a = r_p = r$, the velocity is constant with value $V = \sqrt{\dfrac{\mu}{r}}$.

For the case of a parabola, $E = 0$ and the velocity is $V = \sqrt{\dfrac{2\mu}{r}}$ at all points on the trajectory. The velocity at periapsis is then given by $V_p = \sqrt{\dfrac{2\mu}{r_p}}$. This is termed the *escape velocity*, the minimum velocity at periapsis required for an object to depart from the primary body and not return. It can be seen that the escape velocity is exactly a factor $\sqrt{2}$ greater than the velocity of a circular orbit with the same radius.

The perpendicular and radial components of the velocity vector can be determined as follows. The velocity component perpendicular to the radius vector (see Fig. 2.6), $V_\perp = V\cos\gamma = h/r$, can be combined with $r = \frac{h^2/\mu}{1 + e\cos\theta}$ from Eqn. 2.30 to give:

$$V_\perp = \frac{\mu}{h}(1 + e\cos\theta). \tag{2.43}$$

The velocity component aligned with the radius vector, $V_r = V\sin\gamma$, can be determined from the scalar relation $V_r^2 + V_\perp^2 = V^2 = \frac{2\mu}{r} - \frac{\mu}{a}$ (from Eq. 2.41). Noting that $a = \frac{h^2/\mu}{1-e^2}$ and doing the algebra produces:

$$V_r = \frac{\mu}{h} e\sin\theta. \tag{2.44}$$

Example Problem 2d
Calculate the velocity of a satellite in a circular Earth orbit with an altitude of $h = 300$ km. How different is the velocity if the altitude is $h = 500$ km?

Solution: Since both orbits are circular, the (constant) velocities are given by $V = \sqrt{\frac{\mu}{r}}$. For the first orbit, $r = h + r_\oplus = 300\text{ km} + 6378\text{ km} = 6678\text{ km}$, so the orbital velocity is $V = \sqrt{\frac{398{,}600\frac{\text{km}^3}{\text{s}^2}}{6678\text{ km}}} = 7.73\text{ km/s}$. For the second case, $r = 6778$ km giving $V = \sqrt{\frac{398{,}600\frac{\text{km}^3}{\text{s}^2}}{6778\text{ km}}} = 7.67\text{ km/s}$. This value is less than 1% smaller than the velocity in the 300-km orbit. Both of these are Low Earth Orbits (commonly termed LEO). Most objects in LEO are designed to orbit higher than the dense Earth's atmosphere, but lower than the limit of the Van Allen radiation belts.

Example Problem 2e
Determine the spacecraft velocities in Example Problem 2c, for these points in the orbit: (i) at perigee, (ii) at apogee, and (iii) at the point of Van Allen belt entry. What are the corresponding flight path angles?

Solution: All velocities are given by Eq. 2.42. At perigee, $r = r_p$, so the expression simplifies to

$$V_p = \sqrt{2\mu\left(\frac{1}{r} - \frac{1}{r_a + r_p}\right)} = \sqrt{\frac{2\mu r_a}{r_p(r_a + r_p)}} = \sqrt{\frac{2\left(398{,}600\frac{\text{km}^3}{s^2}\right)(45{,}000\text{ km})}{6500\text{ km}\,(45{,}000 + 6500\text{ km})}} = 10.35\text{ km/s}.$$

Similarly at apogee: $V_a = \sqrt{\frac{2\mu r_p}{r_a(r_a + r_p)}} = \sqrt{\frac{2\left(398{,}600\frac{\text{km}^3}{\text{s}^2}\right)(6500\text{ km})}{45{,}000\text{ km}\,(45{,}000 + 6500\text{ km})}} = 1.50\frac{\text{km}}{\text{s}}$.

Alternatively, using the conservation of angular momentum as before $V_a = \frac{r_p}{r_a}V_p = \frac{6500\text{ km}}{45{,}000\text{ km}}10.35\text{ km/s} = 1.50\text{ km/s}$ gives the same result. The radius at the point of entry to the radiation belts can be determined from the Orbit Equation in the form $r = \frac{r_p(1+e)}{1 + e\cos\theta}$, with $e = 0.748$ and $\cos\theta = 0.619$ from Example Problem 2b. This gives for the radius at that point

$$r = \frac{r_p\,(1+e)}{1+e\,\cos\theta} = \frac{6500\,\text{km}\,(1+0.748)}{1+0.748(0.619)} = 7766.12\,\text{km}.$$

The velocity then again follows from Eq. 2.42:

$$V = \sqrt{2\mu\left(\frac{1}{r} - \frac{1}{r_a + r_p}\right)} = \sqrt{2\mu\left(\frac{1}{7766.12\,km} - \frac{1}{45{,}000 + 6500\,\text{km}}\right)} = 9.34\,\text{km/s}.$$

The flight path angles at both perigee and apogee are exactly zero. The conservation of angular momentum can be applied so

$$h = r_p V_p = rV\cos\gamma = constant,$$

and the flight path angle at the point of belt entry is $\gamma = \cos^{-1}\left(\frac{r_p V_p}{rV}\right) = \cos^{-1}\left[\frac{(6500\,\text{km})\left(10.35\,\frac{\text{km}}{\text{s}}\right)}{(7766.12\,\text{km})(9.34\,\text{km/s})}\right] = +21.96°$. The positive value of the flight path angle is consistent with the fact that the spacecraft at that point is "climbing," as can be seen in Fig. 2c. Given the symmetry of the orbit, the other solution to the inverse cosine, $\gamma = -21.96°$, corresponds to the point where the spacecraft exits the radiation belts and is "falling" in its orbit.

2.6.4 Hyperbolic Trajectories

Along with elliptical orbits, hyperbolic trajectories are of major importance in orbital mechanics. Spacecraft departing Earth on an interplanetary mission, as well as spacecraft approaching planets and during planetary flybys are all on hyperbolic trajectories. The features of a hyperbolic trajectory are illustrated in Fig. 2.11.

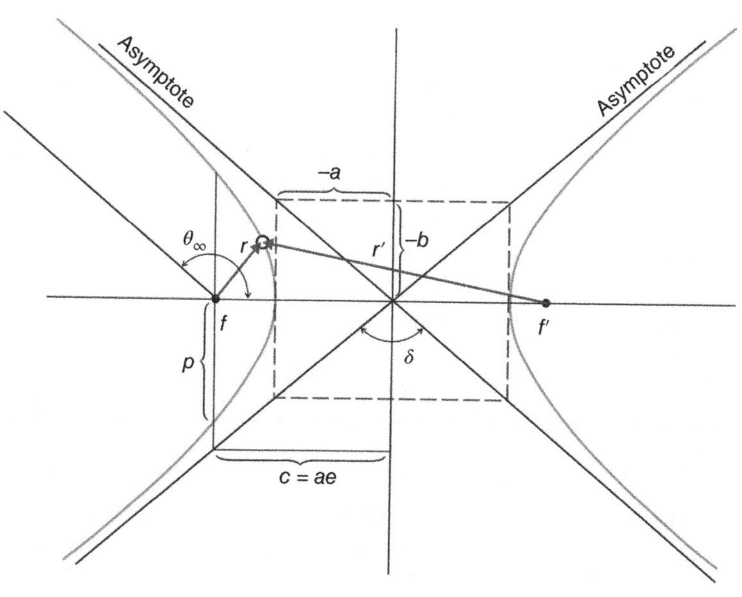

Figure 2.11 Characteristics of a hyperbolic trajectory.

The hyperbola is a class of conic sections as illustrated in Fig. 2.7. Unlike the other conic sections (circle, ellipse, parabola), the hyperbola consists of two distinct branches, each with its own focus. The object of interest moves about the primary body only around the primary focus shown at the left of the figure. The "empty" focus is outside of that trajectory, on the opposite side of the point of periapsis from the primary focus, corresponding to the second, or "imaginary," trajectory.

An important feature of hyperbolic trajectories are the *asymptotes* indicated by the straight lines in Fig. 2.11. The hyperbolic trajectory approaches the asymptotes as the distance from the primary body becomes very large, as shown. Another important characteristic of hyperbolic trajectories is the *turning angle*, δ, which is the angle through which the object is turned during, for example, a flyby maneuver around a planet.

Similar to the ellipse, the hyperbola has a semi-minor axis, a, here defined as half of the distance between the two periapses (those of the primary and imaginary trajectories). The distance from the center of the system to either focus is again c, as for the case of the ellipse. For the hyperbola, this distance exceeds the length of the semi-minor axis, thus the relation $c = ea$ indicates that for the hyperbola $e > 1$. As for the ellipse, the semi-minor axis, b, for the hyperbola follows the relation $a^2 + b^2 = c^2$. For the case of the hyperbola, the parameters a, b, and c are all actually *negative* values.

Mathematically, ellipses, circles, and hyperbolae are described using Cartesian coordinates by these formulas:

$$\begin{aligned}
&\text{Ellipse} && \frac{x^2}{a^2} + \frac{y^2}{b^2} = 1 \\
&\text{Circle} && \frac{x^2}{r^2} + \frac{y^2}{r^2} = 1 \\
&\text{Hyperbola} && \frac{x^2}{a^2} - \frac{y^2}{b^2} = 1.
\end{aligned} \tag{2.45}$$

Recall that for the ellipse, the distances from the orbiting object to both the primary and empty foci can be summed to give $r + r' = 2a$. For the hyperbola, the corresponding relation is $r - r' = 2a$, where a is a negative quantity. It is thus not possible to construct a hyperbola using the "string method" previously described for an ellipse.

The development of a hyperbola might also be imagined using the concept presented in Fig. 2.12. In the first rendering, Case (i), the primary and empty foci are taken to be coincident, corresponding to a circular orbit. In the second Case (ii), the empty focus is displaced a finite distance to the left, resulting in an elliptical orbit. Moving the empty focus further left to infinity then opens up the shape to give the parabolic trajectory shown in Case (iii). Finally, imagining that the empty focus "reappears" on the right side of the figure and then moves to the left results in the creation of an "imaginary" branch and a pair of hyperbolae as shown in Case (iv).

These geometric differences can be summarized in the table below, recalling that the energy is directly related to the semi-major axis, a (which is negative for a hyperbola):

Trajectory	a	$2c$	E
Circle	Finite	Zero	Negative
Ellipse	Finite	Finite	Negative
Parabola	Infinite	Infinite	Zero
Hyperbola	Finite, negative	Finite, negative	Positive

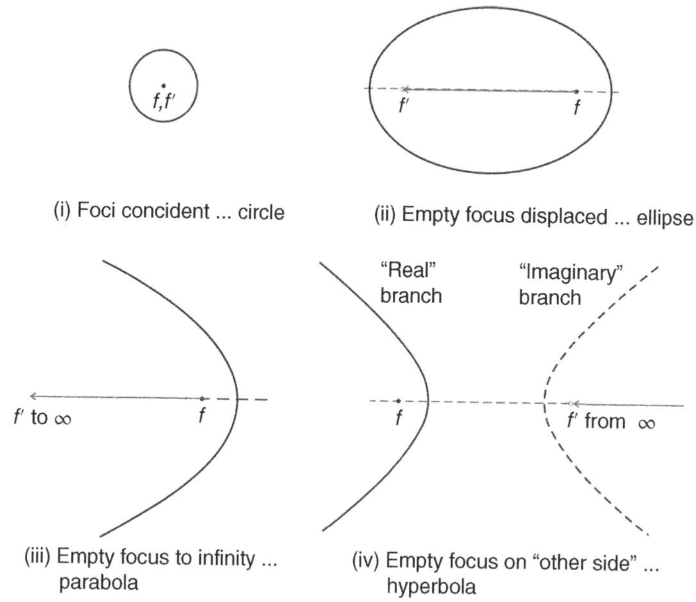

Figure 2.12 Changing a trajectory by moving the empty focus.

Given that the hyperbola, like the ellipse and parabola, is also a conic section, it can be described by the same Orbit Equation (Eq. 2.30). For a hyperbola, this is commonly written as:

$$r = \frac{a(e^2 - 1)}{1 + e\cos\theta}. \tag{2.46}$$

Note that here the semi-major axis, a, is here taken to be *positive*, so the expression is valid for a hyperbola, for which $e > 1$. If Eq. 2.33 is used instead, the semi-major axis must be taken to be *negative* in value. Both approaches ($a > 0$ and $a < 0$) are used in practice. The true anomaly for a hyperbolic trajectory can be determined in a similar fashion as in Example Problem 2c:

$$\cos\theta = \left[\frac{a(e^2 - 1)}{r} - 1\right]\frac{1}{e}. \tag{2.47}$$

This indicates that for very large distances, $r \to \infty$, $\cos\theta \to -\frac{1}{e}$. This gives the limiting value of the true anomaly:

$$\theta_\infty = \cos^{-1}\left(-\frac{1}{e}\right). \tag{2.48}$$

Since the eccentricity $e > 1$ on a hyperbola, Eq. 2.48 implies that the final, asymptotic value of the true anomaly will be less than 180°; only for the case of a parabolic trajectory can a true anomaly of 180° be attained far from the primary body. The periapsis radius for a hyperbola can be written as $r_p = c - a = a(e - 1)$, where a and c are written as being positive, which allows writing the Orbit Equation as $r = \frac{r_p\,(1 + e)}{1 + e\,\cos\theta}$, just as for an ellipse.

A very important consideration in hyperbolic trajectories is the turning angle, δ. This is the angular change in the direction of the trajectory during a flyby maneuver as sketched in Fig. 2.13. Since

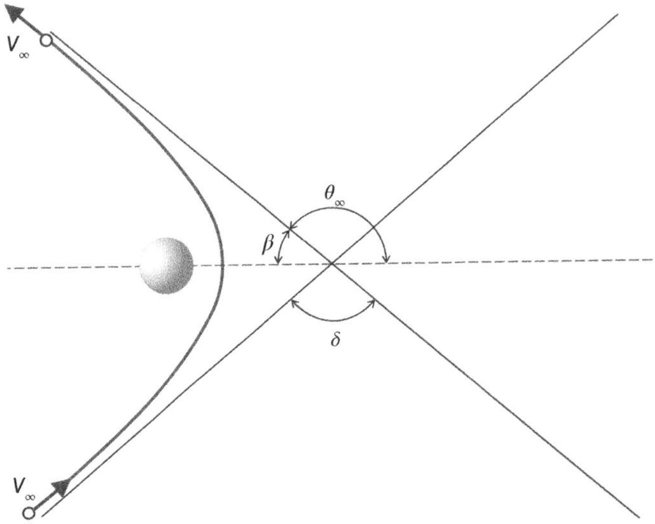

Figure 2.13 Turning on a hyperbolic trajectory.

the energy is conserved during this encounter, the incoming velocity far from the primary object, V_∞, will be the same as the final departure velocity. This velocity is termed the *hyperbolic excess velocity*.

Comparing the angles above and below the axis shown, the symmetry of the geometry can be employed to write for the asymptote angle, β:

$$\beta = \pi - \theta_\infty = \frac{\pi}{2} - \frac{\delta}{2} \tag{2.49}$$

Taking the cosine of both sides then allows the use of the limiting true anomaly (Eq. 2.48) to give:

$$\cos\left(\frac{\pi}{2} - \frac{\delta}{2}\right) = \cos(\pi - \theta_\infty) = -\cos\theta_\infty = \frac{1}{e}. \tag{2.50}$$

Using the trigonometric identity $\cos\left(\frac{\pi}{2} - \frac{\delta}{2}\right) = \sin\frac{\delta}{2}$ then gives the relation for the turning angle as:

$$\delta = 2\sin^{-1}\left(\frac{1}{e}\right). \tag{2.51}$$

This indicates that the higher the value of the eccentricity, the less turning will occur during a hyperbolic trajectory encounter.

Example Problem 2f

What value of eccentricity of a hyperbolic trajectory will be required to give a turning of 90°?

Solution: Taking $\delta = 90° = \frac{\pi}{2} = 2\sin^{-1}\left(\frac{1}{e}\right)$ gives $= \frac{1}{\sin\left(\pi/4\right)} = \sqrt{2}.$

Alternatively, it can be noted that for 90° turning, $a = b$ so $c^2 = a^2 + b^2 = 2a^2 = (ae)^2$ since $c = ae$, so again $e = \sqrt{2}$.

The velocity on a hyperbolic trajectory is the same as that given by Eq. 2.41, which can be written as $V\sqrt{\frac{2\mu}{r}+\frac{\mu}{a}}$, taking a to be positive. The corresponding equation for energy becomes $E=\frac{V^2}{2}-\frac{\mu}{r}=\frac{+\mu}{2a}$. Far from the primary body ($r\rightarrow\infty$), this implies that $V_\infty^2=\mu/a$. Thus, the velocity on a hyperbolic trajectory can be expressed as:

$$V=\sqrt{\frac{2\mu}{r}+V_\infty^2}\,. \tag{2.52}$$

It might be noted that the first term in the radical, $\frac{2\mu}{r}$, is simply the square of the escape velocity for that particular trajectory.

Example Problem 2g

A spacecraft departs low Earth orbit (h = 300 km) on an interplanetary mission that requires a hyperbolic excess velocity of V_∞ = 3.0 km/s. What is the required velocity at injection (the perigee of the departure hyperbola), V_0? How does this compare with the escape velocity for that altitude?

Solution: Using Eq. 2.52 and with the radius $r=h+r_\oplus$ = 300 km + 6378 km = 6678 km gives:

$$V_0=\sqrt{\frac{2\left(398{,}600\frac{\text{km}^3}{\text{s}^2}\right)}{6678\text{ km}}+\left(3.0\frac{\text{km}}{\text{s}}\right)^2}=\sqrt{119.38\frac{\text{km}^2}{\text{s}^2}+9.0\frac{\text{km}^2}{\text{s}^2}}=11.33\text{ km/s}.$$

The escape velocity is simply $V_{esc}=\sqrt{\frac{2\mu}{r}}=\sqrt{119.38\frac{\text{km}^2}{\text{s}^2}}$ = 10.93 km/s. Notice that a comparatively small increase in the injection velocity over the escape velocity, V_0-V_{esc} = 0.40 km/s results in the substantial hyperbolic excess velocity of 3.0 km/s.

Example Problem 2h

The *Viking 2* Mars lander mission was launched in August 1975. The published elements of the departure trajectory from low Earth orbit were a = (−)18,850 km and e = 1.348. What were the: (i) altitude of the initial Earth orbit, (ii) hyperbolic excess velocity, and (iii) the true anomaly far from the Earth? Sketch the departure trajectory, showing the direction of the Earth's orbit and the asymptote angle, β. The departure was in the same direction as the Earth's orbit.

Solution: (i) The periapsis radius of the departure hyperbola is that of the initial Earth orbit, with

$$r_p=a(e-1)=18{,}850\text{ km }(0.348)=6563.5\text{ km}.$$

The altitude of the parking orbit was then $h=r_p-r_\oplus$ = 6563.5 km − 6378 km = 185.5 km. (ii) The hyperbolic excess velocity can be determined from:

$$V_\infty^2=\mu/a\rightarrow V_\infty=\sqrt{\frac{\mu}{a}}=\sqrt{\frac{398{,}600\frac{\text{km}^3}{\text{s}^2}}{18{,}850\text{ km}}}=4.60\text{ km/s}.$$

(iii) The limiting value of true anomaly ($r\rightarrow\infty$) is then from Eq. 2.48:

$$\theta_\infty=\cos^{-1}\left(-\frac{1}{e}\right)=\cos^{-1}\left(-\frac{1}{1.348}\right)=137.9°.$$

The geometry of the trajectory is sketched in Fig. 2f, with the directions of the Earth's orbit and the Sun as shown, along with the asymptote angle $\beta=\pi-\theta_\infty$ = 42.1°. The hyperbolic excess

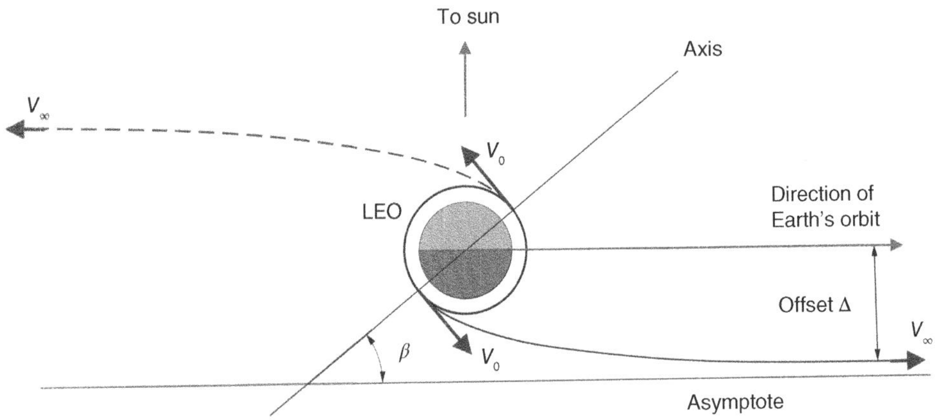

Figure 2f Departure hyperbolae to a superior planet (solid line) and an inferior planet (dashed line). The injection and hyperbolic excess velocities are here the same in both cases.

velocity is added to the orbital velocity of Earth, leading to a trajectory that "climbs" to take the spacecraft on a course to an outer planet (Mars in this case). An interesting extension is suppose the excess velocity is applied in the direction *opposite* to the Earth's motion, as shown in the figure by the dashed line. This results in a velocity, relative to the Sun, less than that of the Earth's orbit and will cause the spacecraft to "fall" toward an inner planet (say Venus). The departure velocity, relative to the Earth, is however the same in both cases considered here – only the direction is different. Both maneuvers also require the exact same injection velocity.

The quantity Δ shown in the figure is the distance between the departure trajectory and the Earth's orbit, termed the "offset." This can be determined noting that the angular momentum for $r \to \infty$ is $h = V_\infty \Delta$, so:

$$\frac{h^2}{\mu} = p = \frac{V_\infty{}^2 \Delta^2}{\mu} = a\left(e^2 - 1\right), \tag{2.53}$$

where a is again taken to be positive. The energy definition gives for $r \to \infty$:

$$E = \frac{V^2}{2} - \frac{\mu}{r} = \frac{V_\infty{}^2}{2} = \frac{+\mu}{2a}. \tag{2.54}$$

Combining Eqs. 2.53 and 2.54 allows solving for the offset distance:

$$\Delta = a\sqrt{e^2 - 1}. \tag{2.55}$$

For the Viking 2 mission, the offset distance was equal to $\Delta = (18{,}850\,\text{km})\sqrt{(1.348)^2 - 1} =$ 17,040 km. This distance is negligible compared to the mean radius of the Earth's orbit, $R_\oplus = 150 \times 10^6$ km, a fact that will prove useful in defining interplanetary trajectories in Chapter 6.

2.6.5 Degenerate Cases

It was stated previously that *all* 2-body trajectories are conic sections, using the two-cone system as shown in Fig. 2.7. Consider an extreme case, where the cone system is intersected only along an edge, with the resulting "shape" being a straight line. Are such "degenerate" cases also conic

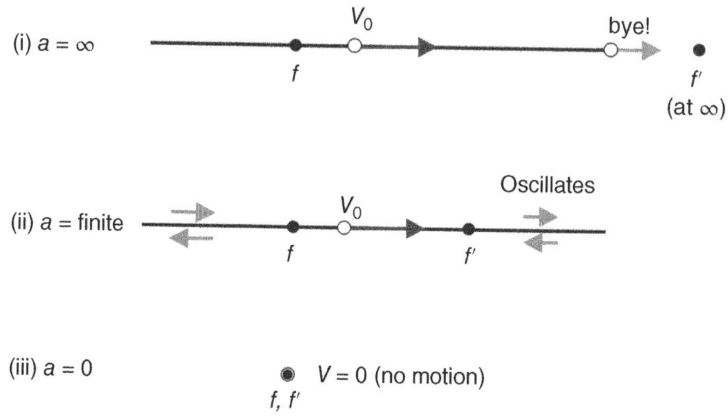

Figure 2.14 "Degenerate" trajectories.

sections? Three such degenerate trajectories are shown in Fig. 2.14, along with the primary and empty foci f and f', respectively. In the first example, (i), a spacecraft departs the vicinity of f with sufficient velocity to escape the system. In this case $a = \infty$.

That the angular momentum is zero in this case, $p = h^2/\mu = a(1-e^2) = 0$, implies that $e = 1$ and the line can be considered a degenerate parabola. In the second case, (ii), the initial velocity, V, is less than that in case i), so the spacecraft will reach a finite apoapsis distance, then return, leading to a linear oscillation about the primary focus (recall that in the 2-body model all mass is concentrated at a point, which can be taken to be of negligible size). In this case, the periapsis radius is $r_p = a(1 - e) = 0$, again implying that $e = 1$ since a is finite. The apoapsis is then $r_a = a(1 + e) = 2a$. This trajectory might also be considered a degenerate ellipse, since a is finite. In the last case, (iii), there is no motion and the spacecraft is fixed in position at the focus f. In this case $r_p = r_a = a = 0$, so e is arbitrary and the point could be considered a degenerate circle, ellipse, parabola, or hyperbola – all conic sections (if completely impractical).

2.7 Problems

2.1 Suppose two stars of identical mass m separated by a distance d. In this case, the stars have a constant speed and the system rotates around the center of mass (situated halfway between the stars) with a constant angular velocity, ω. Taking the origin of the system to be at the center of mass, use Newton's second "law" to find the expression for ω.

2.2 A satellite with mass m is in a circular orbit with radius r around the Earth. Using Newtonian gravitation (Eq. 2.1) and dynamics *only* verify that the orbital velocity is consistent with Eq. 2.42 for a circular orbit. The gravitational constant is G. Hint: What is the radial acceleration of a body moving in uniform circular motion with velocity V and radius r?

2.3 The orbital radius of a planet around a star can be used to determine the mass of the star. Take as an example the earth, which takes 365.25 days to orbit the Sun with an orbital radius of

149.6×10^6 km, and use this information to calculate the mass of the Sun (kg). Begin your analysis by using the relation for the orbital velocity in a *circular* orbit to get the period, T.

2.4 If the radius of the earth is $r = 1$ DU "distance units," what is then the circular orbital velocity of a low Earth orbit in the units of DU, TU (TU = "time unit")? Using these units, the gravitational parameter μ is simply equal to 1 DU^3/TU^2.

2.5 The position and velocity vectors of a satellite at a given instant are given by:

$$r = 2\hat{i} - 2\hat{j} + 4\hat{k} \text{ (in DU)}$$
$$v = -0.4\hat{i} + 1.0\hat{j} - 0.5\hat{k} \text{ (in DU per TU)}$$

where the unit vectors $\hat{i}, \hat{j}, \hat{k}$ represent a nonrotating, geocentric coordinate system. Find the angular momentum and total energy of the satellite. The *Distance* and *Time* units are defined in Problem 2.4.

2.6 An early warning system detects an unidentified object with the following parameters: Altitude = 0.5 DU, Speed = $\sqrt{2/3}$DU/TU, and Flight-path angle = 30°.
Based on these data, is the object a space probe designed to escape the earth, an earth satellite, or a ballistic missile? The DU and TU are defined in Problem 4. Humorous side-note: Be "quick" with this one – if it's a ballistic missile, that conclusion will be desperately needed in less than half an hour!

2.7 Derive the expression for the eccentricity of an orbit/trajectory, e, in terms of a given initial speed, V_0, radius, r_0, and flight path angle, γ_0 (these all at the same instant). Evaluate (i.e. do algebra) and discuss these special cases: (i) $r_0 V_0^2/\mu = 2$, and (ii) $r_0 V_0^2/\mu = 1$.

2.8 Make a plot of elliptical, parabolic, and hyperbolic trajectories starting with $e = 0$ and increasing by 0.2 up to a final value of $e = 3.0$. The result should look similar to Fig. 2.9.

2.9 For the case of a parabolic trajectory, what is the relationship between the flight path angle, γ, and the true anomaly, θ? Hint: Use a standard trigonometric identity or two – your solution need not require complicated algebra!

2.10 A spacecraft with mass m deploys a large "solar sail" in its orbit around the Sun. The sail faces the Sun at all times, and the radiation pressure on the sail produces an acceleration, $\boldsymbol{a}_s = \frac{2SA}{mc}\left(\frac{\boldsymbol{r}}{r^3}\right)$, where S is the solar constant (solar radiation per unit solid angle, W/steradian), A is the sail area, r is the distance from the Sun, and c is the speed of light. Of course, gravity acts on the spacecraft as well.

(a) Show that the equation of motion for this spacecraft is $\ddot{\boldsymbol{r}} = -\left(\mu - \frac{2SA}{mc}\right)\left(\frac{\boldsymbol{r}}{r^3}\right)$.

(b) What types of conic sections are possible when μ is greater than $2SA/mc$? When they are equal? When μ is less than $2SA/mc$?

(c) In one of the cases in part (b), the vehicle will *always* escape the solar system. Which case is that? Show that the spacecraft will arrive (eventually) at $r = \infty$ with residual speed $v_\infty = \left(v_0^2 - 2\left(\mu - \frac{2SA}{mc}\right)\frac{1}{r_0}\right)^{1/2}$, where r_0 and v_0 are the initial distance and speed when the sail is deployed.

2.11 A satellite in a *geostationary* orbit remains precisely over the same spot on the earth's equator, with an orbital period that matches the period of the earth's rotation.[9]
(a) What will be the velocity (m/s) and orbital radius (km) of a satellite in such an orbit?
(b) Suppose that the satellite is first launched into a circular, *low earth orbit* with an altitude of 250 km. What will be the required increase in velocity (m/s) to boost the satellite up to the altitude of the geostationary orbit? Hint: The resulting "transfer orbit" should be an elliptical orbit, with the apogee exactly equal to the geostationary orbit altitude.
(c) To remain in geostationary orbit, it will be necessary to give the satellite arriving on the orbit in part (b) an additional "boost." What increase in velocity (m/s) will be required to properly insert the satellite into the geostationary orbit?

2.12 Find the minimum additional speed needed to escape the earth from a geostationary orbit (GEO), in km/s. Compare this with the increase in speed needed to escape from low earth orbit (LEO). In practice, interplanetary missions are initiated from LEO, not GEO – give reason(s) for this.

2.13 What departure velocity relative to Earth (km/s) will be required to escape the solar system? What will be the needed injection velocity from a low Earth orbit (LEO)? Consider also a spacecraft being launched from Mars – what will be the departure velocity and injection velocity in that case (assuming a "low Mars orbit")? State clearly any approximations and assumptions made in your analysis.

2.14 Compute and plot the velocity versus radial distance from Earth of a spacecraft with hyperbolic excess velocities of $V_\infty = 2, 4, 6, 8$, and 10 km/s. Assume the departure is from LEO and consider also the case of a parabolic departure. Plot the velocities out to a distance from Earth of 1.5×10^6 km. At what approximate distance from the Earth does the velocity become essentially constant? This point is the *sphere of influence*, the distance beyond which the Earth no longer has any significant impact on the spacecraft's velocity.

9 Detail: Is the period of the earth's rotation exactly 24 hours, relative to the stars?

3

Elapsed Time and Position on Orbits and Trajectories

The fundamental geometric descriptions of orbits and trajectories were developed in the previous chapter for the case of 2-body motion. These descriptions relate consider key parameters such as semi-major axis, parameter, and eccentricity, and can be extended to give the velocity of an object at any point. However, those formulations do not directly provide information on the *times* involved, such as the position of an object at a specific time, or how much time is required to traverse from one point on a trajectory to another. Determining the corresponding times on orbits and trajectories is the focus of this chapter. The chapter outline is as follows:

- The simplest case – time on circular orbits
- Kepler's second law and the periods of elliptical orbits
- Development of the time equation for elliptical orbits
  - Geometric method
  - Mathematical method
- Orbits with single/multiple passage(s) through periapsis
- The inverse time problem for elliptical orbits and numerical solution approach
- Parabolic time equation (mathematical derivation)
- The time equation for hyperbolic orbits
  - Geometric method
  - Mathematical method

3.1 Circular Orbits

Determining the time versus position on a circular orbit is straightforward, given that both the orbital radius and the orbital speed are constant (Fig. 3.1). From Eq. 2.42, the (constant) circular orbital speed is simply $V_c = \sqrt{\frac{\mu}{r}}$. Relating this to the circumference of the orbit then gives the period directly:

$$T = \frac{2\pi r}{\sqrt{\mu/r}} = 2\pi\sqrt{\frac{r^3}{\mu}}. \tag{3.1}$$

Introduction to Orbital Mechanics: A Concise and Practical Approach, First Edition. James C. Hermanson.

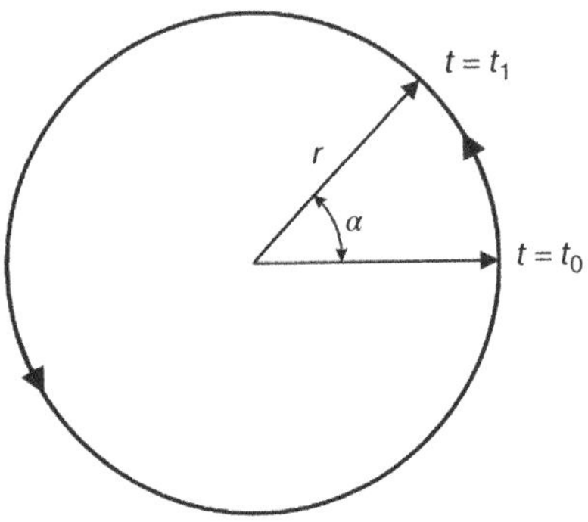

Figure 3.1 A circular orbit.

The angular distance around the orbit from any reference point with time $t = t_0$ (there is no periapsis in a circular orbit, so the true anomaly is undefined) is then simply given by

$$\alpha = \frac{2\pi(t - t_0)}{T}, \tag{3.2}$$

where $t - t_0$ is the time elapsed from the reference point at time t_0.

3.2 Elliptical Orbits

Determining the time-varying position on an elliptical orbit is more complicated than for the circular orbit, given that both the velocity and radius on an elliptical orbit are not constant with time or angular position. This problem can be addressed in either a geometric or mathematical fashion, and both are presented in the following sections.

3.2.1 Geometric Method - Elliptical Orbits

The geometric method considers the area "swept" by an object in an elliptical orbit, as shown, for example, in Fig. 3.2.

For a differential time increment dt, an object sweeps a differential area dA, as shown in Fig. 3.3. As dt becomes differentially small, the area is simply that of the triangle $A = V_\perp \frac{r}{2} dt$. The quantity $rV_\perp$ is the (constant) angular momentum, h, so the rate of swept area change is:

$$\frac{dA}{dt} = \frac{1}{2}h = \text{constant}. \tag{3.3}$$

This is the essence of Kepler's second law, that equal areas are swept in equal times for objects in an elliptical orbit, as shown in Fig. 3.2. This principle can then be used to determine the period of an

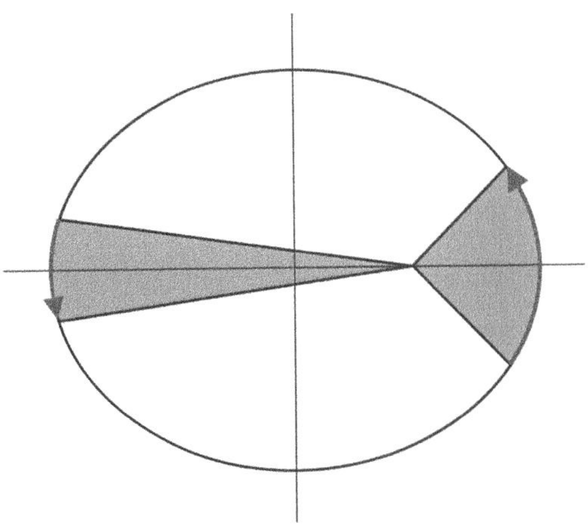

Figure 3.2 Swept areas in an elliptical orbit.

elliptical orbit. Noting that the area of an entire ellipse is $A_e = \pi ab$, this can be related to the rate of area change to give the orbital period

$$T = \frac{A_e}{\frac{dA}{dt}} = \frac{\pi ab}{\frac{1}{2}h}. \tag{3.4}$$

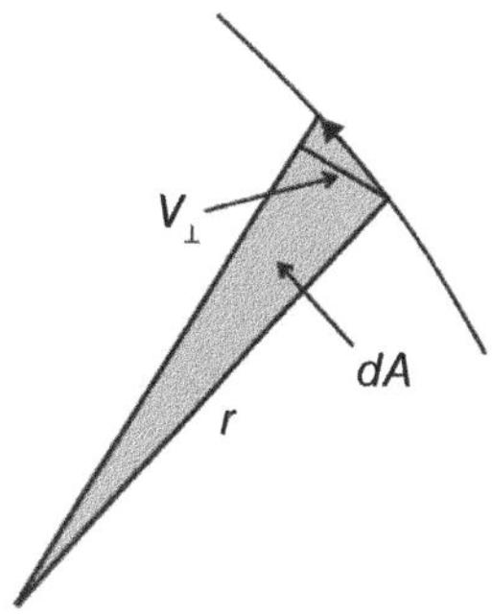

Figure 3.3 Differential swept area.

From the Orbit Equation (Eq. 2.33) $\frac{h^2}{\mu} = p = a(1-e^2)$, so $h = \sqrt{\mu a(1-e^2)}$. Since $a^2 + b^2 = c^2$ and $c = ae$, $b = a\sqrt{1-e^2}$. Inserting a, b, and h into Eq. 2.33 then gives the expression for the period of an elliptical orbit:

$$T = 2\pi\sqrt{\frac{a^3}{\mu}} \tag{3.5}$$

This is a formal demonstration of Kepler's third law, that the period of an orbit squared is proportional to the semi-major axis squared, or $T^2 \sim a^3$. It also follows from Eq. 3.5 that the time from periapsis to apoapsis (or the reverse) is simply $T/2$ given the symmetry of the elliptical orbit about its major axis.

That the rate at which area is swept out in an elliptical orbit is constant allows for the determination of the time at any point by comparing the area swept to the total area of the ellipse. Consider the area A_1 swept in going from periapsis to some arbitrary point S as shown in Fig. 3.4. Designating the elapsed time to reach point S as $t - t_0$, where t_0 would be the time at periapsis (which could be set to zero for convenience), then gives:

$$\frac{t-t_0}{T} = \frac{A_1}{A_e} \text{ or } t - t_0 = \frac{A_1}{A_e}T. \tag{3.6}$$

The challenge is determining the swept area A_1. This was accomplished by Kepler by the ingenious use of an *auxiliary circle* with radius a that encloses the elliptical orbit, as shown in Fig. 3.4.

The needed area A_1 can then be determined by comparing the various areas that can be inferred from Fig. 3.4. The points F and O correspond to the focus and center of the orbit, respectively; the periapsis location is point P. The projections of point S (the location of the orbiting object) onto the auxiliary circle and the ellipse are indicated respectfully as points Y^+ and Y^-. The angle between the

Figure 3.4 Definition of the auxiliary circle. The area A_1 is shaded.

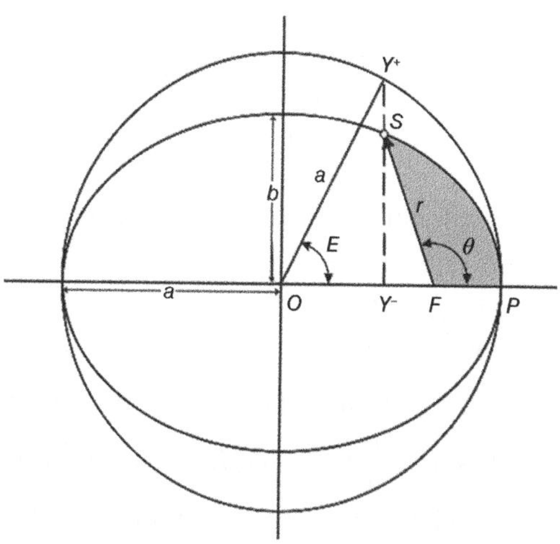

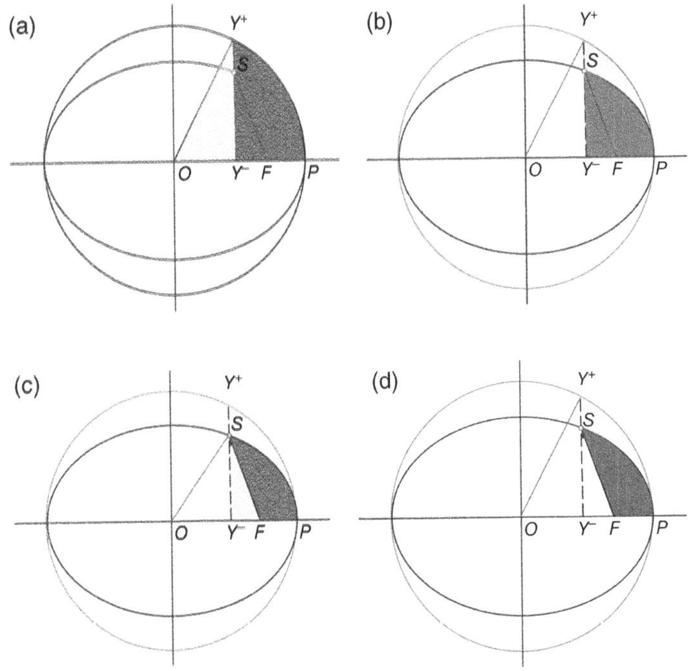

Figure 3.5 Geometric argument progression. Steps (a) through (d) are explained in the following text.

radius from the center to the auxiliary circle is termed the *eccentric anomaly*, E, which is distinct from the *true anomaly*, θ. The progression of the geometric argument is shown in Fig. 3.5.

The first step, shown in Fig. 3.5a, is to recognize that the region bounded by the points $OY^{+}P$ is simply the area of a circular sector, or $A_C = a^2E$, where the eccentric anomaly E is in radians and a, the radius of the auxiliary circle, is is equal to the semi-major axis of the orbit. Next, subtracting the area of the triangle bounded by $OY^{+}Y$ from A_C gives a new area $A_2 = a^2E - \frac{1}{2}(a\sin E)(a\cos E) = a^2\left(E - \frac{1}{2}\sin E\cos E\right)$, which is then the area $Y^{-}Y^{+}P$ as shown in Fig. 3.5b. This area can then be related to the corresponding area directly under the ellipse using the functional definitions of an ellipse and a circle:

$$\text{On an ellipse (orbit)}\ \frac{x^2}{a^2} + \frac{y^2}{b^2} = 1 \Rightarrow y_{ellipse} = \frac{b}{a}\left(a^2 - x^2\right)^{1/2} \tag{3.7}$$

$$\text{On a circle (auxiliary circle)}\ \frac{x^2}{a^2} + \frac{y^2}{a^2} = 1 \Rightarrow y_{circle} = \left(a^2 - x^2\right)^{1/2} \tag{3.8}$$

Thus, the ratio of all vertical distances from the major axis to the ellipse to those of the auxiliary circle is:

$$\frac{y_{ellipse}}{y_{circle}} = \frac{b}{a}. \tag{3.9}$$

This expression holds for all values of x, including when integrating, so the area under the ellipse ($Y^{-}SP$) is $A_3 = \frac{b}{a}A_2 = \frac{1}{2}ab\left(E - \frac{1}{2}\sin E\cos E\right)$, as shown in Fig. 3.5c. Lastly, the smaller triangle $Y^{-}SF$ is subtracted from A_3 to give the needed area A_1. Noting that the distance between points O

and F is $c = ae$ for the ellipse, the base of triangle ΥSF is given by $ae{-}a\cos E = a(e{-}\cos E)$. Using again Eq. 3.9, the height of that triangle is then $r\sin\theta = \frac{b}{a}(a\sin E) = b\sin E$ so the area of triangle ΥSF is $1/2\ ab\sin E(e{-}\cos E)$. Subtracting this from the area A_3 then gives the desired area, A_1 as:

$$A_1 = A_3 - \frac{1}{2}ab\sin E(e - \cos E) = \frac{1}{2}ab\left(E - \frac{1}{2}\sin E\cos E\right) - \frac{1}{2}ab\sin E(e - \cos E),$$

which simplifies to

$$A_1 = \frac{1}{2}ab(E - e\sin E). \tag{3.10}$$

Combining this result with Eq. 3.6 then gives the expression for the elapsed time from periapsis to any point on an elliptical orbit as:

$$t - t_0 = \frac{A_1}{A_e}T = \frac{\frac{1}{2}ab(E - e\sin E)}{\pi ab}\left[2\pi\left(\frac{a^3}{\mu}\right)^{\frac{1}{2}}\right],$$

which simplifies to

$$t - t_0 = \left(\frac{a^3}{\mu}\right)^{\frac{1}{2}}(E - e\sin E). \tag{3.11}$$

Noting that the period of the orbit is $T = 2\pi\left(\frac{a^3}{\mu}\right)^{\frac{1}{2}}$ points to the quantity $\left(\frac{a^3}{\mu}\right)^{\frac{1}{2}}$ as being the mean time (averaged over the entire orbit) for the object to sweep 1 radian of arc on the auxiliary circle. Thus, the *mean angular rate* can be defined as follows:

$$n = \frac{2\pi}{T} = \left(\frac{\mu}{a^3}\right)^{\frac{1}{2}}. \tag{3.12}$$

The actual angular rate varies over an elliptical orbit, with the mean angular rate only being identical to the actual angular rate for the case of a circular orbit. Introducing the *mean anomaly* $M = E - e\sin E$ then allows writing simply:

$$\boxed{M = n(t - t_0).} \tag{3.13}$$

This is *Kepler's time equation* that directly gives the mean anomaly, and hence the eccentric anomaly, at any point in time for an elliptical orbit with eccentricity e. What is needed however is how the *true anomaly* varies with time to give the actual position on the orbit. The relationship between the true anomaly and the eccentric anomaly can be determined using the distance along the x-axis between point O and point Υ. Noting again that the distance between points O and F is $c = ae$ gives the $O{-}\Upsilon$ distance

$$a\cos E = ae + r\cos\theta \text{ (noting that } \cos\theta < 0 \text{ for } \theta > 90°)$$

Inserting the $r = \frac{a(1-e^2)}{1 + e\cos\theta}$ variation of the Orbit Equation (Eq. 2.33) and simplifying then gives:

$$\cos E = \frac{e + \cos\theta}{1 + e\cos\theta}, \tag{3.14}$$

or, rearranging:

$$\cos\theta = \frac{e - \cos E}{e\cos E - 1}. \tag{3.15}$$

Knowing the eccentric anomaly from Kepler's time equation (Eq. 3.13) thus allows determination of the true anomaly, which combined with the orbit equation, specifies the object's position on the orbit. There is however the potential for quadrant ambiguity, given that $\cos(\theta) = \cos(-\theta)$. This can be resolved by making a careful sketch of the orbit and positions in question. Alternatively, the trigonometric identity $\tan^2\left(\frac{\alpha}{2}\right) = \frac{1-\cos\alpha}{1+\cos\alpha}$ can be employed, where α is any angle (such as the eccentric anomaly, E) to get:

$$\tan^2\left(\frac{E}{2}\right) = \frac{1-\cos E}{1+\cos E}.$$

Inserting $\cos E$ from Eq. 3.14 and simplifying gives:

$$\tan\left(\frac{E}{2}\right) = \left(\frac{1-e}{1+e}\right)^{\frac{1}{2}} \tan\left(\frac{\theta}{2}\right). \tag{3.16}$$

This useful relationship has no quadrant ambiguity, as it can be seen by inspecting Fig. 3.4 to verify that $E/2$ and $\theta/2$ are necessarily always in the same quadrant.

3.2.2 Analytical Method - Elliptical Orbits

The change in position with time on an elliptical orbit can also be derived analytically, though that involves some rather complicated integration. Referring back to Fig. 3.3, it can be seen that the time rate of change in the true anomaly is given by $\frac{d\theta}{dt} = \frac{V_\perp}{r} = \frac{h}{r^2}$ since $h = rV_\perp$. Thus $dt = (r^2/h)d\theta$ and taking $r = (h^2/\mu)/(1 + e\cos\theta)$ from the Orbit Equation gives:

$$dt = \frac{h^3}{\mu^2}\frac{d\theta}{(1+e\cos\theta)^2}.$$

This can then be integrated to give the elapsed time to reach a given true anomaly:

$$t - t_0 = \frac{h^3}{\mu^2}\int_0^\theta \frac{d\theta}{(1+e\cos\theta)^2}, \tag{3.17}$$

where t_0 is the (arbitrary) time at periapsis ($\theta = 0$). The integral can be evaluated using this result from integral tables[1]:

$$\int \frac{dx}{(a+b\cos x)^2} = \frac{b\sin x}{(b^2-a^2)(a+b\text{cos}x)} - \frac{2}{(b^2-a^2)\sqrt{a^2-b^2}}\tan^{-1}\left(\frac{\sqrt{a^2-b^2}\tan\frac{x}{2}}{a+b}\right), \tag{3.18}$$

which is valid for $b < a$. This is suitable for the elliptical orbit, where $b = e$ is less than $a = 1$, so $b^2 - a^2$ is negative. The integral in Eq. 3.18 evaluated from 0 to θ can then be rearranged to give:

$$\int_0^\theta \frac{d\theta}{(1+e\cos\theta)^2} = \frac{1}{(1-e^2)^{3/2}}\left(2\tan^{-1}\left(\sqrt{\frac{1-e}{1+e}}\tan\frac{\theta}{2}\right) - \frac{e\sqrt{1-e^2}\sin\theta}{1+e\cos\theta}\right). \tag{3.19}$$

The first term in the bracket on the right side can be seen from Eq. 3.16 to be simply $2\tan^{-1}\left(\sqrt{\frac{1-e}{1+e}}\tan\frac{\theta}{2}\right) = E$, the eccentric anomaly defined previously. A little algebra using

1 CRC Standard Mathematical Tables and Formulae, 30th Edition 30th ed. Edition.

Eq. 3.15 shows the second term to be $\frac{e\sqrt{1-e^2}\sin\theta}{1+e\cos\theta} = e\sin E$. This gives for the elapsed time in Eq. 3.17:

$$t - t_0 = \frac{h^3}{\mu^2}\frac{1}{(1-e^2)^{\frac{3}{2}}}(E - e\sin E).$$

This can be simplified by noting from Eq. 2.33 taking $\theta = \pi/2$ that $p = {h^2}/{\mu} = a(1-e^2)$ to give:

$$t - t_0 = \left(\frac{a^3}{\mu}\right)^{\frac{1}{2}}(E - e\sin E),$$

which is exactly the result of the geometric approach earlier (Eqs. 3.10–3.13).

Example Problem 3a

An Earth-orbiting satellite is in an elliptical orbit with perigee radius $r_p = 6900$ km and eccentricity $e = 0.9$. What would be the time to fly from point A ($\theta_A = 135°$) to point B ($\theta_B = 220°$), as shown in Fig. 3a.

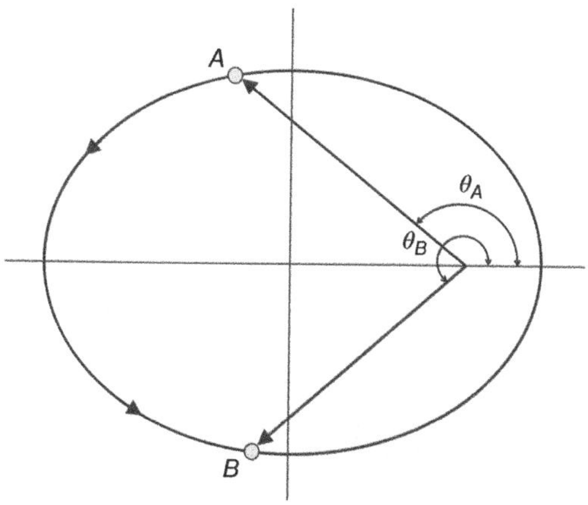

Figure 3a Orbit for Example Problem 3a (not to scale).

Solution:

The eccentric anomalies corresponding to points A and B can be determined from Eq. 3.14, which can be rearranged to give:

$$E_A = \cos^{-1}\left[\frac{e + \cos(\theta_A)}{1 + \cos(\theta_A)}\right] = \cos^{-1}\left[\frac{0.9 + \cos(135°)}{1 + \cos(135°)}\right] = 57.96° = 1.012\text{ rad}$$

and similarly

$$E_B = \cos^{-1}\left[\frac{e + \cos(\theta_B)}{1 + \cos(\theta_B)}\right] = \cos^{-1}\left[\frac{0.9 + \cos(220°)}{1 + \cos(220°)}\right] = 295.5° = 5.159\text{ rad},$$

where the other value from the inverse cosine, $E_B = 64.45°$, is rejected as it is apparent from the figure that E_B must be greater than 180° (π radians). The elapsed time between points A and B can be determined by referencing each point to the corresponding time since perigee passage,

$t_B - t_A = (t_B - t_0) - (t_B - t_0)$, where t_0 is the (here arbitrary) time at perigee. Incorporating Kepler's time equation (Eq. 3.13) then gives: $t_B - t_A = \frac{1}{n}[(E_B - e\sin E_B) - (E_A - e\sin E_A)]$. The mean angular rate is (Eq. 3.12) $n = \left(\frac{\mu}{a^3}\right)^{\frac{1}{2}}$, where $a = r_p/(1-e) = \frac{6900}{0.1} = 69{,}000\,km$ giving $n = \left[\frac{398{,}600\,km^3/s^2}{(69{,}000\,km)^3}\right]^{1/2} = (28{,}708s)^{-1}$. Inserting these values gives the desired elapsed time:

$$t_B - t_A = 28{,}708\text{ s}[(5.159 - 0.9\sin(5.159)) - (1.012 - 0.9\sin(1.012))] = 164{,}238\text{ s} = 45.62\text{ hr},$$

where the calculation is properly performed using the eccentric anomaly in radians. The elapsed time can be seen to be comparable to the period of the entire orbit, $T = 2\pi n = 50.1$ hr. The time from perigee to point A is much shorter, $t_A - t_0 = \frac{1}{n}(E_A - e\sin E_A) = 1.99$ hr. The spacecraft at point B will return to perigee in the time difference $t_0' - t_B = T - (t_B - t_A) - (t_A - T_0) = 50.1 - 45.62 - 1.99 = 2.49$ hr, where t_0' would be the time of the *next* perigee passage (note also that point B is farther from perigee than point A).

3.2.3 Orbits with Periapsis Passage

Since the above analysis is based on the time after periapsis passage, t_0, a modified approach is needed where passage through periapsis occurs. In fact all solutions to the elliptical time problem contain a *period ambiguity*, since a given position/true anomaly will occur repeatedly on subsequent orbits. These situations are discussed below.

For the case without periapsis passage (Case I in Fig. 3.6), the result from Example Problem 3a can be applied directly, so that

$$(t - t^*)_I = (t - t_0) - (t^* - t_0) = \frac{1}{n}[(E - e\sin E) - (E^* - e\sin E^*)], \tag{3.20}$$

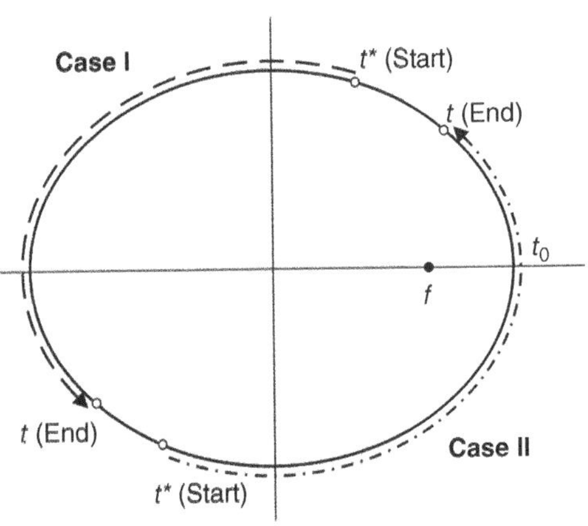

Figure 3.6 Orbits without (Case I) and with (Case II) periapsis passage.

where t is the time at epoch (the final position), and t^* and E^* are, respectively, the time and the eccentric anomaly at the initial position on the orbit. For the case with periapsis passage (Case II in Fig. 3.6), the elapsed time can be determined by subtracting the time for Case I from the total period of the orbit, to give $(t - t^*)_{II} = T - (t^* - t)_I = T + (t - t^*)_I$. The sign reversal in the last term reflects that $t^* < t$ if Case I is employed here. Since the period of the orbit is $T = 2\pi/n$, the elapsed time for Case II is then given by:

$$(t - t^*)_{II} = \frac{1}{n}[2\pi + (E^* - e\sin E^*) - (E - e\sin E)].$$

This is the elapsed time for a *single* passage through periapsis. To account for the possibility of multiple perigee passages in general, a time increment of $\Delta t = 2\pi k/n$ needs to be incorporated into the expression above to give:

$$(t - t^*)_{II} = \frac{1}{n}[2k\pi + (E^* - e\sin E^*) - (E - e\sin E)], \tag{3.21}$$

which allows for any k number of periapsis passages ($k = 1, 2, 3, \ldots$) for Case II. For Case I, the time associated with additional orbits can be incorporated in similar fashion to give:

$$(t - t^*)_{I} = \frac{1}{n}[2k\pi + (E - e\sin E) - (E^* - e\sin E^*)], \tag{3.22}$$

which, unlike Case II, allows for the possibility of no periapsis passage at all ($k = 0$).

3.2.4 Inverse Problem

The approach described earlier allows for the direct solution of the elapsed time between any two points on an elliptical orbit, provided that the true anomalies of those two positions are known. This can be termed the "direct" time problem. Another practical situation is where an initial position on the orbit is known, and the position after a specified elapsed time must be determined. This can be termed the "inverse" time problem. In this case, the Kepler time equation (Eq. 3.13) cannot be solved directly for the eccentric anomaly (and hence the true anomaly) since the mean anomaly, $M = E - e\sin E$, is a transcendental function, so the eccentric anomaly, E, cannot be written as an exact, analytical function of time. In this case, another solution method is needed. These could include simply solving by "trial and error" using a spreadsheet, for example, or by solving graphically by plotting the mean anomaly against the eccentric anomaly for the known eccentricity of the orbit, then finding the point where Eq. 3.13 is satisfied.

A practical numerical approach to this inverse problem is the Newton-Rhapson method, which involves iteratively improving an initial estimate until the precise solution is reached. The method proceeds as shown in Fig. 3.7.

The goal of this numerical method is to determine the zero point (or root) of a transcendental function, $f(x)$. In this case, the function is $f(E) = E - e\sin E - M$, where $M = n(t - t_0)$ is known. The first step in the solution is to assume (or "guess") a value of the independent variable, $x = x_1$. The slope of the function at that point, $df(x)/dx = f'(x_1)$, is then used to determine the next iteration on x:

$$\frac{f(x_1) - 0}{x_1 - x_2} = f'(x_1),$$

to give: $x_2 = x_1 - \frac{f(x_1)}{f'(x_1)}$. The iteration can be applied as often as needed, so that in general for step k

$$x_{k+1} = x_k - \frac{f(x_k)}{f'(x_k)}. \tag{3.23}$$

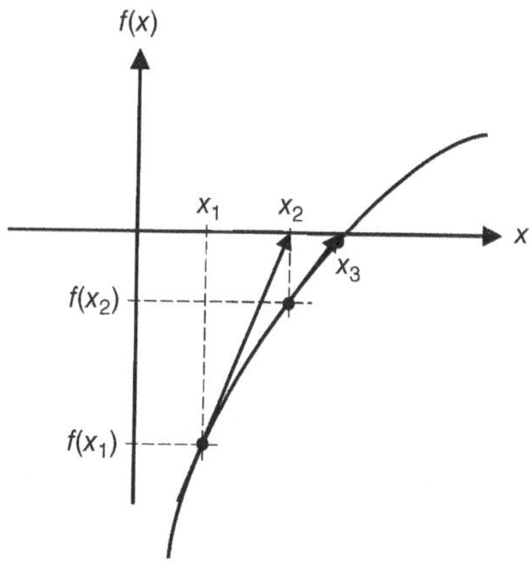

Figure 3.7 Illustration of the Newton-Rhapson method.

The derivative of the needed function in the time problem is $f'(E) = 1 - e\cos E$, so the iteration formula becomes in this case:

$$E_{k+1} = E_k - \frac{(E_k - e\sin E_k) - M}{1 - e\cos E_k}. \tag{3.24}$$

This is then simply applied repeatedly until the increment in E becomes sufficiently small, or $\left|\frac{f(E_k)}{f'(E_k)}\right| < |\Delta E_k| < \varepsilon$, where ε is an arbitrarily small number. The method described here works well because the function $f(E) = E - e\sin E - M$ is "well-behaved" in that it is not oscillatory and does not have multiple roots over the interval $0 \le E \le 2\pi$. The final consideration is how to choose the initial value, E_0, to start the iterative process that will best lead to the solution. The following recipe can prove useful:

- For moderate values of the eccentricity, e, take $E_0 = M = n(t - t_0)$;
- For high values of e, try $E_0 = M + e/2$;
- As a "last resort," simply use $E_0 = \pi$.

Should the solution procedure not converge, a different initial guess can then be taken.

Example Problem 3b

An Earth-orbiting satellite has the parameters $a = 4R_\oplus$ and $r_p = 1.5R_\oplus$, where $R_\oplus = 6378$ km is the mean radius of the Earth. What is the true anomaly four hours after perigee passage?

Solution:

The eccentricity of the orbit is $e = 1 - r_p/a = 1 - 1.5/4 = 0.625$. For this value of eccentricity, a good initial guess for the eccentric anomaly could be $E_0 = M = n(t - t_0)$. Then

$$E_0 = \left(\frac{\mu}{a^3}\right)^{\frac{1}{2}}(t - t_0) = \left[\frac{398{,}600\ km^3/s^2}{(25{,}512\ km)^3}\right]^{\frac{1}{2}}(4\ hr)\left(\frac{3600\ s}{hr}\right) = 2.2311\ \text{rad} = M.$$

The first iteration using Eq. 3.24 gives:

$$E_1 = E_0 - \frac{(E_0 - e\sin E_0) - M}{1 - e\cos E_0} = 2.2311 - \frac{(2.2311 - 0.625\sin(2.2311)) - 2.2311}{1 - 0.625\cos(2.2311)} = 2.5879.$$

Repeating this gives the next iteration

$$E_2 = E_1 - \frac{(E_1 - e\sin E_1) - M}{1 - e\cos E_1} = 2.5879 - \frac{(2.5879 - 0.625\sin(2.5879)) - 2.2311}{1 - 0.625\cos(2.5879)} = 2.5695.$$

This is within 10^{-4} of the exact value. Such rapid convergence in the solution is common, given the well-behaved nature of the function $E - e\sin E - M$. The true anomaly is then given by rearranging Eq. 3.16 to give:

$$\tan\left(\frac{\theta}{2}\right) = \left(\frac{1+e}{1-e}\right)^{\frac{1}{2}} \tan\left(\frac{E}{2}\right) = 1.285 \rightarrow \theta = 2.861\ \text{rad} = 163.9°.$$

The solution can be checked by taking the computed value of the true anomaly and using Kepler's equation (Eq. 3.13) and calculating the elapsed time $t - t_0 = \left(\frac{a^3}{\mu}\right)^{\frac{1}{2}}[E - e\sin E] = (1.549 \times 10^{-4}/s)^{-1}[2.570 - 0.625\sin(2.570)] = 14{,}408\,s = 4\,hr$, and the solution is verified.

3.3 Parabolic Trajectories

For the case of a parabolic trajectory $e = 1$, which allows Eq. 3.17 to be simply reformulated to give:

$$t - t_0 = \frac{h^3}{\mu^2}\int_0^\theta \frac{d\theta}{(1 + \cos\theta)^2}. \tag{3.25}$$

A standard trigonometric identity can be used to write $1 + \cos\theta = 2\cos^2(\theta/2)$. Eq. 3.25 then becomes:

$$t - t_0 = \frac{h^3}{4\mu^2}\int_0^\theta \sec^4\left(\frac{\theta}{2}\right)d\theta,$$

since $\cos^2\left(\frac{\theta}{2}\right) = \frac{1}{\sec^2(\theta/2)}$. Changing variables in the integrand to $\theta/2 = \theta'$, $d\theta = 2\theta'$ gives the simplification:

$$t - t_0 = \frac{h^3}{2\mu^2}\int_0^\theta \sec^4(\theta)d\theta.$$

From integral tables $\int_0^\theta \sec^4(\theta)d\theta = \tan\theta + \frac{1}{3}\tan^3\theta$, so

$$t - t_0 = \frac{h^3}{2\mu^2}\left[\tan\theta'\Big|_0^{\frac{\theta}{2}} + \frac{1}{3}\tan^3\theta'\Big|_0^{\frac{\theta}{2}}\right].$$

Evaluating the limits of integration then gives the final result:

$$t - t_0 = \frac{h^3}{2\mu^2}\left[\tan\frac{\theta}{2} + \frac{1}{3}\tan^3\frac{\theta}{2}\right]. \tag{3.26}$$

This is commonly termed *Barker's Equation*, the analytical solution of the time problem for parabolic trajectories. Since $h = \sqrt{\mu p} = \sqrt{2\mu r_p}$ for a parabolic trajectory, the prefactor in Eq. 3.26 can also be expressed as $\frac{h^3}{2\mu^2} = \left(\frac{2r_p}{\mu}\right)^{1/2}$. Equation 3.25 can be rearranged to write an equation similar in form to Kepler's time equation (Eq. 3.13):

$$M_p = \frac{\mu^2}{h^3}(t - t_0), \tag{3.27}$$

where the *parabolic anomaly* $M_p = \frac{1}{2}\tan\frac{\theta}{2} + \frac{1}{6}\tan^3\frac{\theta}{2}$.

The parabolic time equation (Eq. 3.27) can then be solved in the same fashion as Kepler's time equation for elliptical orbits, namely: (i) if the true anomaly(ies) are known, the equation is solved directly, and (ii) if the elapsed time is known, the iterative method described above can be applied.

3.4 Hyperbolic Trajectories

Determining the time-varying position on a hyperbolic trajectory is further complicated by the fact that, unlike an elliptical orbit, the trajectory is open-ended and therefore has no period. However, the similarity of the hyperbola to the ellipse (the hyperbola can be thought of as an "inside out" ellipse, as discussed in Section 2.6) suggests that similar geometric and analytical approaches can be applied. Both a geometric and a mathematical approach are thus presented here.

3.4.1 Geometric Method - Hyperbolic Trajectories

One approach to developing the time equation for hyperbolic trajectories is to recognize the difference in the defining equation for a hyperbola, $\frac{x^2}{a^2} - \frac{y^2}{b^2} = 1$, versus that for an ellipse, $\frac{x^2}{a^2} + \frac{y^2}{b^2} = 1$. While the unit circle ($x^2 + y^2 = 1$) gives the circular sine and cosine functions, the unit hyperbola ($x^2 - y^2 = 1$) leads to the hyperbolic functions $\sinh(F) = \frac{e^F - e^{-F}}{2}$ and $\cosh(F) = \frac{e^F + e^{-F}}{2}$, where F is the *hyperbolic anomaly*. Unlike the circular trigonometric functions, the hyperbolic functions are nonperiodic and are unbounded (as is a hyperbola). The hyperbolic functions are analogous to the circular trigonometric functions in several ways, for example, that the hyperbolic tangent is $\tanh(F) = \frac{\sinh(F)}{\cosh(F)}$ and $\cosh^2(F) - \sinh^2(F) = 1$. The relationship of the hyperbolic anomaly to the parameters of a hyperbolic trajectory are illustrated in Fig. 3.8, where for given parameters a and b the horizontal and vertical distances from the center of the hyperbola are $x = a\cosh F$ and $y = b\cosh F$, where a and b are taken to be positive. The hyperbolic anomaly F, unlike the eccentric anomaly E, is not an angle but relates to the area $A = ½a^2F$ shown in the figure, which can be termed the *hyperbolic sector.*

Substituting the hyperbolic sine for the circular sine into the definition of the mean anomaly for an ellipse, $M = E - e\sin E$, suggests this form for the hyperbolic time equation:

$$t - t_0 = \left(\frac{-a^3}{\mu}\right)^{\frac{1}{2}}(e\sinh F - F) = \left(\frac{-a^3}{\mu}\right)^{\frac{1}{2}} M_h, \tag{3.28}$$

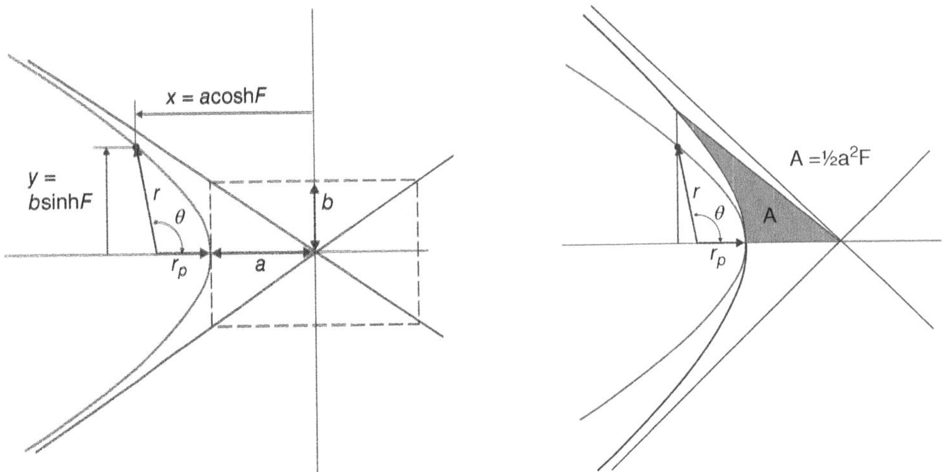

Figure 3.8 Hyperbolic geometry illustrating the hyperbolic sector.

where $M_h = \sinh F - F$ is the *hyperbolic mean anomaly*. The sign change in the first factor, from a to $-a$, reflects that the semi-major axis, a, is strictly negative for a hyperbola. The hyperbolic time equation can be written in the form:

$$\boxed{M_h = n(t - t_0),} \tag{3.29}$$

which is analogous to the eccentric mean anomaly defined for an ellipse, where $n = \left(\frac{\mu}{-a^3}\right)^{\frac{1}{2}}$ is the mean angular rate for a hyperbolic trajectory with the semi-major axis again taken as being negative.

The relationship between the true anomaly, θ, and the hyperbolic anomaly F can be determined noting that $\cosh F = a + r_p - r\cosh F$, where r is the radius from the focus of the hyperbola to the object. Noting that and $r = \dfrac{a(e^2-1)}{1+e\cos\theta}$ (Eq. 2.46) and $r_p = a(e-1)$ (taking a to positive) and some algebra leads to:

$$\cosh F = \frac{e + \cos\theta}{1 + e\cos\theta}, \tag{3.30}$$

or, rearranging:

$$\cos\theta = \frac{e - \cosh F}{e\cosh F - 1}, \tag{3.31}$$

which are very similar to the expressions given in Eqs. 3.14 and 3.15 for an ellipse. Knowing the hyperbolic anomaly from the hyperbolic time equation (Eq. 3.29) thus allows the determination of the true anomaly, which combined with the orbit equation specifies the object's position on the orbit, as for the case of the ellipse. There is no quadrant ambiguity in this case as the hyperbolic functions are nonperiodic. Alternatively, the identity $\tanh^2\left(\frac{F}{2}\right) = \dfrac{\cosh F - 1}{\cosh F + 1}$ can be combined with Eq. 3.30 to give:

$$\tanh\left(\frac{F}{2}\right) = \left(\frac{e-1}{e+1}\right)^{\frac{1}{2}} \tan\left(\frac{\theta}{2}\right) \tag{3.32}$$

3.4.2 Analytical Method - Hyperbolic Trajectories

The analytical derivation of the hyperbolic time equation is based on the same equation used at the start of the elliptical time equation development, Eq. 3.17. The needed integral in the hyperbolic case is:

$$\int \frac{dx}{(a + b\cos x)^2} = \frac{b\sin x}{(b^2 - a^2)(a + b\cos x)} - \frac{a}{(b^2 - a^2)} \frac{1}{\sqrt{b^2 - a^2}} \ln\left(- \frac{\sqrt{b^2 - a^2}\tan\frac{x}{2} + a + b}{\sqrt{b^2 - a^2}\tan\frac{x}{2} - a - b} \right), \tag{3.33}$$

which is valid for $b > a$. In the case of the hyperbolic trajectory, $b = e > a = 1$. The integral in Eq. 3.33 rearranged and evaluated from 0 to θ becomes:

$$\int_0^\theta \frac{d\theta}{(1 + e\cos\theta)^2} = \frac{1}{(e^2 - 1)^{3/2}} \left(\frac{e\sqrt{e^2 - 1}\sin\theta}{1 + e\cos\theta} - \ln\left(\frac{\sqrt{e+1} + \sqrt{e-1}\tan\frac{\theta}{2}}{\sqrt{e+1} - \sqrt{e-1}\tan\frac{\theta}{2}} \right) \right). \tag{3.34}$$

The first term in the bracket on the right side can be shown by using Eq. 3.32 and some algebra to be $\dfrac{e\sqrt{e^2-1}\sin\theta}{1+e\cos\theta} = e\sinh F$. Considerable additional algebra then shows the second term to be $\ln\left(\dfrac{\sqrt{e+1}+\sqrt{e-1}\tan\frac{\theta}{2}}{\sqrt{e+1}-\sqrt{e-1}\tan\frac{\theta}{2}}\right) = F$. This gives for the elapsed time, going back to Eq. 3.17:

$$t - t_0 = \frac{h^3}{\mu^2}\frac{1}{(e^2-1)^{\frac{3}{2}}}(e\sin F - F),$$

or:

$$t - t_0 = \left(\frac{-a^3}{\mu}\right)^{\frac{1}{2}}(e\sin F - F),$$

using $h^2/_\mu = a(1-e^2)$. This is identical to the result of the geometric approach earlier (Eq. 3.28).

The hyperbolic time equation (Eq. 3.29) can also be solved in the same fashion as Kepler's time equation for elliptical orbits, directly if the true anomaly(ies) is/are known, and iteratively if the elapsed time is given.

Example Problem 3c

The *Voyager 2* spacecraft had a flyby of Neptune in 1989 (this was the only time a spacecraft visited that planet). How long after periapsis passage at Neptune was needed to fly out to the orbit of Neptune's largest moon, Triton? The orbit of Triton is at a radius of $r_t = 354{,}600$ km from Neptune. The trajectory elements for the hyperbolic flyby were $a = -19{,}985$ km, $e = 2.459$, and the geometry is sketched in Fig. 3c.

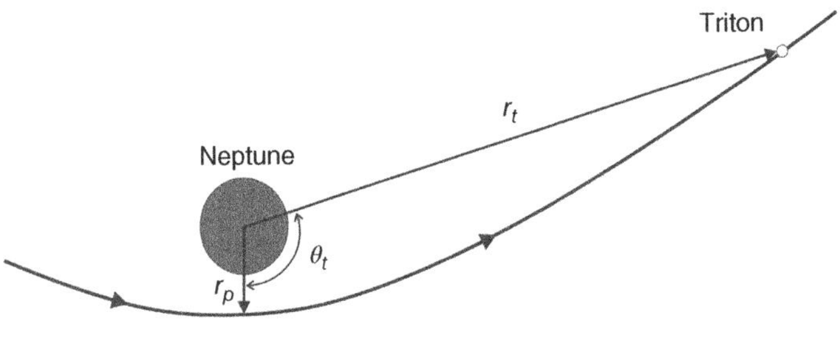

Figure 3c Neptune flyby/Triton encounter.

Solution:

Taking the Orbit Equation in hyperbolic form (a positive), $r = \dfrac{a(e^2-1)}{1+e\cos\theta}$, can be solved to get the true anomaly at Triton's orbit,

$$\cos\theta_t = \frac{1}{e}\left[\frac{a(e^2-1)}{r_t} - 1\right] = \frac{1}{2.459}\left[\frac{+19{,}985\text{ km}\left((2.459)^2-1\right)}{354{,}600\text{ km}} - 1\right] = -0.291,$$

so $\theta_t = 106.9^\circ$. Next, determine the hyperbolic anomaly using Eq. 3.30:

$$\cosh F = \frac{e + \cos\theta_t}{1+\cos\theta_t} = \frac{2.459 - 0.291}{1 + 2.459(-0.291)} = 7.622,$$

which gives $F = 2.7199$, and $\sinh F = 7.5563$. The hyperbolic mean anomaly is

$$M_h = \sinh F - F = (2.459)(7.5563) - 2.7199 = 15.8611.$$

The time to Triton's orbit after periapsis passage is then $t-t_0 = M_h/n$, where the mean angular rate is $n = \left(\frac{\mu_*}{-a^3}\right)^{\frac{1}{2}}$, where μ_* is the gravitational parameter for Neptune and the semi-major axis is taken to be negative. The time from periapsis to Triton's orbit is then:

$$t-t_0 = \frac{M_h}{\left(\frac{\mu_*}{-a^3}\right)^{\frac{1}{2}}} = \frac{15.8611}{\left(\frac{6.896 \times \frac{10^6\text{km}^3}{\text{s}^2}}{-(19{,}985\text{ km})^3}\right)^{\frac{1}{2}}} = 17{,}064\text{ s} = 4.74\text{ hr}.$$

3.5 Problems

3.1 A hypervelocity gun located at sea level launches a projectile at a velocity of 6000 m/s at an angle of 45° relative to the horizontal.

How far from the launch site does the projectile hit the ground (surface distance, km)?
What will be the maximum altitude of the projectile (km)?
What will be the flight time of the projectile (minutes)?

3.2 Halley's comet last passed perihelion on February 9, 1986. The famous comet's orbit has semi-major axis $a = 17.8$ AU and eccentricity $e = 0.967$. Calculate the period of the comet's orbit (years) and predict the date of the next return (i.e. when the comet again reaches perihelion). Compare your result with a published value (cite your source) and comment on any reason(s) for the difference.

Next, solve Kepler's equation and calculate the quantities E (radians), θ (degrees), and the length of the radius vector r (in AU) for the current date. For simplicity, take the time of day for all dates to be exactly noon, local time. How does your value of r compare with the actual current position of the comet on that date?

Finally, make a to-scale sketch of the orbit of the comet, including the location of the sun, the current location of the comet, and the orbits of earth and Neptune. Note: The orbit of Halley's comet is in fact quite inclined compared to that of the planets, but it is acceptable for this problem to sketch all orbits in the same plane.

3.3 The Space Shuttle, in a circular low earth orbit, launches a deep-space probe that departs earth orbit on a parabolic trajectory, as shown. The perigee radius of the probe (and the Shuttle's orbital radius) is $r_p = 7000$ km.

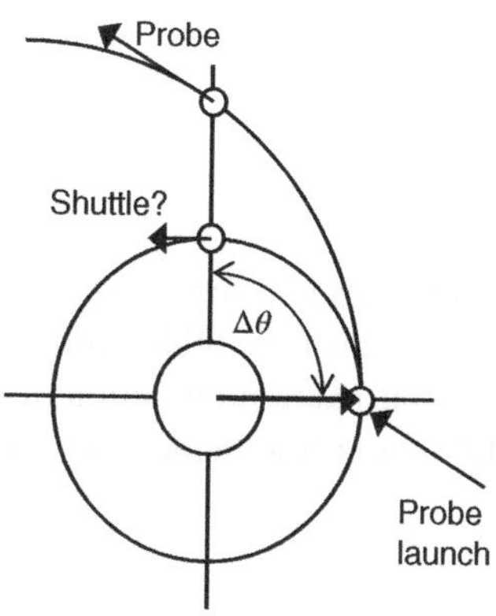

(a) How long (hours) does it take the departing probe to reach a true anomaly of $\theta = 90°$?
(b) How far will the Shuttle have gone around its circular orbit during that time? Express your answer in terms of its angle from the point of probe launch, $\Delta\theta$. Will the Shuttle be "ahead" or "behind" the probe (i.e. how does $\Delta\theta$ compare to 90°)?

3.4 A spacecraft is at point Q on the elliptical orbit shown. The elliptical orbit has semi-major axis a and eccentricity e. Express the answers for each part below in terms of a and e (and μ as required) only.

(a) What is the orbital radius at point Q? What is the radius at perigee, r_p?
(b) How long will it take for the spacecraft to move from point Q to perigee?

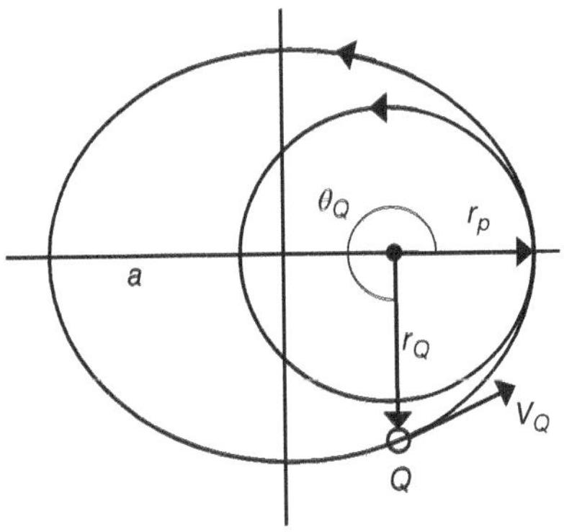

3.5 A space probe approaching Mars is at a distance $r_1 = 98{,}000$ km relative to the planet, with a true anomaly of $\theta_1 = 130°$. The trajectory has periapsis radius $r_p = 7000$ km. $\mu_{\text{Mars}} = 43{,}050$ km^3/s^2.

How long (hours) will it take to probe to reach periapsis from the point $r_1 = 98{,}000$ km, $\theta_1 = 130$?

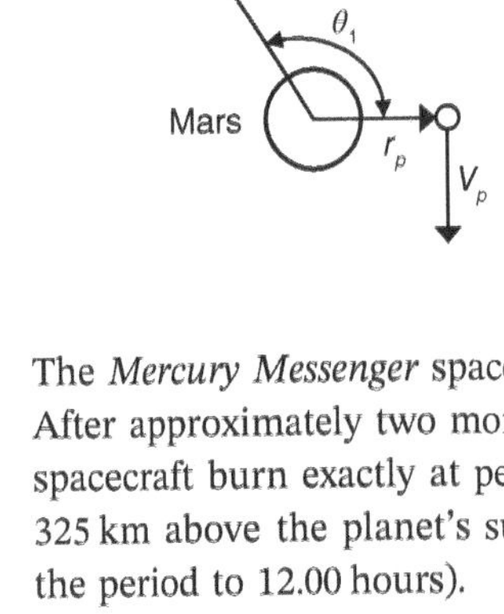

3.6 The *Mercury Messenger* spacecraft was inserted into an orbit around the planet Mercury. After approximately two months in orbit, it was decided to adjust the orbit period with a spacecraft burn exactly at periapsis. The preadjustment orbit had a minimum altitude of 325 km above the planet's surface, and a period of 11.80 hours (the maneuver increased the period to 12.00 hours).

i) What was the eccentricity, e, and semi-major axis, a (km), of the initial orbit (period $T = 11.80$ hours)?

ii) The spacecraft entered Mercury's shadow at a true anomaly of $\theta = 155°$. How long after shadow entry did the maneuvering burn occur (you can assume here that the burn duration is infinitely short)? As part of this calculation, give the values of the eccentric anomaly, the mean anomaly (radians), and the mean motion (inverse seconds) for the true anomaly of $\theta = 155°$.

3.7 In Example Problem 2c, the true anomalies of a satellite where it entered and left the Van Allen radiation belts were determined. The perigee and apogee radii of the satellite were 6500 and 45,000 km, respectively. It was assumed that the lower limit of the radiation belts was at an altitude of $h = 500$ km (no upper limit was considered).

For how long (minutes) will the satellite be *inside* the radiation belts? How long (minutes) will it be *outside* the belts?

3.8 For the *ARES* system, it is proposed to conduct a short-duration test of the Lunar Surface Access Module (*LSAM*) in earth orbit. The *LSAM* is first to be launched, together with the *Orion* spacecraft, into a circular orbit with an altitude of 300 km.

The *LSAM* is then to be tested by boosting it into an elliptical orbit where the apogee is *at least* twice the radius of the circular parking orbit, as sketched below. After an unspecified number of orbits, the two spacecraft (the *LSAM* and *Orion*) will rendezvous (this would be at the point of perigee in the *LSAM*'s test orbit). It is desired to conduct the entire test in the shortest possible time.

(a) How many orbits will have been completed by the *LSAM*, and the *Orion* spacecraft, when the rendezvous occurs?

(b) What would be the apogee of the *LSAM* test orbit (km)?

(c) What will be the total time required for the test (hours)? Neglect any time associated with docking/undocking, orbital insertion, and rendezvous maneuvers.

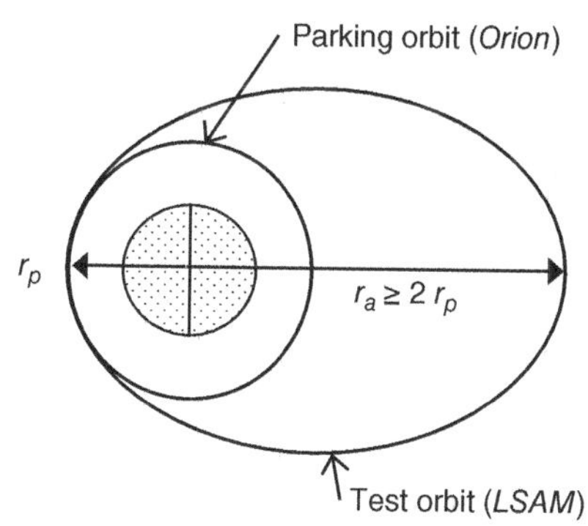

3.9 A hostile country has launched a spy satellite (object ①) into a circular orbit with radius $r_1 = 20{,}000$ km. It is planned to destroy the spy satellite by hitting it with a spacecraft (object ②) launched from the earth's surface in a retrograde, highly elliptical trajectory. See the sketch and orbital parameters given below (all orbits are in the same plane), and neglect the atmosphere.

(a) How long (minutes), from the moment of interceptor launch will it take for the intercept to occur?
(b) What will be the magnitude of the relative velocity, $|\boldsymbol{V}_2 - \boldsymbol{V}_1|$, in km/s, between the two spacecraft at the moment of collision?
$a_2 = 31{,}890$ km
$e_2 = 0.8$
$r_{p_2} = 6378$ km
$r_1 = 20{,}000$ km
$e_1 = 0.0$
$V_1 = 4.46$ km/s

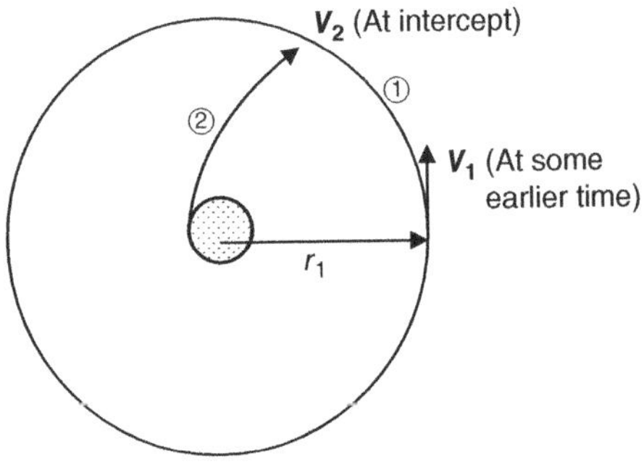

3.10 Determine the time required to fly from point B_1 to point B_2 (both on minor axis) in terms of the orbital period T and eccentricity e. Identify three different ways this problem can be solved.

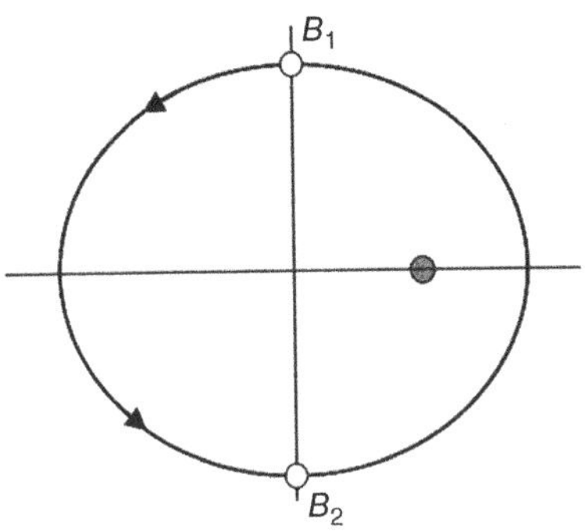

3.11 How long does it take for a comet on a parabolic trajectory to go from perihelion (periapsis for the sun) out to the distance of the parabola's parameter, p?

3.12 If the eccentricity of the elliptical orbit shown is 0.5, calculate, in terms of the period T, the time required to fly from periapsis to the point where the orbital radius is triple that of the periapsis.

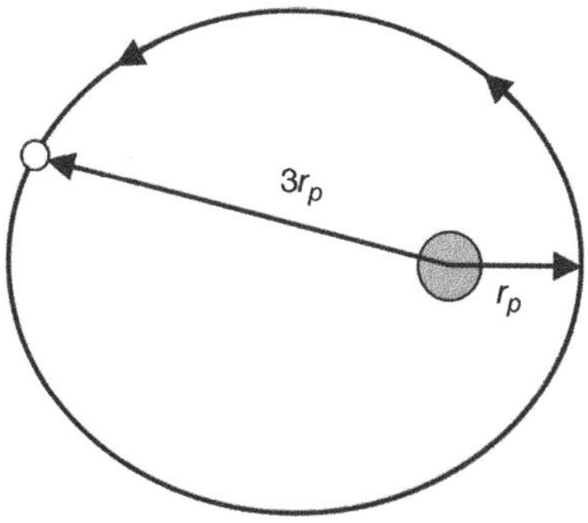

3.13 A satellite in the Earth's orbit has a perigee radius of $r_p = 8000$ km and $e = 0.30$. What is its true anomaly 60 minutes after passing the true anomaly of 90°.

3.14 Calculate the time required for a spacecraft launched into a parabolic trajectory at a perigee altitude of 200 km to leave the earth's sphere of influence[2] ($r = 294{,}000$ km). Perform the same calculation for the case where the injection into the parabolic escape trajectory occurs from geosynchronous orbit (altitude $z = 35{,}786$ km). For which case will the location of the sphere of influence be reached more quickly? Provide a (brief) physical explanation for the difference.

2 The sphere of influence is the point where the spacecraft has essentially escaped Earth's gravity and its motion becomes dominated by the sun.

4

Orbits in Three Dimensions

The discussion of orbital motion thus far has completely described the trajectory and velocity of an object in an orbit for the case of 2-body motion. In the absence of any perturbations, such trajectories and the corresponding velocity vectors necessarily remain in a single plane. What has not yet been considered is the *orientation* of trajectories relative to the primary body. For example, two orbits with identical two-dimensional parameters (such as semi-major axis and eccentricity) could have very different inclinations. In addition to the orientation of an orbit, the three-dimensional reference frame on which the orbit is defined needs to be specified (for example, is the orbit that of a satellite about the Earth, or a planet about the Sun). Developing complete, three-dimensional descriptions of orbits in different reference frames is discussed here. The chapter outline is:

- The elements of three-dimensional orbits
- Reference frames
  - Heliocentric (sun-centered)
  - Geocentric (Earth-centered)
  - Topocentric (Earth-surface-centered)
  - Perifocal (orbit-centered)
- Transformations
  - Rotation and translation
  - Transformations between reference frames

4.1 Three-Dimensional Orbital Elements

In the previous two chapters, it was stated that any two scalar parameters (such as, r_a and r_p, or a and e) are sufficient to describe any 2-body trajectory. The addition of the elapsed time from periapsis passage then specifies the location of the object. This, however, describes the motion only in the plane of the orbit. Relative to the primary body (the Earth, Sun, or a target planet), the plane of an orbit (or other trajectory) can be oriented in any arbitrary fashion. Examples include an equatorial orbit, exactly over the plane of a planet's equator, or a polar orbit that exactly passes over both of the planet's poles, or anything in between.

The description of an orbit in three dimensions requires the addition of three more parameters to the 2-body orbital elements presented previously. This description will require returning to the

Introduction to Orbital Mechanics: A Concise and Practical Approach, First Edition. James C. Hermanson.

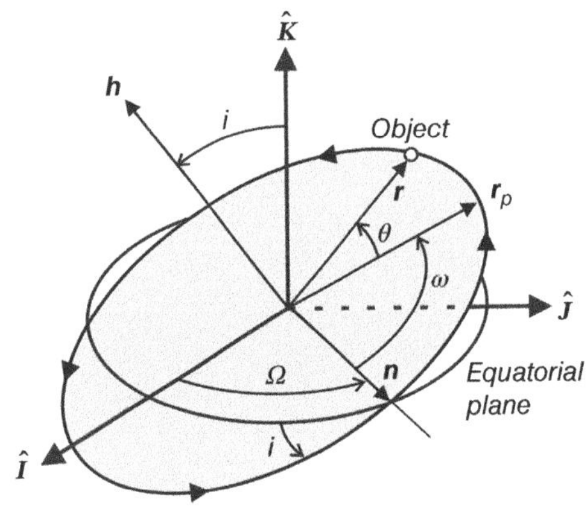

Figure 4.1 Three-dimensional orbital elements.

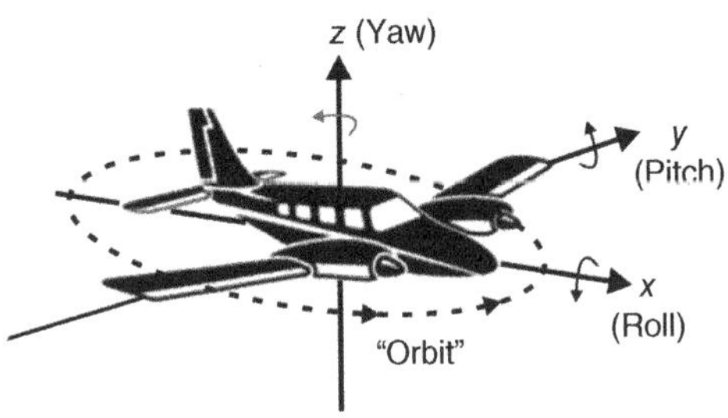

Figure 4.2 An aircraft "orbit."

vector descriptions of an object's motion (e.g. $\boldsymbol{r}$, $\boldsymbol{V}$) used in developing the 2-body trajectories in Chapter 2. The additional orbital elements are shown in Fig. 4.1. The three-dimensional coordinate system can be based on, for example, the direction of the planet's north pole (the $\hat{\boldsymbol{K}}$ axis) and the plane of the planet's equator (the $\hat{\boldsymbol{I}}$ and $\hat{\boldsymbol{J}}$ axes). The three additional elements needed are the *inclination*, i, the *longitude of the ascending node*, Ω, and the *argument of periapsis*, ω. The vector $\boldsymbol{h}$ is the angular momentum vector; the vector $\boldsymbol{n}$ points to the location of the ascending node.

The inclination reflects how the orbit is tilted relative to the equatorial plane. The longitude of the ascending node is the point, relative to the selected coordinate system, where the object in an inclined orbit crosses the equator "upward," or in the direction of the north pole (here the $\hat{\boldsymbol{K}}$ axis). Given that the orbit is centered on the primary body's center, the descending node, where the object crossed the equator "downward," is necessarily at a true anomaly 180° greater than the location of the ascending node. Finally, the argument of periapsis refers to the location of the periapsis relative to the ascending node. For an equatorial orbit (zero inclination), both the longitude of the ascending node and the argument of periapsis are undefined.

The various orbital orientations relative to a planet can be visualized by an analogy using the motion of an aircraft in flight (Fig. 4.2), where the orbital plane is modeled as being aligned with the planform of the aircraft and the periapsis located at the nose, and the axes x, y, and z are aligned with the aircraft's centerline, wings, and vertically, as shown.

A change in orbital inclination is analogous to a change in aircraft pitch, or rotation about the y axis. This change is shown relative to a planet in Fig. 4.3a. For fixed values of the inclination and the location of the ascending node, rotating the orbit in yaw (about the z axis) moves the location of periapsis, changing the argument of periapsis (Fig. 4.3b). Finally, moving the orbit in roll (rotation about the x axis) causes the location of the ascending node to change, as shown in Fig. 4.3c. This last example is only strictly true for a polar orbit; in general changing only the ascending node would require two rotations, a combination of roll and pitch. The effects of combined rotations are discussed in Section 4.3.

The angular momentum vector, $\boldsymbol{h} = \boldsymbol{r} \times \boldsymbol{V}$, is always perpendicular to the plane of the orbit, as shown in Fig. 4.1. The vector $\boldsymbol{n}$ is normal to the plane defined by the vectors $\boldsymbol{h}$ and $\hat{\boldsymbol{K}}$ and can be defined as the unit vector $\hat{\boldsymbol{K}} \times \left(\dfrac{\boldsymbol{h}}{h}\right)$. The three-dimensional orbital elements can be determined from the angles between the various vectors, as the angle α between any two vectors $\boldsymbol{a}$ and $\boldsymbol{b}$ is

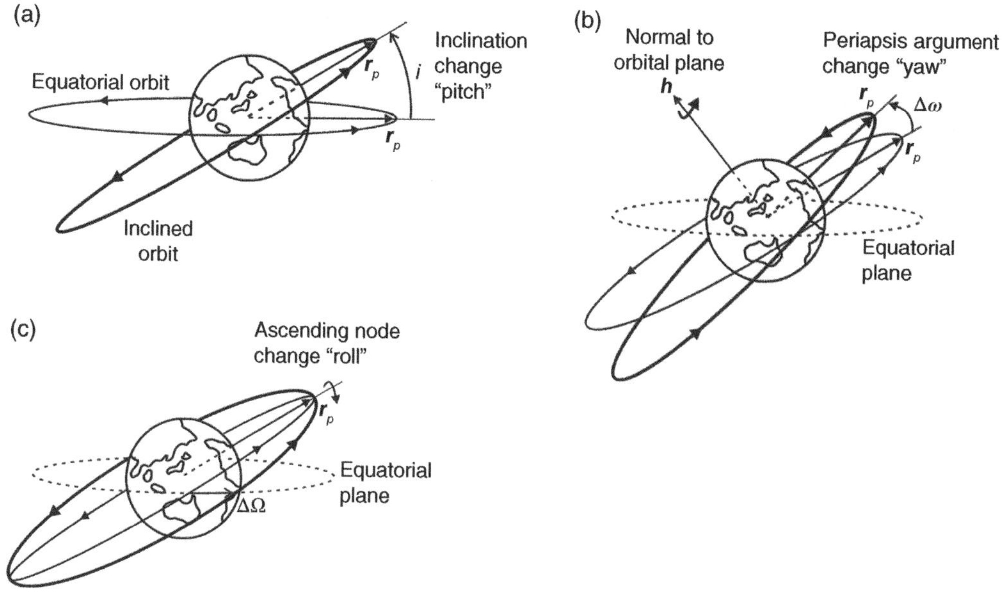

Figure 4.3 Changes in three-dimensional orbital elements. (a) Change in inclination; (b) change in argument of periapsis; and (c) change in ascending node.

given by $\alpha = \cos^{-1}\left[\frac{\boldsymbol{a}\cdot\boldsymbol{b}}{ab}\right]$. Thus, the inclination, the angle between the angular momentum vector $\boldsymbol{h}$ and the unit vector $\hat{\boldsymbol{K}}$, is given by:

$$i = \cos^{-1}\left[\frac{\hat{\boldsymbol{K}}\cdot\boldsymbol{h}}{h}\right]. \tag{4.1}$$

By definition, the inclination is only specified over the range $0 \le i < 180°$. An equatorial orbit has $i = 0$; a polar orbit, $i = 90°$. An orbit with $0 \le i < 90°$ is termed *prograde* or *direct*, which would correspond to motion eastward for an Earth satellite (this is the most common situation). Orbits with $90° < i < 180°$ are termed *retrograde*, with westward motion for an Earth satellite.

The longitude of the ascending node is the angle between the unit vectors $\boldsymbol{n}$ and $\hat{\boldsymbol{I}}$, which is similarly given by:

$$\Omega = \cos^{-1}\left[\frac{\hat{\boldsymbol{I}}\cdot\boldsymbol{n}}{n}\right]. \tag{4.2}$$

The double-valued nature of Eq. 4.2 can be resolved by considering a "bird's eye" view of the equatorial plane, looking down the $\hat{\boldsymbol{K}}$ axis, as shown in Fig. 4.4. Denoting the component of the unit vector $\boldsymbol{n}$ in the $\hat{\boldsymbol{J}}$ direction as n_J suggests the convention:

$\Omega < 180°$ for $n_J > 0$ and $\Omega > 180°$ for $n_J < 0$.

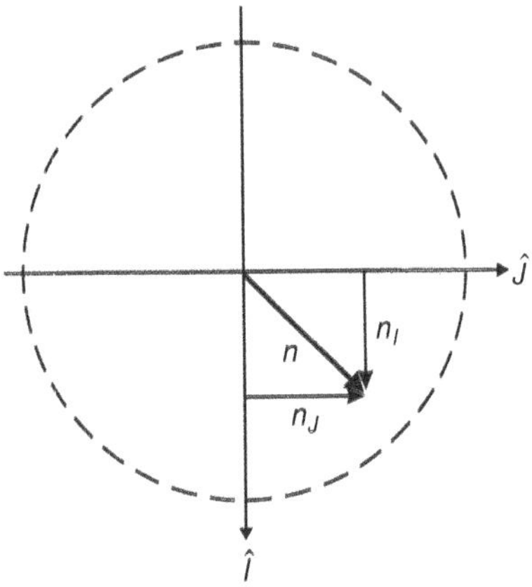

Figure 4.4 Definition convention for the longitude of the ascending node.

The argument of periapsis can be similarly determined by using the eccentricity vector, **e**, which points to periapsis as shown in Chapter 2. This then implies that:

$$\omega = \cos^{-1}\left[\frac{\boldsymbol{n}\cdot\boldsymbol{e}}{ne}\right], \tag{4.3}$$

where ω can have any value in the range $0 \le \omega < 360°$. The eccentricity and position radius vectors, **e** and **r**, using similar reasoning give the true anomaly as:

$$\theta = \cos^{-1}\left[\frac{\boldsymbol{e}\cdot\boldsymbol{r}}{er}\right], \tag{4.4}$$

with $0 \le \theta < 360°$.

Example Problem 4a

The radius and velocity vectors of an Earth-orbiting satellite are $\boldsymbol{r} = -0.8\hat{\boldsymbol{I}} + 0.6\hat{\boldsymbol{J}} + 0.5\hat{\boldsymbol{K}}$ DU and $\boldsymbol{V} = -0.4\hat{\boldsymbol{I}} - 0.8\hat{\boldsymbol{J}} + 0.6\hat{\boldsymbol{K}}$ DU/TU, using the DU and TU units introduced in Chapter 2. Find all of the orbital elements (a e, i, Ω, ω, θ), expressed in DU and TU units as needed.

Solution:

The semi-major axis can be determined from the energy equation as described previously: $E = \frac{V^2}{2} - \frac{\mu}{r} = -\frac{\mu}{2a}$. The scalar magnitudes of the radius and velocity vectors are, respectively, $r = \sqrt{(-0.8)^2 + (0.6)^2 + (0.5)^2} = 1.118$ DU and $V = \sqrt{(-0.4)^2 + (-0.8)^2 + (0.6)^2} = 1.077$ DU/TU. Taking $\mu = 1$ DU^3/TU^2 gives $a = 1.59$ DU.

The angular momentum vector is $\boldsymbol{h} = \boldsymbol{r}\times\boldsymbol{V} = (-0.8\hat{\boldsymbol{I}} + 0.6\hat{\boldsymbol{J}} + 0.5\hat{\boldsymbol{K}}) \times (-0.4\hat{\boldsymbol{I}} - 0.8\hat{\boldsymbol{J}} + 0.6\hat{\boldsymbol{K}}) = 0.76\hat{\boldsymbol{I}} + 0.28\hat{\boldsymbol{J}} + 0.88\hat{\boldsymbol{K}}$ DU^2/TU with the scalar value $h = \sqrt{(0.76)^2 + (0.28)^2 + (0.88)^2} = 1.196$ DU/TU. This then gives the eccentricity using Eq. 2.40 as follows:

$$e = \sqrt{1 - \frac{h^2}{\mu}\frac{1}{a}} = \sqrt{1 - \frac{(1.196)^2}{(1)}\frac{1}{1.59}} = 0.317$$

The orbital inclination follows from Eq. 4.1: $i = \cos^{-1}\left[\frac{\hat{\boldsymbol{K}}\cdot\boldsymbol{h}}{h}\right] = \cos^{-1}\left[\frac{\hat{\boldsymbol{K}}\cdot(-0.8\hat{\boldsymbol{I}} + 0.6\hat{\boldsymbol{J}} + 0.5\hat{\boldsymbol{K}})\,\text{DU/TU}}{1.196\,\text{DU/TU}}\right] = 42.6°$ (the inclination is always taken to be positive).

To determine the longitude of the ascending node requires first the unit vector pointing in that direction, $\boldsymbol{n} = \hat{\boldsymbol{K}} \times \left(\frac{\boldsymbol{h}}{h}\right) = \hat{\boldsymbol{K}} \times \left(\frac{-0.8\hat{\boldsymbol{I}} + 0.6\hat{\boldsymbol{J}} + 0.5\hat{\boldsymbol{K}}}{1.196\,h}\right) = -0.234\hat{\boldsymbol{I}} + 0.635\hat{\boldsymbol{J}}$ with the scalar value $n = 0.6772$. The vector gives for the longitude of the ascending node from Eq. 4.2: $\Omega = \cos^{-1}\left[\frac{\hat{\boldsymbol{I}}\cdot\boldsymbol{n}}{n}\right] = \cos^{-1}\left[\frac{\hat{\boldsymbol{K}}\cdot(-0.234\hat{\boldsymbol{I}} + 0.635\hat{\boldsymbol{J}})}{0.6772}\right] = 110.2°$. This is the correct quadrant since $n_J > 0$, as seen in Fig. 4.4.

The eccentricity vector is determined from Eq. 2.25: $\boldsymbol{e}=\boldsymbol{B}/\mu \quad (\boldsymbol{V}\times\boldsymbol{h}-\mu\frac{\boldsymbol{r}}{r})=\frac{1}{(1)}$ $\left[(-0.4\hat{\boldsymbol{I}}-0.8\hat{\boldsymbol{J}}+0.6\hat{\boldsymbol{K}})\times(-0.8\hat{\boldsymbol{I}}+0.6\hat{\boldsymbol{J}}+0.5\hat{\boldsymbol{K}})-(1)\frac{-0.8\hat{\boldsymbol{I}}+0.6\hat{\boldsymbol{J}}+0.5\hat{\boldsymbol{K}}}{1.118}\right]=-0.4\hat{\boldsymbol{I}}-0.8\hat{\boldsymbol{J}}+0.6\hat{\boldsymbol{K}}$. The argument of periapsis is then from Eq. 4.3: $\omega=\cos^{-1}\left[\frac{\boldsymbol{n}\cdot\boldsymbol{e}}{ne}\right]=\cos^{-1}$ $\left[\frac{(-0.234\hat{\boldsymbol{I}}+0.635\hat{\boldsymbol{J}})\cdot(-0.4\hat{\boldsymbol{I}}-0.8\hat{\boldsymbol{J}}+0.6\hat{\boldsymbol{K}})}{(0.6772)(0.317)}\right]=13.1°$. Finally, the true anomaly given by Eq. 4.4 is: $\theta=\cos^{-1}\left[\frac{\boldsymbol{e}\cdot\boldsymbol{r}}{er}\right]=\cos^{-1}\left[\frac{(-0.4\hat{\boldsymbol{I}}-0.8\hat{\boldsymbol{J}}+0.6\hat{\boldsymbol{K}})\cdot(-0.8\hat{\boldsymbol{I}}+0.6\hat{\boldsymbol{J}}+0.5\hat{\boldsymbol{K}})}{(0.317)(1.118)}\right]=28.2°$. Given that the inverse cosine is double-valued another possible solution is $\theta=-28.2°=331.8°$. If the quantity $\boldsymbol{e}\cdot\boldsymbol{r}>0$, the object must be in the first or fourth quadrant of its orbital plane. In addition, if $\boldsymbol{r}\cdot\boldsymbol{V}>0$, the object is "climbing" in its orbit, so it must be in either the first or second quadrant. In this case, $\boldsymbol{e}\cdot\boldsymbol{r}=0.312>0$ and $\boldsymbol{r}\cdot\boldsymbol{V}=0.14>0$, so the solution is in the first quadrant and $\theta=+28.2°$ is correct.

4.2 Reference Frames

The definition of orbital elements presented in the previous section is dependent on the choice of the *reference frame* that specifies the three-dimensional coordinate system in which the motion is defined. The choice of reference frame is guided by the orbital motion being considered, for example, an object in Earth orbit, a probe orbiting the Sun, or a spacecraft approaching a planet. A particular reference frame will be needed for observations taken from the Earth's surface. In complex missions, for example, an interplanetary mission involving Earth departure, transit to the target planet, and the planetary encounter, it will be necessary to use multiple reference frames, transitioning from one to another, as discussed in Section 4.3. The various reference frames are described as follows.

4.2.1 Heliocentric Reference Frame

The motions of objects about the Sun, such as planets and interplanetary spacecraft, are best described using the *heliocentric* (Sun-centered) reference frame, as shown in Fig. 4.5.

In the heliocentric reference frame, the $\hat{\boldsymbol{Z}}$ or "north" direction can be taken to be perpendicular to the plane of the Earth's orbit (termed the *ecliptic plane*). The $\hat{\boldsymbol{Z}}$ vector does not align with the north pole of the sun, which is tilted from the ecliptic plane by 7.25°. The $\hat{\boldsymbol{X}}$ direction can be defined by the position of the sun as seen by the Earth at the moment that the northern hemisphere spring begins (the *vernal equinox*) when the sun is seen to cross the Earth's equator. The direction corresponding to the vernal equinox is commonly termed the *first point of Aries* and designated using the symbol ♈, as shown in Fig. 4.5. Using Cartesian coordinates, the definitions of the $\hat{\boldsymbol{X}}$ and $\hat{\boldsymbol{Z}}$ directions then define the $\hat{\boldsymbol{Y}}$ direction, which corresponds to the position of the Earth on the first day of the northern hemisphere winter, as shown.

The first point of Aries[1] is currently situated in the zodiacal constellation Pisces. The shift from Aries is due to the *precession of the equinoxes*, which is a continual change in the direction of the Earth's north pole due to the torque exerted on the (slightly nonspherical) Earth by the Moon

1 The first point of Aries was defined by Hipparchus in the year 130 BC, when the location of the vernal equinox was near the star γ Arietis.

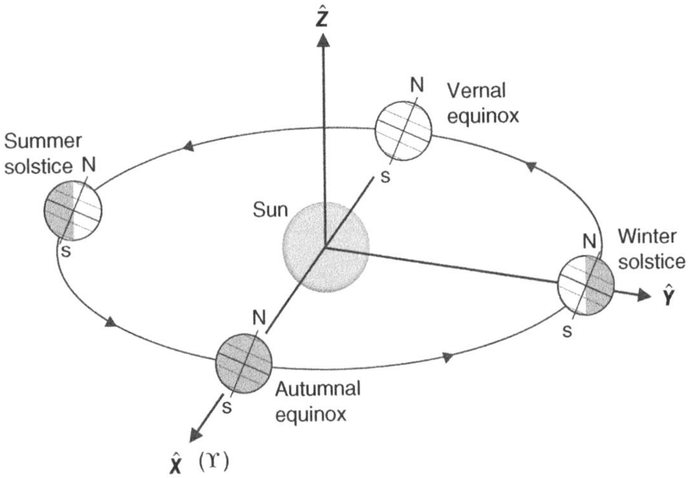

Figure 4.5 Heliocentric reference frame.

(see Chapter 7). This perturbation does not affect the inclination of the north pole relative to the ecliptic, which remains a constant 23.5°. The procession of the equinoxes causes the location of the *north celestial pole* (the point in space exactly above the Earth's north pole) to move in a circle over a period of 25,772 years. This apparent motion of the Earth's celestial poles is shown in Fig. 4.6.

The north celestial pole at present points very nearly to Polaris, the familiar and bright "North Star". In the year 3000 BC, the North Star would have been the fainter star Thuban (α Draconis) in the constellation Draco. The current situation in the southern hemisphere is unfavorable, as the south celestial pole is the near faint star σ Octanis, in the faint constellation Octans. The centerpoint in Fig. 4.6 is the *ecliptic pole*, which is exactly perpendicular to the plane of the Earth's orbit.

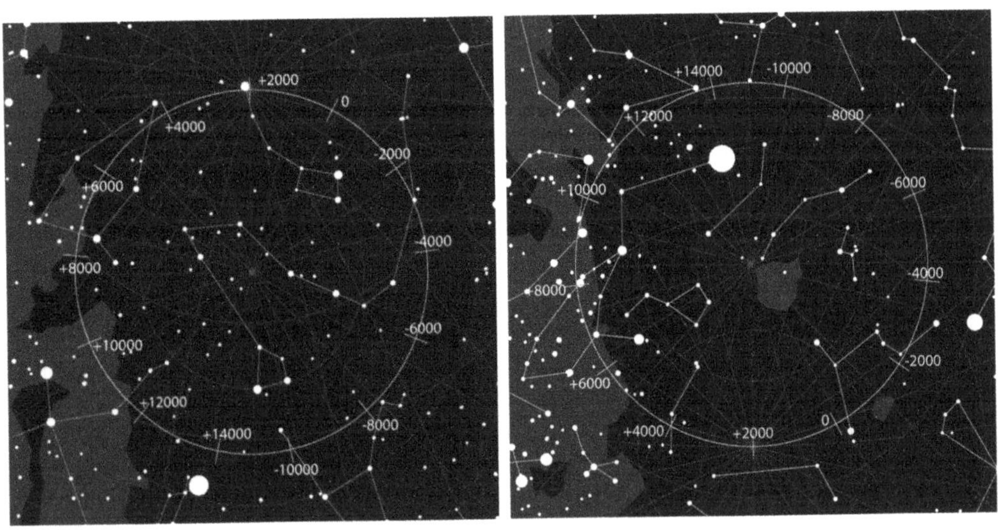

Figure 4.6 Precession of the equinoxes. Left, northern sky; right, southern sky. *Source:* Taúolunga / Wikimedia Commons / Public domain / CC BY-SA 4.0.

Example Problem 4b

The elements of the Earth's orbit are[2] $a = 1$ AU, $e = 0.01673$, and $\omega = 102.93°$ ($i = 0$ by definition, so Ω is undefined). What is the seasonal position of the Earth, in heliocentric coordinates, at the times of the autumnal equinox, the winter solstice, the vernal equinox, and the summer solstice? What are the positions at perihelion and aphelion?

Solution:

From the Orbit Equation (in Eq. 2.33 form), the Earth's radial distance from the sun is $r = \frac{a(1-e^2)}{1 + e\cos\theta}$. At the time of the autumnal equinox, the Earth is exactly aligned with the $\hat{\boldsymbol{X}}$ direction, as seen in Fig. 4.5. There is no component in the $\hat{\boldsymbol{Y}}$ direction (and never one in the $\hat{\boldsymbol{Z}}$ direction, by the definition of the ecliptic plane). At that point, the true anomaly $\theta = -\omega = -102.93° = 257.07°$. Expressed in astronomical units, the radius is $r = \frac{(1)\left[1-(0.01673)^2\right]}{1 + (0.01673)\cos(257.07°)} = 1.0035$ AU. The radius vector in heliocentric coordinates is then $\boldsymbol{r} = 1.0035\hat{\boldsymbol{X}}$ AU. At the winter solstice, the true anomaly increases by 90°, to give $\theta = 90° - \omega = -12.93° = 347.07°$. This gives $\boldsymbol{r} = 0.9837\,\hat{\boldsymbol{Y}}$ AU, as shown in Fig. 4.5. The radii for the vernal equinox ($\theta = 77.07°$) and summer solstice ($\theta = 167.07°$) are determined similarly.

At perihelion, $\theta = 0$, which gives a radial distance of $r = 0.9997$ AU. The radius vector is $\boldsymbol{r} = r\left(\cos\omega\hat{\boldsymbol{X}} + \sin\omega\,\hat{\boldsymbol{Y}}\right) = 0.9997\left[\cos(102.93°)\hat{\boldsymbol{X}} + \sin(102.93°)\hat{\boldsymbol{Y}}\right] = -0.2237\hat{\boldsymbol{X}} + 0.9746\hat{\boldsymbol{Y}}$ AU. At aphelion, $\theta = 180°$ and $r = 1.0003$ AU. In this case, $\boldsymbol{r} = r\left[\cos(\omega + 180°)\hat{\boldsymbol{X}} + \sin(\omega + 180°)\,\hat{\boldsymbol{Y}}\right]$. The complete results are tabulated as follows:

Earth position	True anomaly, θ (deg.)	Angle from $\hat{X}$ (deg.)	r (AU)	$\boldsymbol{r}$ (AU)
Autumnal equinox	257.07	0	1.0035	$1.0035\,\hat{\boldsymbol{X}}$
Winter solstice	347.07	90	0.9837	$0.9837\,\hat{\boldsymbol{Y}}$
Vernal equinox	77.07	180	0.9960	$-0.9960\,\hat{\boldsymbol{X}}$
Summer solstice	167.07	270	1.0163	$-1.0163\,\hat{\boldsymbol{Y}}$
Perihelion	0	102.93	0.9997	$-0.2237\hat{\boldsymbol{X}} + 0.9746\hat{\boldsymbol{Y}}$
Aphelion	180	282.93	1.0003	$0.2238\hat{\boldsymbol{X}} - 0.9749\hat{\boldsymbol{Y}}$

The orbit of the Earth is nearly circular, but it can be seen that perihelion occurs slightly after the northern hemisphere winter solstice. Aphelion occurs near the summer solstice in the northern hemisphere.

4.2.2 Geocentric Reference Frame

For objects in motion about the Earth (the Moon, satellites, orbital debris, etc.), it is convenient to use the *geocentric reference frame* shown in Fig. 4.7. In this case, the direction $\hat{\boldsymbol{Z}}$ (or $\hat{\boldsymbol{K}}$) is taken to align with the Earth's geographic north pole, normal to the equator as the reference plane. The $\hat{\boldsymbol{X}}$ (or $\hat{\boldsymbol{I}}$) direction is then the first point of Aries as defined in the heliocentric frame. $\hat{\boldsymbol{Z}}$ and $\hat{\boldsymbol{X}}$ then define the third coordinate axis, $\hat{\boldsymbol{Y}}$ (or $\hat{\boldsymbol{J}}$).

2 The Astronomical Almanac for the Year 2015 (United States Naval Observatory/Nautical Almanac Office, 2014).

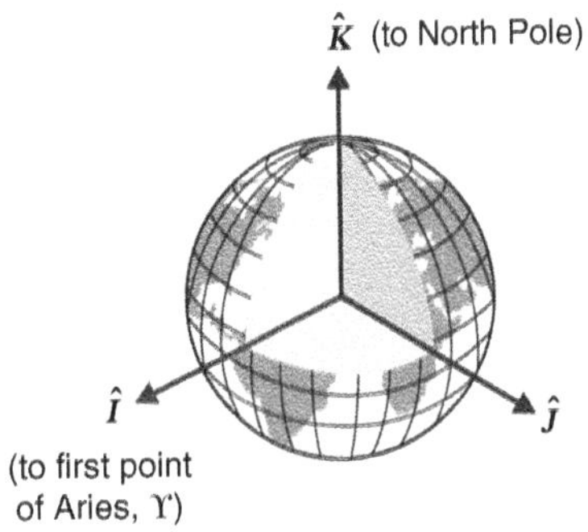

Figure 4.7 Geocentric reference frame.

4.2.3 Perifocal Reference Frame

It is often convenient to describe an orbit in terms of *perifocal coordinates*, which are defined by the plane of the orbit itself, as shown in Fig. 4.8. Defining the coordinate $\hat{\boldsymbol{p}}$ as pointing to the direction of the periapsis and the coordinate $\hat{\boldsymbol{q}}$ in the direction of the parameter of the orbit gives the third axis, $\hat{\boldsymbol{w}}$, as normal to the plane of the orbit.

The utility of the perifocal reference frame is that for an unperturbed, 2-body orbit, the entire motion can be described in two dimensions using the coordinates $\hat{\boldsymbol{p}}$ and $\hat{\boldsymbol{q}}$ only (there is no out-of-plane motion in the $\hat{\boldsymbol{w}}$ direction). This allows the direct application of the 2-body orbit described in Chapter 2 as follows.

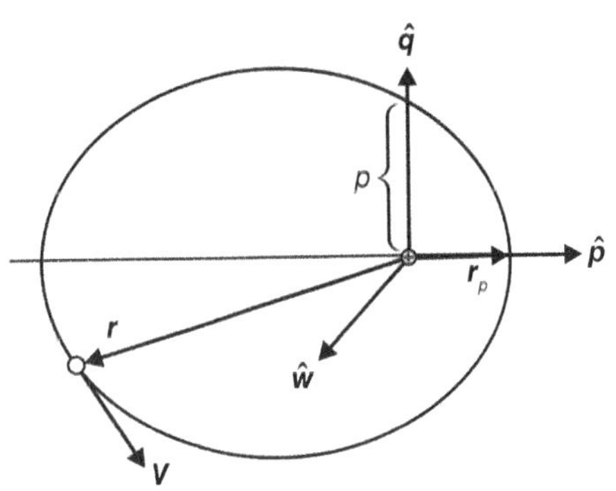

Figure 4.8 Perifocal reference frame.

The radius and velocity vectors of an object are shown in perifocal coordinates in Fig. 4.8, where θ is the true anomaly. Since there is no $\hat{\boldsymbol{w}}$ component, the radius vector can be written as follows:

$$\boldsymbol{r} = r\cos\theta\,\hat{\boldsymbol{p}} + r\sin\theta\,\hat{\boldsymbol{q}}. \tag{4.5}$$

Similarly, expressing the velocity vector in terms of its radial and tangential components gives:

$$\boldsymbol{V} = (V_r\cos\theta - V_\perp\sin\theta)\,\hat{\boldsymbol{p}} + (V_r\cos\theta - V_\perp\sin\theta)\,\hat{\boldsymbol{q}}. \tag{4.6}$$

Noting that $V_r = \sqrt{\frac{\mu}{p}}e\sin\theta$ and $V_\perp = \sqrt{\frac{\mu}{p}}(1 + e\cos\theta)$ from Eqs. 2.43 and 2.44 allows rewriting Eq. 4.6 to give:

$$\begin{aligned}\boldsymbol{V} &= \sqrt{\mu/p}\{[e\sin\theta\,\cos\theta - (1 + e\cos\theta)\sin\theta]\hat{\boldsymbol{p}} + [e\sin\theta\,\cos\theta + (1 + e\cos\theta\cos\theta)]\hat{\boldsymbol{q}}\} \\ &= \sqrt{\mu/p}[-\sin\theta\hat{\boldsymbol{p}} + (e + \cos\theta)\hat{\boldsymbol{q}}].\end{aligned} \tag{4.7}$$

Equations 4.6 and 4.7 give the complete description of the orbital position and velocity in perifocal coordinates, given the planar orbital elements e and p. The true anomaly θ can be determined for any given time by using the Kepler time equation discussed in Chapter 3. It should be pointed out that perifocal coordinates are only of use in the plane of the orbit; to be practical, those coordinates will need to be converted to geocentric, heliocentric, or other reference frames.

Example Problem 4c

An earth satellite has the orbital elements $a = 10{,}000$ km and $e = 0.3$. What are the radius and velocity vectors in perifocal coordinates at a true anomaly of $\theta = 145°$?

Solution:

The radius at that point is $r = \dfrac{a(1-e^2)}{1 + e\cos\theta} = \dfrac{10{,}000\left[1-(0.3)^2\right]}{1 + 0.3\cos(145)} = 12{,}065$ km. From Eq. 4.6, the radius vector is $\boldsymbol{r} = r\cos\theta\,\hat{\boldsymbol{p}} + r\sin\theta\,\hat{\boldsymbol{q}} = 12{,}065[\cos(145)\hat{\boldsymbol{p}} + \sin(145)\hat{\boldsymbol{q}}] = -9883\hat{\boldsymbol{p}} + 6920\hat{\boldsymbol{q}}$ km.

Equation 4.7, noting that $p = a(1-e^2)$, gives the velocity vector as

$$V = \sqrt{\mu/a(1-e^2)}\left[-\sin\theta\hat{\boldsymbol{p}} + (e+\cos\theta)\hat{\boldsymbol{q}}\right]$$

$$= \sqrt{398{,}600/10{,}000(1-(0.3)^2)}\left[-\sin(145)\hat{\boldsymbol{p}} + (0.3+\cos(145))\hat{\boldsymbol{q}}\right]$$

$$= -3.80\hat{\boldsymbol{p}} - 3.44\hat{\boldsymbol{q}}\frac{\text{km}}{\text{s}}.$$

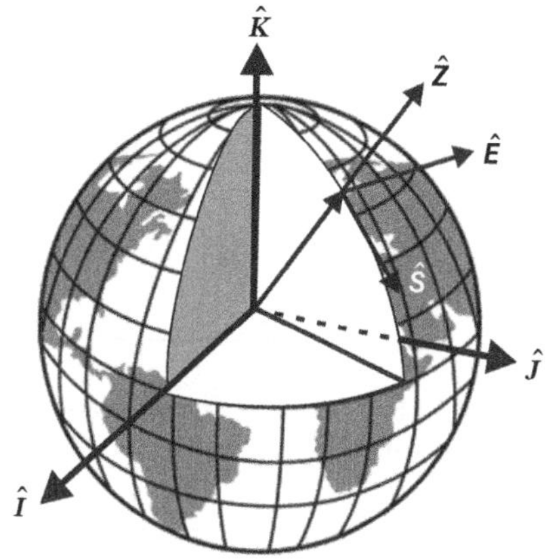

Figure 4.9 Topocentric reference frame. The topographic unit vectors $\hat{\boldsymbol{S}}, \hat{\boldsymbol{E}}, \hat{\boldsymbol{Z}}$ and geocentric unit vectors $\hat{\boldsymbol{I}}, \hat{\boldsymbol{J}}, \hat{\boldsymbol{K}}$ are shown.

4.2.4 Topocentric Reference Frame

Observations taken from the Earth's surface, for example, astronomical sightings or satellite tracking, require the use of the *topocentric reference frame*, which is based on the Earth's surface. The vertical coordinate, "up," normal to the horizon is designated $\hat{\boldsymbol{Z}}$. The other two coordinate directions are taken to be due east ($\hat{\boldsymbol{E}}$) and due south ($\hat{\boldsymbol{S}}$). This *SEZ* coordinate system is particularly convenient for tracking rocket launches, which normally proceed on a southeasterly trajectory for launches in the northern hemisphere. The relationship of an *SEZ* coordinate system to geocentric (*IJK*) coordinates is shown in Fig. 4.9. Unlike the geocentric coordinate system, the reference directions in the topographic *SEZ* coordinate system vary with the latitude and longitude of the observation site.

A different rendering of the topocentric (*SEZ*) system is provided in Fig. 4.10. The term *El* is the observed *elevation* of an object; *Az* is termed the *azimuth*.

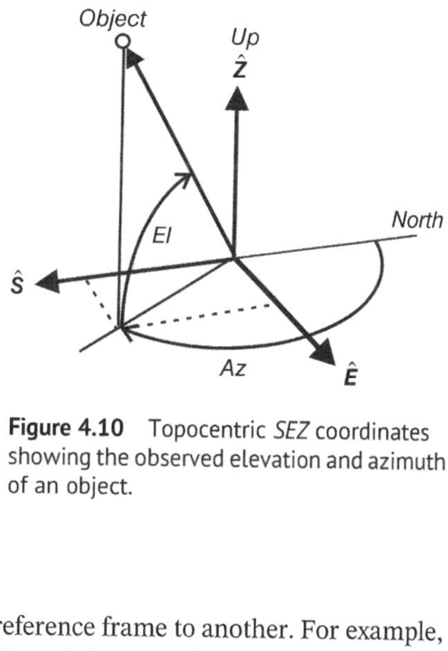

Figure 4.10 Topocentric *SEZ* coordinates showing the observed elevation and azimuth of an object.

4.3 Coordinate Frame Transformations

In practice, it will often be necessary to change from one reference frame to another. For example, the motion of a deep-space probe in an Earth parking orbit could be described in geocentric coordinates, but the transit to the target planet after Earth departure would require the use of heliocentric coordinates. Another example would be the motion of an Earth satellite, which can easily be described using perifocal coordinates, but tracking the motion over the Earth's surface would need to be in geocentric coordinates.

The geometry associated with changing coordinate frames can be first illustrated in two dimensions, as shown in Fig. 4.11. Consider a transformation between the $\hat{\boldsymbol{I}} - \hat{\boldsymbol{J}}$ coordinate system (this could be geocentric coordinates) and the $\hat{\boldsymbol{i}} - \hat{\boldsymbol{j}}$ system (this might be the perifocal coordinates $\hat{\boldsymbol{p}}$ and $\hat{\boldsymbol{q}}$).

Suppose that the $\hat{\boldsymbol{i}} - \hat{\boldsymbol{j}}$ coordinate system is rotated by the angle α relative to the $\hat{\boldsymbol{I}} - \hat{\boldsymbol{J}}$ system ($\alpha = \alpha_1$ in Fig. 4.11). The radius vector at a specific time expressed in $\hat{\boldsymbol{i}} - \hat{\boldsymbol{j}}$ coordinates would be

$$\boldsymbol{r} = r_i\hat{\boldsymbol{i}} + r_j\hat{\boldsymbol{j}}.$$

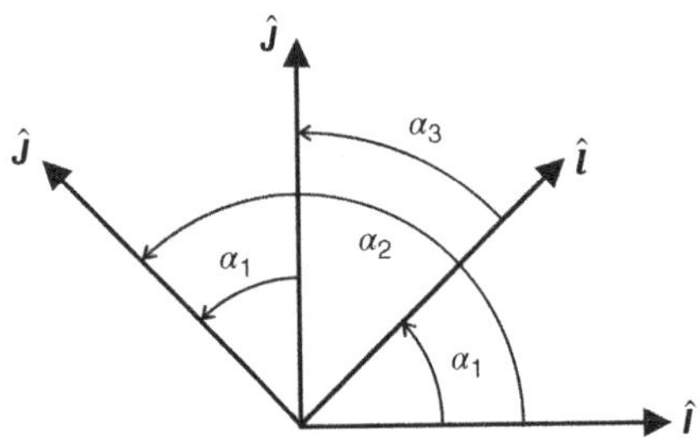

Figure 4.11 Coordinate transformation in two dimensions.

where r_i and r_j are the two components of the vector $\boldsymbol{r}$. The same position vector in the $\hat{\boldsymbol{I}}-\hat{\boldsymbol{J}}$ reference frame is $\boldsymbol{r} = r_I\hat{\boldsymbol{I}} - r_J\hat{\boldsymbol{J}}$, with the different components r_I and r_J. The unit vectors can be transformed geometrically from Fig. 4.11 to give:

$$\hat{\boldsymbol{I}} = \hat{\boldsymbol{\imath}}\cos\alpha - \hat{\boldsymbol{j}}\sin\alpha$$

and

$$\hat{\boldsymbol{J}} = \hat{\boldsymbol{\imath}}\sin\alpha + \hat{\boldsymbol{j}}\cos\alpha.$$

The transformation between the two systems can be conveniently described using *directional cosines*, which are the cosines of the angles between each coordinate axis pairs, as shown in Fig. 4.11 by the angles α_1, α_2, and α_3 . Denoting the $\hat{\boldsymbol{I}}$ and $\hat{\boldsymbol{J}}$ axes as "1" and the $\hat{\boldsymbol{\imath}}$ and $\hat{\boldsymbol{j}}$ axes as "2" then allows definition of the transformations in the unit vectors as follows:

from $\hat{\boldsymbol{\imath}}$ to $\hat{\boldsymbol{I}}$: $Q_{11} = \cos\alpha = \hat{\boldsymbol{\imath}}\cdot\hat{\boldsymbol{I}}$
from $\hat{\boldsymbol{j}}$ to $\hat{\boldsymbol{J}}$: $Q_{22} = \cos\alpha = \hat{\boldsymbol{j}}\cdot\hat{\boldsymbol{J}}$
from $\hat{\boldsymbol{j}}$ to $\hat{\boldsymbol{I}}$: $Q_{21} = -\sin\alpha = \hat{\boldsymbol{j}}\cdot\hat{\boldsymbol{I}}$
from $\hat{\boldsymbol{\imath}}$ to $\hat{\boldsymbol{J}}$: $Q_{12} = \sin\alpha = \hat{\boldsymbol{\imath}}\cdot\hat{\boldsymbol{J}}$.

Using these definitions and referring back to Fig. 4.11 allows relating the radius vector between the two reference frames:

$$\begin{aligned} r_i\hat{\boldsymbol{\imath}} + r_j\hat{\boldsymbol{j}} &= r_I\left[Q_{11}\hat{\boldsymbol{\imath}} + Q_{21}\hat{\boldsymbol{j}}\right] + r_J\left[Q_{12}\hat{\boldsymbol{\imath}} + Q_{22}\hat{\boldsymbol{j}}\right] \\ &= [r_IQ_{11} + r_JQ_{12}]\hat{\boldsymbol{\imath}} + [r_IQ_{21} + r_JQ_{22}]\hat{\boldsymbol{j}}. \end{aligned} \tag{4.8}$$

Equating the components of the vectors in Eq. 4.8 then gives:

$$r_i = Q_{11}r_I + Q_{21}r_J \text{ and } r_j = Q_{21}r_I + Q_{22}r_J. \tag{4.9}$$

This can be expressed in matrix form as

$$\begin{bmatrix} r_i \\ r_i \end{bmatrix} = \begin{bmatrix} Q_{11} & Q_{12} \\ Q_{21} & Q_{22} \end{bmatrix}\begin{bmatrix} r_I \\ r_J \end{bmatrix}, \text{ or more compactly as}$$

$$\boldsymbol{r}_{ij} = [\boldsymbol{Q}]\boldsymbol{r}_{IJ} \tag{4.10}$$

where $[\boldsymbol{Q}] = \begin{bmatrix} Q_{11} & Q_{12} \\ Q_{21} & Q_{22} \end{bmatrix}$ is the *transformation matrix* for this rotational transformation from the *IJ* to the *ij* coordinate frame. Similarly, the transformation from the *ij* to the *IJ* coordinate frame is accomplished by employing the inverse $[\boldsymbol{Q}]^{-1}$ of the transformation matrix $[\boldsymbol{Q}]$:

$$\boldsymbol{r}_{IJ} = [\boldsymbol{Q}]^{-1}\boldsymbol{r}_{ij} \tag{4.11}$$

The methodology just described can readily be extended to three dimensions (see Fig. 4.12). Adding the third component to each coordinate axis (i.e. $\hat{\boldsymbol{K}}$ to the $\hat{\boldsymbol{I}}, \hat{\boldsymbol{J}}$ axes and $\hat{\boldsymbol{k}}$ to the $\hat{\boldsymbol{\imath}}, \hat{\boldsymbol{j}}$ axes) gives the three-dimensional direction cosines/transformation matrix

$$[\boldsymbol{Q}] = \begin{bmatrix} \hat{\boldsymbol{I}}\cdot\hat{\boldsymbol{\imath}} & \hat{\boldsymbol{I}}\cdot\hat{\boldsymbol{j}} & \hat{\boldsymbol{I}}\cdot\hat{\boldsymbol{k}} \\ \hat{\boldsymbol{J}}\cdot\hat{\boldsymbol{\imath}} & \hat{\boldsymbol{J}}\cdot\hat{\boldsymbol{j}} & \hat{\boldsymbol{J}}\cdot\hat{\boldsymbol{k}} \\ \hat{\boldsymbol{K}}\cdot\hat{\boldsymbol{\imath}} & \hat{\boldsymbol{K}}\cdot\hat{\boldsymbol{j}} & \hat{\boldsymbol{K}}\cdot\hat{\boldsymbol{k}} \end{bmatrix} = \begin{bmatrix} Q_{11} & Q_{12} & Q_{13} \\ Q_{21} & Q_{22} & Q_{23} \\ Q_{31} & Q_{32} & Q_{33} \end{bmatrix}. \tag{4.12}$$

The complete three-dimensional transformation matrix $[Q]$ can be derived by considering rotation about one axis at a time to transform a coordinate system, as resulting in the transformation shown in Fig. 4.12. The order in which the rotations are applied is important and needs to be specified. Consider the airplane example shown in Fig. 4.1. The final orientation resulting from first an increase in pitch of 90°, followed by a roll to the right of 90°, is very different than the attitude resulting from a 90° roll, followed by the 90° pitch-up.

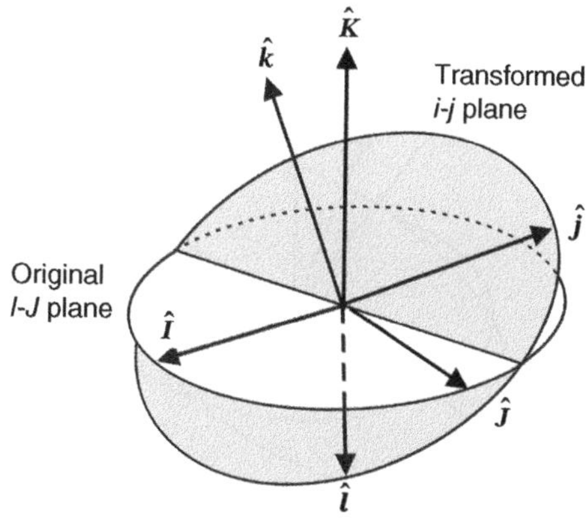

Figure 4.12 Three-dimensional coordinate frame transformation.

One convention to bring clarity to coordinate axes rotations is to use the *Euler angles* shown in Fig. 4.13. Consider an initial coordinate system with coordinate axes x_1, y_1, and z_1 as shown. It is desired to transform this system to the system labeled x_4, y_4, z_4. Starting with the two coordinate axes exactly aligned, the x-y plane is first rotated by the Euler angle α about the z_1-axis, as shown. The new set of axes, indicated by x_2, y_2, and z_2, is now rotated about the new y axis, denoted in Fig. 4.13 as y_2, by the Euler angle β. Finally, a rotation about the new z-axis (z_3) by the Euler angle γ is performed.

The first rotation, about the z_1 axis by the amount α, results in the corresponding three-dimensional transformation:

$$[Q]_1 = \begin{bmatrix} \cos\alpha & \sin\alpha & 0 \\ -\sin\alpha & \cos\alpha & 0 \\ 0 & 0 & 1 \end{bmatrix}. \tag{4.13}$$

The corresponding conversion for any vector would be $\boldsymbol{d}_{ijk} = [\boldsymbol{Q}]_1\boldsymbol{d}_{IJK}$. A rotation about the new y-axis (y_2) by the angle β gives the transformation matrix

$$[Q]_2 = \begin{bmatrix} 1 & 0 & 0 \\ 0 & \cos\beta & \sin\beta \\ 0 & -\sin\beta & \cos\beta \end{bmatrix}. \tag{4.14}$$

Finally, rotation about the new z-axis (z_3) by the angle γ produces the transformation matrix

$$[Q]_3 = \begin{bmatrix} \cos\gamma & 0 & -\sin\gamma \\ 0 & 1 & 0 \\ \sin\gamma & 0 & \cos\gamma \end{bmatrix}. \tag{4.15}$$

Applying each rotational transformation in turn gives the complete transformation matrix

$$[Q] = [Q]_3[Q]_2[Q]_1, \tag{4.16}$$

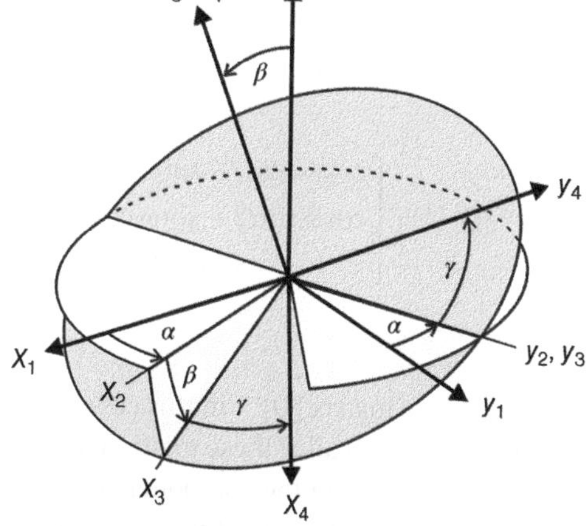

Figure 4.13 Coordinate frame rotation from initial axes x_1, y_1, z_1 to final axes x_4, y_4, z_1 using Euler angles.

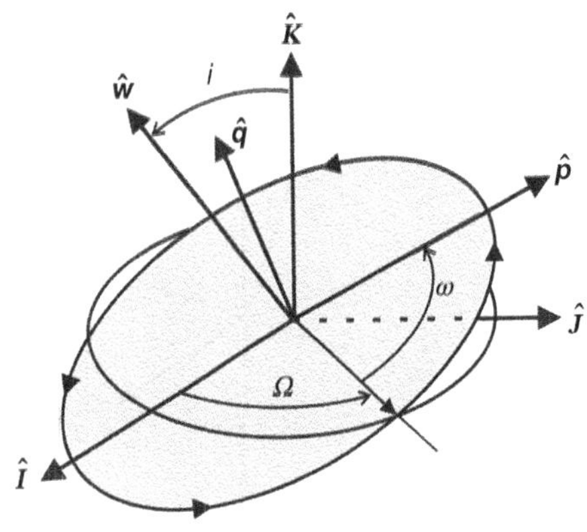

Figure 4.14 Transformation from geocentric (*IJK*) to perifocal (*pqw*) coordinates.

so that

$$\mathbf{d}_{IJK} = [\mathbf{Q}]_3[\mathbf{Q}]_2[\mathbf{Q}]_1\mathbf{d}_{SEZ} = [\mathbf{Q}]\mathbf{d}_{SEZ} \tag{4.17}$$

4.3.1 Transformations Between Perifocal, Geocentric, and Topocentric Coordinates

The use of the Euler angles shown earlier directly provides for conversion between perifocal coordinates and geocentric coordinates. It may be noted that the first angular rotation in Fig. 4.13, α, is equivalent to a change in the longitude of the ascending node, Ω. Similarly, the second angular rotation, β, is akin to a change in orbital inclination, i. The third angle, γ, is analogous to a change in the argument of periapsis, ω. The resulting transformation is shown in Fig. 4.14, where the geocentric system (*IJK*) is transformed to the perifocal axes *pqw*.

The corresponding transformation matrix $[\mathbf{Q}]$ is the matrix product of the three rotations shown in Fig. 4.13, taken now in the order of rotations Ω, i, then ω. This gives the transformation matrix:

$$[\mathbf{Q}] = [\mathbf{Q}]_3[\mathbf{Q}]_2[\mathbf{Q}]_1 = \begin{bmatrix} \cos\omega & \sin\omega & 0 \\ -\sin\omega & \cos\omega & 0 \\ 0 & 0 & 1 \end{bmatrix} \begin{bmatrix} 1 & 0 & 0 \\ 0 & \cos i & \sin i \\ 0 & -\sin i & \cos i \end{bmatrix} \begin{bmatrix} \cos\Omega & \sin\Omega & 0 \\ -\sin\Omega & \cos\Omega & 0 \\ 0 & 0 & 1 \end{bmatrix}$$

$$= \begin{bmatrix} \cos\omega\cos\Omega - \sin\omega\cos i\sin\Omega & \cos\omega\sin\Omega + \sin\omega\cos i\cos\Omega & \sin\omega\sin i \\ -\sin\omega\cos\Omega - \cos\omega\cos i\sin\Omega & -\sin\omega\sin\Omega + \cos\omega\cos i\cos\Omega & \cos\omega\sin i \\ \sin i\sin\Omega & -\sin i\cos\Omega & \cos i \end{bmatrix}. \tag{4.18}$$

This is the transformation matrix in going from the *IJK* to the *pqw* system, that is, $\mathbf{d}_{pqw} = [\mathbf{Q}]\mathbf{d}_{IJK}$. To convert from *pqw* to *IJK* coordinates requires the inverse of this transformation matrix, $[\mathbf{Q}]^{-1}$. The inverse matrix is:

$$[\mathbf{Q}]^{-1} = \begin{bmatrix} \cos\omega\cos\Omega - \sin\omega\cos i\sin\Omega & -\sin\omega\cos\Omega - \cos\omega\cos i\sin\Omega & \sin i\sin\Omega \\ \cos\omega\sin\Omega + \sin\omega\cos i\cos\Omega & -\sin\omega\sin\Omega + \cos\omega\cos i\cos\Omega & -\sin i\cos\Omega \\ \sin\omega\sin i & \cos\omega\sin i & \cos i \end{bmatrix}. \tag{4.19}$$

The inverse matrix $[\mathbf{Q}]^{-1}$ in this case is also the transpose $[\mathbf{Q}]^T$ of the original matrix $[\mathbf{Q}]$. This is due to the fact that all of these rotational transformation matrices are *orthogonal*, with each row and column representing the components of orthogonal unit vectors. Thus, the inverse matrix simply consists of each row having the components of the corresponding column in the original matrix, in the proper order, as can be seen by comparing Eqs. 4.18 and 4.19.

Example Problem 4d

Suppose that the orbit considered in Example Problem 4c has the orbital elements i = 28.4, Ω = 120°, and ω = 90°. Determine the vector position of the point of perigee in geocentric coordinates. What is the corresponding position at a true anomaly of θ = 145°?

Solution:

Converting the positions given in perifocal coordinates to geocentric coordinates requires the inverse transformation matrix given in Eq. 4.19, so that $\boldsymbol{r}_{IJK} = [\boldsymbol{Q}]^{-1}\boldsymbol{r}_{pqw}$. The orbital radius at perigee is $r = a(1-e) = 10{,}000(1-0.3) = 7000$ km so $\boldsymbol{r}_{pqw} = 7000\,\hat{\boldsymbol{p}}$ km. Multiplying the components of this vector by those of the inverse transformation (Eq. 4.19) for the given values of i, Ω, and ω gives:

$$\begin{bmatrix} r_I \\ r_J \\ r_K \end{bmatrix} = \begin{bmatrix} -0.7618 & 0.5000 & 0.4119 \\ -0.4398 & -0.8660 & 0.2378 \\ 0.4756 & 0 & 0.8796 \end{bmatrix} \begin{bmatrix} 7000 \\ 0 \\ 0 \end{bmatrix} = \begin{bmatrix} -5333 \\ -3079 \\ 3329 \end{bmatrix} \text{km.}$$

Thus, at perigee $\boldsymbol{r}_{IJK} = -5333\hat{\boldsymbol{I}} - 3079\,\hat{\boldsymbol{J}} + 3329\,\hat{\boldsymbol{K}}$ km. The radius vector at $\theta = 145°$ is from the Orbit Equation (2.33) and Eq. 4.5 $\boldsymbol{r} = -9{,}883\hat{\boldsymbol{p}} + 6920\,\hat{\boldsymbol{q}}$ to give:

$$\begin{bmatrix} r_I \\ r_J \\ r_K \end{bmatrix} = \begin{bmatrix} -0.7618 & 0.5000 & 0.4119 \\ -0.4398 & -0.8660 & 0.2378 \\ 0.4756 & 0 & 0.8796 \end{bmatrix} \begin{bmatrix} -9883 \\ 6920 \\ 0 \end{bmatrix} = \begin{bmatrix} 10{,}989 \\ -1646 \\ -4700 \end{bmatrix}$$

or $\boldsymbol{r}_{IJK} = 10{,}989\hat{\boldsymbol{I}} - 1646\hat{\boldsymbol{J}} - 4700\hat{\boldsymbol{K}}$ km.

The scalar distances are necessarily the same in both coordinate frames: at perigee, $r_{IJK} = r_{pqw} = 7000$ km and for $\theta = 145°$ $r_{IJK} = r_{pqw} = 12{,}065$ km.

In some cases, the necessary coordinate transformations can be simply performed by rotating the reference frames, as shown in Figs. 4.13 and 4.14 and in the discussion and example earlier. Oftentimes, however, the reference frames are offset in their origins, in addition to undergoing the rotations explained earlier. This would be the case for converting from the topocentric to the geocentric reference frame. In this case, in addition to the offset in the origin of the topocentric frame from the center of the geocentric frame, it is also necessary to take into account that the Earth, and the topocentric reference frame along with it, are rotating relative to the fixed geocentric frame. This situation is shown in Fig. 4.15.

The radius vector to an object (such as an orbiting satellite) in the topocentric, or *SEZ* coordinate system, is indicated by the vector $\boldsymbol{\rho}$. This is different from the radius vector $\boldsymbol{r}$ from the Earth's center. The quantity θ is here the eastward longitude relative to the first point of Aries, ϒ. This would be the geographic longitude relative to the Greenwich meridian plus the angle of the Greenwich meridian relative to the first point of Aries. The quantity ϕ is the latitude north from the equator (the $\hat{\boldsymbol{I}} - \hat{\boldsymbol{J}}$ plane). The two radii are simply related by

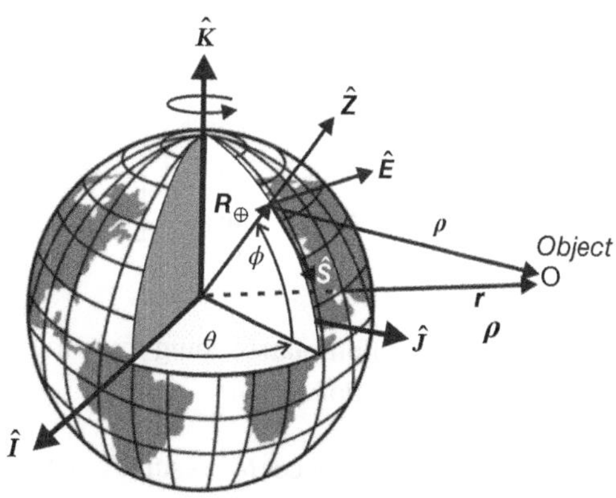

Figure 4.15 Topocentric (*SEZ*) and geocentric (*IJK*) reference frames.

$$\boldsymbol{r} = \boldsymbol{\rho} + \boldsymbol{R}_{\oplus}, \tag{4.20}$$

where $\boldsymbol{R}_{\oplus}$ is the radius vector from the Earth's center to the origin of the topocentric *SEZ* system. In this case, the Earth is taken to be a perfect sphere with radius $R_{\oplus}$ (The oblateness of the Earth due to its rotation is discussed in Chapter 7). To determine the velocity in geocentric coordinates, the motion of the *SEZ* frame due to the rotation of the Earth needs to be taken into account. The true velocity of the object in the fixed *IJK* frame is the sum of the velocity in the moving *SEZ* frame, plus the velocity of the point in the *SEZ* frame due to the Earth's rotation. The *SEZ* frame moves with the vector velocity $\boldsymbol{\omega}_{\oplus} \times \boldsymbol{R}_{\oplus}$ due to the Earth's rotation, where the angular vector velocity of the Earth's rotation is $\boldsymbol{\omega}_{\oplus} = \omega_{\oplus}\hat{\boldsymbol{K}}$. The rotation of the frame results in the added vector velocity for the object of $\boldsymbol{\omega}_{\oplus} \times \boldsymbol{\rho}$. Thus, the total contribution to the measured velocity in the *IJK* frame is $\boldsymbol{\omega}_{\oplus} \times \boldsymbol{R}_{\oplus} + \boldsymbol{\omega}_{\oplus} \times \boldsymbol{\rho} = \boldsymbol{\omega}_{\oplus} \times (\boldsymbol{R}_{\oplus} + \boldsymbol{\rho}) = \boldsymbol{\omega}_{\oplus} \times \boldsymbol{r}$. The latter is the vector product of the angular rotation rate of the Earth (also the angular rotation rate of all points in the *SEZ* reference frame) and the radial distance to the object in the *IJK* frame. Taking the angular velocity of the Earth's rotation to be $\boldsymbol{\omega}_{\oplus} = \omega_{\oplus}\hat{\boldsymbol{K}}$, the velocity in the *IJK* frame is thus given by:

$$\boldsymbol{V} = \dot{\rho} + \boldsymbol{\omega}_{\oplus} \times \boldsymbol{r} \tag{4.21}$$

where $\dot{\boldsymbol{\rho}}$ is the velocity observed in the *SEZ* frame. The vectors $\boldsymbol{\rho}$, $\boldsymbol{r}$, $\dot{\rho}$, and $\boldsymbol{V}$ have different values depending on the reference frame and need to be reconciled by performing the transformation described later.

The rotations between the *SEZ* and *IJK* frames can be taken into account as follows. First, the *SEZ* frame is rotated by the difference in longitude, θ, as shown in Fig. 4.15. The components of the transformation matrix due to this single rotation are similar to Eq. 4.13 and are:

$$[\boldsymbol{Q}]_1 = \begin{bmatrix} \cos\theta & \sin\theta & 0 \\ -\sin\theta & \cos\theta & 0 \\ 0 & 0 & 1 \end{bmatrix}. \tag{4.22}$$

This intermediate *SEZ* frame is then rotated around the new $\hat{\boldsymbol{E}}$ axis by the colatitude, $90° - \phi$. This gives the transformation matrix due to that rotation only of

$$[\boldsymbol{Q}]_2 = \begin{bmatrix} \cos(90° - \phi) & 0 & -\sin(90° - \phi) \\ 0 & 1 & 0 \\ \sin(90° - \phi) & 0 & \cos(90° - \phi) \end{bmatrix} = \begin{bmatrix} \sin\phi & 0 & -\cos\phi \\ 0 & 1 & 0 \\ \cos\phi & 0 & \sin\phi \end{bmatrix}. \tag{4.23}$$

The result after both rotations is the transformation matrix $[\boldsymbol{Q}] = [\boldsymbol{Q}]_2[\boldsymbol{Q}]_1$ or

$$[\boldsymbol{Q}] = \begin{bmatrix} \sin\phi & 0 & -\cos\phi \\ 0 & 1 & 0 \\ \cos\phi & 0 & \sin\phi \end{bmatrix} \begin{bmatrix} \cos\theta & \sin\theta & 0 \\ -\sin\theta & \cos\theta & 0 \\ 0 & 0 & 1 \end{bmatrix} = \begin{bmatrix} \sin\phi\cos\theta & \sin\phi\sin\theta & -\cos\phi \\ -\sin\theta & \cos\theta & 0 \\ \cos\phi\cos\theta & \cos\phi\sin\theta & \sin\phi \end{bmatrix}. \tag{4.24}$$

This is the rotation matrix going from the *IJK* to *SEZ* coordinates. The rotation matrix for the reverse transformation is:

$$[\boldsymbol{Q}]^{-1} = \begin{bmatrix} \sin\phi\cos\theta & -\sin\theta & \cos\phi\cos\theta \\ \sin\phi\sin\theta & \cos\theta & \cos\phi\sin\theta \\ -\cos\phi & 0 & \sin\phi \end{bmatrix}. \tag{4.25}$$

These rotations only account for the angular changes between the two reference frames. To complete the conversion between the geocentric and topocentric reference frames, the offset due to the Earth's radius as well as the Earth's rotation need to be considered, as given in Eqs. 4.20 and 4.21.

Example Problem 4e

A ground station in Washington State is at a latitude of 47° N and a longitude of 122° W. At a time when the Greenwich meridian is 92° east of the first point of Aries, the ground station detects an orbiting satellite with the elements:

$$\boldsymbol{\rho}_{SEZ} = 16{,}854\hat{\boldsymbol{S}} + 13{,}305\hat{\boldsymbol{E}} + 9337\hat{\boldsymbol{Z}}\ \text{km and } \dot{\boldsymbol{\rho}}_{SEZ} = -0.706\hat{\boldsymbol{S}} + 1.671\,\hat{\boldsymbol{E}} - 0.658\,\hat{\boldsymbol{Z}}\ \text{km/s}.$$

What are the position and velocity of the object in geocentric *IJK* coordinates? The satellite is suspected to be a GPS[3] satellite in a circular orbit – can this be verified?

Solution:

The measured elements in the *SEZ* frame are first rotated to align with *IJK* coordinates using the inverse transformation matrix (Eq. 4.25). Noting that $\phi = +47°$ and $\theta = 92°–122° = -30°$ gives the matrix components as:

$$[Q]^{-1} = \begin{bmatrix} 0.6334 & 0.5000 & 0.5906 \\ -0.3657 & 0.8660 & -0.3410 \\ -0.6820 & 0 & 0.7314 \end{bmatrix}.$$

The radius vector first needs to be corrected for the offset of the *SEZ* system from the Earth's center, as given in Eq. 4.20:

$$\boldsymbol{r} = \boldsymbol{\rho} + \boldsymbol{R}_{\oplus} = \left(16{,}854\hat{\boldsymbol{S}} + 13{,}305\hat{\boldsymbol{E}} + 9337\hat{\boldsymbol{Z}}\right) + 6378\hat{\boldsymbol{Z}}$$
$$= 16{,}854\hat{\boldsymbol{S}} + 13{,}305\hat{\boldsymbol{E}} + 15{,}715\hat{\boldsymbol{Z}}\ \text{km}.$$

Applying the transformation matrix then gives components in the *IJK* frame:

$$\begin{bmatrix} r_I \\ r_J \\ r_K \end{bmatrix} = \begin{bmatrix} 0.6334 & 0.5000 & 0.5906 \\ -0.3657 & 0.8660 & -0.3410 \\ -0.6820 & 0 & 0.7314 \end{bmatrix} \begin{bmatrix} 16{,}854 \\ 13{,}305 \\ 15{,}715 \end{bmatrix} = \begin{bmatrix} 26{,}612 \\ -0.2 \\ 0 \end{bmatrix} \text{km}.$$

Neglecting the two small terms, $\boldsymbol{r} = 26{,}612\,\hat{\boldsymbol{I}}$ km in geocentric *IJK* coordinates.

The velocity vector in *SEZ* coordinates can similarly be converted to *IJK* coordinates:

$$\begin{bmatrix} V_I \\ V_J \\ V_K \end{bmatrix} = \begin{bmatrix} 0.6334 & 0.5000 & 0.5906 \\ -0.3657 & 0.8660 & -0.3410 \\ -0.6820 & 0 & 0.7314 \end{bmatrix} \begin{bmatrix} -0.706 \\ 1.671 \\ -1.013 \end{bmatrix} = \begin{bmatrix} 0 \\ 1.927 \\ 0 \end{bmatrix}.$$

Finally, this velocity is corrected for the Earth's rotation as given in Eq. 4.21 to give:

$$\boldsymbol{V} = \dot{\boldsymbol{\rho}} + \boldsymbol{\omega}_{\oplus} \times \boldsymbol{r} = 1.927\hat{\boldsymbol{J}} + 0.00007283\hat{\boldsymbol{K}} \times 26{,}612\,\hat{\boldsymbol{I}} = 1.927\hat{\boldsymbol{J}} + 1.938\,\hat{\boldsymbol{J}} = 3.865\hat{\boldsymbol{J}}\ \text{km/s},$$

where $\boldsymbol{\omega}_{\oplus} = 7.283 \times 10^{-5}\,\hat{\boldsymbol{K}}\,\text{s}^{-1}$ is the angular rotation vector of the earth. The circularity of the orbit can be checked by comparing the expected velocity for a circular orbit with that radius.

---

3 GPS is the acronym for Global Positioning System, the American network of 38 navigation satellites for military and civilian applications. The satellites are arranged in six medium-Earth-orbit (MEO) planes.

The scalar values of the measured radius and velocity are $r = 26{,}612$ km and $V = 3.865$ km/s, respectively. The expected scalar velocity for a circular orbit is $V = \sqrt{\frac{\mu}{r}} = \sqrt{\frac{398{,}600}{26{,}612}} = 3.870$ km/s, which agrees well with the measured value. That the vectors $\boldsymbol{V}$ and $\boldsymbol{r}$ are clearly orthogonal indicates a zero flight path angle, thus confirming that that satellite is in a circular orbit. The period of the orbit is $T = 2\pi r/V = 12$ hr, which is typical for a GPS satellite.

4.4 Problems

4.1 Which of these coordinate system transformations can be done by considering rotations only?

(a) Perifocal to geocentric
(b) Perifocal to topocentric
(c) Topocentric to geocentric
(d) Geocentric to heliocentric

4.2 Suppose the angular momentum vector, $\boldsymbol{h}$, of an orbit is known in geocentric coordinates $(\hat{\boldsymbol{I}}, \hat{\boldsymbol{J}}, \hat{\boldsymbol{K}})$. Using this information only, is it possible to determine the following (yes/no)?

(a) The inclination, i
(b) The longitude of the ascending node, Ω
(c) The argument of periapsis, ω
(d) The true anomaly, θ?

4.3 The radius and velocity vectors of an Earth-orbiting satellite are, in the geocentric reference frame, $\boldsymbol{r} = -0.8\hat{\boldsymbol{I}} + 0.6\hat{\boldsymbol{J}} + 0.5\hat{\boldsymbol{K}}$ DU and $\boldsymbol{V} = -0.4\hat{\boldsymbol{I}} - 0.8\hat{\boldsymbol{J}} + 0.6\hat{\boldsymbol{K}}$ DU/TU, using the DU and TU units introduced in Chapter 2.
Find all of the orbital elements (a, e, i, Ω, ω, θ), expressed in DU and TU units as needed.

4.4 A near-earth space object has the following data at a point in time: radius vector $\boldsymbol{r} = 2\hat{\boldsymbol{I}} + \hat{\boldsymbol{J}}$ DU, velocity vector $\boldsymbol{V} = -0.5\hat{\boldsymbol{I}} + 0.2\hat{\boldsymbol{K}}$ DU/TU.
All coordinates in this problem are geocentric and recall that $\mu = 1$ DU^3/TU^2. Determine (i) the inclination of the trajectory, i (degrees); (ii) whether the orbit is direct or retrograde; and (iii) the period of the orbit (in TU).

4.5 The elliptical orbit of an Earth satellite has the following orbital elements at epoch:

$$a = 2\times10^4 \text{ km} \quad e = 0.5 \quad \Omega = 20^\circ \quad \omega = 60^\circ \quad i = 90^\circ \quad \theta = 30^\circ.$$

(a) Sketch the orbit in the geocentric reference frame (draw the portion of the orbit for positive $\hat{\boldsymbol{I}}, \hat{\boldsymbol{J}}, \hat{\boldsymbol{K}}$ only). Clearly identify in your sketch the orbit parameters Ω, ω, i, and θ, the location of perigee, and the satellite's location at epoch;
(b) What is the position radius vector, $\boldsymbol{r}$, at epoch in geocentric coordinates? Express your answer in "distance units" where 1 DU = 6378 km.

4.6 A space object near the earth has the following properties, in geocentric coordinates, at a given time:

Radius vector $\boldsymbol{r} = 25{,}000\,\hat{\boldsymbol{K}}$ (km), velocity vector $\boldsymbol{V} = -4.0\hat{\boldsymbol{J}} - 6.0\hat{\boldsymbol{K}}$ (km/s).

Determine:

(a) The (scalar) eccentricity, e,
(b) The perigee radius, r_p (km),
(c) The inclination, i.
(d) What time after perigee will be required to reach a true anomaly of $\theta = 145°$?

4.7 An object has the coordinates $x = -10, y = +5$ in the x-y plane (arbitrary units). What are the coordinates in the axis rotated counterclockwise by 40° in the same plane? Designate those coordinates at x' and y'. If the object has the velocity components $\dot{x} = 8$ and $\dot{y} = 2$ (again arbitrary units), what are those components in x'-y' coordinates?

4.8 Derive the transformation matrix resulting from rotations about the three axes (see Eqs. 4.13–4.15). Consider first a rotation about the x-axis by the angle α, then about the y-axis by β, then finally about the z-axis by the angle γ. Compare this with the transformation matrix resulting from doing the rotations in reverse order.

4.9 An Earth satellite has, in geocentric coordinates, the radius and velocity vectors:

$\boldsymbol{r}_{IJK} = 8000\hat{\boldsymbol{I}} + 5000\hat{\boldsymbol{J}}$ km and $\boldsymbol{V} = 4\hat{\boldsymbol{J}} + 6\hat{\boldsymbol{K}}$ km/s.

What are the vectors in a reference frame $\hat{\boldsymbol{i}}\hat{\boldsymbol{j}}\hat{\boldsymbol{k}}$, which is rotated by the three angles in order as follows: $\alpha = 30°$, $\beta = 10°$, $\gamma = 60°$ (See Eqs. 4.13–4.15)

4.10 What are the corresponding vectors in perifocal coordinates for the satellite in Problem 9? What will be the position, in geocentric coordinates, of the satellite four hours later?

4.11 A new object is detected from a ground station situated at 34.7° north latitude and 120.6° west longitude at local noon on the date of the winter solstice (December 21). The measured radius and velocity are $\boldsymbol{\rho}_{SEZ} = 12{,}000\hat{\boldsymbol{S}} + 10{,}000\hat{\boldsymbol{E}} + 9000\hat{\boldsymbol{Z}}$ km and $\dot{\boldsymbol{\rho}}_{SEZ} = -5.0\hat{\boldsymbol{S}} + 4.0\hat{\boldsymbol{E}} - 3.0\hat{\boldsymbol{Z}}$ km/s. What are the radius and velocity in geocentric coordinates? Is the trajectory that of a satellite, a departing deep space probe, or a ballistic missile?

4.12 A space probe to Mars is launched from the Earth on a day halfway between the autumnal equinox and the winter solstice. The spacecraft departs Earth with the excess velocity $V_\infty = 4$ km/s on a trajectory tangential to the Earth's orbit. The geometry is as shown in Fig. 2f. Determine the scalar velocity relative to the sun as well as the velocity vector in heliocentric coordinates. The orbital velocity of the Earth is 29.8 km/s.

5

Orbital Maneuvers

Maneuvers are often performed to change the trajectory of a spacecraft. Such maneuvers include raising the altitude of an orbit, intercepting an object in a different orbit, departing the Earth on an interplanetary mission, inserting a spacecraft into orbit around a planet, changing orbital elements such as its inclination, and more. This chapter will discuss the basic orbital maneuvers, the velocity changes required, and the implications for mission duration. The chapter outline is as follows:

- Impulsive maneuvers
- The basic maneuver: The Hohmann Transfer
- Limiting geometries for transfer between two orbits
- Other orbital maneuvers
  - Bi-elliptic Hohmann transfer
  - Phasing maneuver
  - General transfer (noncollinear maneuvers)
  - Chasing maneuver
  - Apse line rotation
- Plane change maneuvers
- Combined maneuvers

5.1 Impulsive Maneuvers

The first consideration in considering orbital maneuvers is the nature of the rocket burn(s) required. The maneuvers discussed here are assumed to be impulsive – essentially instantaneous relative to the duration of the mission. This can be appreciated by considering a practical example. The Trans-Lunar Insertion (TLI) maneuver in the *Apollo 16* mission required a rocket-burn duration in Low Earth Orbit (LEO) of 340 seconds. The subsequent flight time to Lunar Orbit Insertion (LOI) on the transfer trajectory to the moon was 71.817 hours, or 258,540 seconds. This gives the ratio of the burn time to transfer time of $\frac{\Delta t_{TLI}}{\Delta t_{xfer}} = \frac{340}{258{,}540} = 0.0013 = 0.13\%$. Beyond the comparatively short duration, the impulsive-maneuver assumption is also practical in light of the fact that there is often a need for subsequent corrections to the trajectory. These mid-course corrections can be required to compensate for inaccuracies in a maneuver, such as the actual burn time or the flight

Introduction to Orbital Mechanics: A Concise and Practical Approach, First Edition. James C. Hermanson.

path angle being slightly different from nominal. Long-duration interplanetary missions will be subject to perturbations from other planets. In addition, insertion maneuvers are sometimes designed to be intentionally "off course" so that spent rocket stages will not follow a spacecraft to its destination. Practical exceptions to the impulse-maneuver scenario include solar sails and continuous-thrust electric propulsion systems; the latter are discussed in Chapter 10.

5.2 The Hohmann Transfer

The most basic orbital maneuver is the *Hohmann Transfer*, shown in Fig. 5.1 for a transfer between two circular, coplanar orbits. As seen in the figure, the Hohmann transfer has "just enough" change in radial distance to reach the target orbit. This transfer could be from a lower orbit to a higher one, or the reverse (the former situation is shown in Fig. 5.1). This is also the maneuver that normally requires the least total change in velocity. Minimizing the total velocity change is highly desirable, as the required velocity changes for many maneuvers are of the order of kilometers/second, which can place very large demands on the propulsion required. Rocket propulsion is discussed in Chapter 10.

The Hohmann Transfer maneuver shown in Fig. 5.1 consists of three elements: the initial parking orbit, ①, the target/destination orbit, ②, and the transfer trajectory, denoted as ⓣ. The thrusting maneuvers to depart orbit ① and for insertion into orbit ② are both exactly tangent to the corresponding circular orbits. The velocities in circular orbits ① and ② are:

$$V_{c1} = \sqrt{\frac{\mu}{r_1}} \text{ and } V_{c2} = \sqrt{\frac{\mu}{r_2}}. \tag{5.1}$$

The velocity on the transfer trajectory at any point was given by Eq. 2.42, which in this case can be written as:

$$V = \sqrt{2\mu\left(\frac{1}{r} - \frac{1}{r_1 + r_2}\right)}.$$

This gives the velocities on the transfer trajectory at departure from orbit ① and for the arrival at orbit ② as:

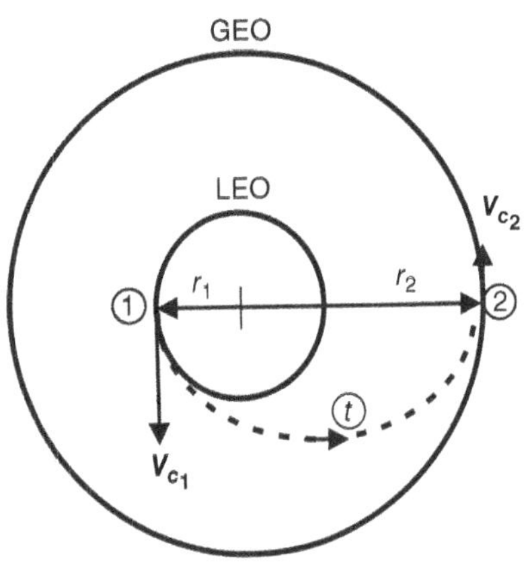

Figure 5.1 Hohmann Transfer between two circular, coplanar orbits.

$$V_1 = \sqrt{2\mu\left(\frac{1}{r_1} - \frac{1}{r_1 + r_2}\right)} \text{ and } V_2 = \sqrt{2\mu\left(\frac{1}{r_2} - \frac{1}{r_1 + r_2}\right)}. \tag{5.2}$$

The required change in velocity to initiate the transfer from orbit ① will be:

$$V_1 - V_{c1} = \Delta V_1 = \sqrt{2\mu\left(\frac{1}{r_1} - \frac{1}{r_1 + r_2}\right)} - \sqrt{\frac{\mu}{r_1}}, \tag{5.3}$$

which can be seen to be positive, i.e. a boost maneuver. If nothing is done upon arrival at target orbit ②, the satellite will continue on the transfer orbit and fall back to the original insertion point in orbit ①. Insertion into the target orbit ② thus requires a second, smaller velocity change in the amount of:

$$V_{c2} - V_2 = \Delta V_2 = \sqrt{\frac{\mu}{r_2}} - \sqrt{2\mu\left(\frac{1}{r_2} - \frac{1}{r_1 + r_2}\right)}, \tag{5.4}$$

which is also a boost maneuver. The total velocity change required for the transfer will be $\Delta V_{tot} = \Delta V_1 + \Delta V_2$.

The flight time on the transfer trajectory is simply that of a half-ellipse (see Eq. 3.5):

$$T_t = \pi\sqrt{\frac{a^3}{\mu}} = \pi\sqrt{\frac{\left(\frac{r_1 + r_2}{2}\right)^3}{\mu}}. \tag{5.5}$$

Example Problem 5a

A GPS satellite in LEO with an altitude of 300 km is to be transferred to a 12-hour circular orbit. What are the required velocity changes to depart LEO, and to insert the satellite into the 12-hour orbit, using a Hohman Transfer?

Solution:

The radius of the LEO is $r_{LEO} = 300 + 6378 = 6678$ km, with the orbital velocity $V_{LEO} = \sqrt{\frac{\mu}{r}} = \sqrt{\frac{398{,}600}{6{,}678}} = 7.726$ km/s. The radius of the 12-hour orbit was seen in Example Problem 4e to be $r_{GPS} = 26{,}612$ km, with an orbital velocity of $V_{GPS} = 3.870$ km/s. The velocity required at the start of the transfer maneuver from Eq. 5.2 is $V_1 = \sqrt{2\mu\left(\frac{1}{r_{LEO}} - \frac{1}{r_{GPS} + r_{LEO}}\right)} = \sqrt{2(398{,}600)\left(\frac{1}{6{,}678} - \frac{1}{6{,}678 + 26{,}612}\right)} = 9.769$ km/s. Thus, the required increase in velocity to depart LEO is $V_1 - V_{LEO} = 9.769 - 7.726 = 2.043$ km/s. The velocity of the satellite when it reaches the target orbit, V_2, is similarly found to be:

$$V_2 = \sqrt{2\mu\left(\frac{1}{r_{GPS}} - \frac{1}{r_{GPS} + r_{LEO}}\right)} = \sqrt{2(398{,}600)\left(\frac{1}{26{,}612} - \frac{1}{6{,}678 + 26{,}612}\right)} = 2.451 \text{ km/s}.$$

Insertion into the target GPS orbit requires a second, smaller velocity change in the amount $V_{GPS} - V_2 = 3.870 - 2.451 = 1.419$ km/s. As seen earlier, both required orbital maneuvers are boosts. The total velocity change requirement is $\Delta V_{tot} = \Delta V_1 + \Delta V_2 = (V_1 - V_{LEO}) + (V_{GPS} - V_2) = 2.043 + 1.419 = 3.462$ km/s. The transfer time will be from Eq. 5.5

$$T_t = \pi\sqrt{\frac{\left(\frac{r_{LEO} + r_{GPS}}{2}\right)^3}{\mu}} = \pi\sqrt{\frac{\left(\frac{6678 + 26{,}612}{2}\right)^3}{398{,}600}} = 10{,}686 \text{ s} = 2.97 \text{ hr}.$$

Other examples of Hohmann transfers include:

1) A transfer from LEO ($r_1 = 6900$ km) to GEO ($r_2 = 42{,}164$ km):

$\rightarrow \Delta V_1 = 2.42$ km/s, $\Delta V_2 = 1.47$ km/s, $T_t = 19{,}120$ s $= 5.28$ hr.

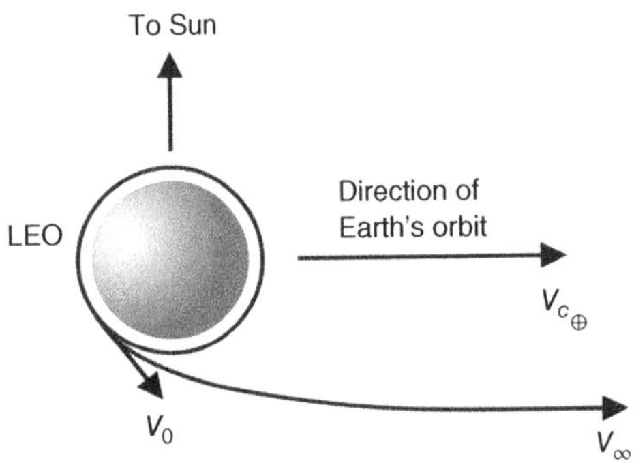

Figure 5.2 Earth departure on Hohmann transfer to Mars.

2) A transfer from LEO ($r_1 = 6900$ km) to the orbit of the Moon ($r_2 = 385{,}000$ km):

$$\rightarrow \Delta V_1 = 3.05 \text{ km/s}, \; T_t = 5.00 \text{ d}.$$

Note that the increase in injection velocity required is comparatively small compared to the injection velocity in case (1).

3) A transfer from LEO to Mars. The orbital radii of the Earth and Mars are respectively $r_1 = 149.5 \times 10^6$ km and $r_2 = 227.8 \times 10^6$ km, and the velocity of the Earth in its orbit is $V_{c\oplus} = 29.79$ km/s. This gives for this Hohmann transfer (in this case using the gravitational parameter of the sun):

$$\rightarrow \Delta V_1 = V_1 - V_{c\oplus} = 2.95 \text{ km/s}, \; T_t = 258.6 \text{ d} = 8\tfrac{1}{2} \text{ m}.$$

The calculated velocity change in this case curiously appears to be less than that for transfer (2) to the Moon. The explanation is that this is the velocity difference relative to the *Sun*, which sees the spacecraft departing Earth with the velocity V_1. ΔV_1 is thus the velocity after departing the Earth, V_∞, as shown in Fig. 5.2 (this is the same hyperbolic-departure situation as Example Problem 2f).

The actual velocity at the point of departure from LEO, V_0, can be determined using Eq. 2.52 to give:

$$V_0 = \sqrt{\frac{2\mu_\oplus}{r_{LEO}} + V_\infty^2} = \sqrt{\frac{2(398{,}600)}{6900} + (2.95)^2} = 11.15 \text{ km/s}.$$

Subtracting from this, the velocity in LEO $V_{LEO} = \sqrt{\frac{\mu_\oplus}{r}} = \sqrt{\frac{398{,}600}{6900}} = 7.60$ km/s gives the actual required boost to depart LEO for Mars as $\Delta V_0 = V_0 - V_{LEO} = 3.55$ km/s, which exceeds the velocity increase needed to reach the Moon. The cited Earth departure velocity of $V_\infty = 4.60$ km/s in Example Problem 2f indicates that the Viking spacecraft's transit to Mars was on a more energetic (and faster) trajectory than a Hohmann transfer.

5.3 Limiting Geometries for Transfer Between Two Orbits

Many other transfer trajectories are possible than the Hohmann transfer. Consider a general two-impulse transfer between two circular orbits as shown in Fig. 5.3. Each transfer involves an elliptical transfer trajectory with periapsis radius r_p and apoapsis radius r_a. Take the radii of the two circular orbits to be respectively r_1 and r_2. The desired transfer will perform one impulsive maneuver to depart orbit ① and a second impulsive maneuver to enter orbit ②.

In the first example, Fig. 5.3a, the periapsis radius of the transfer trajectory is inside the inner orbit ($r_p < r_1$) and the apoapsis is outside the outer orbit ($r_a > r_2$), so the transfer is possible. In the second case, Fig. 5.3b, $r_a > r_2$ but $r_p > r_1$, so departing the inner orbit is not possible. In the third case, Fig. 5.3c, $r_p < r_1$ but $r_a < r_2$, so the transfer trajectory will not reach the outer target orbit.

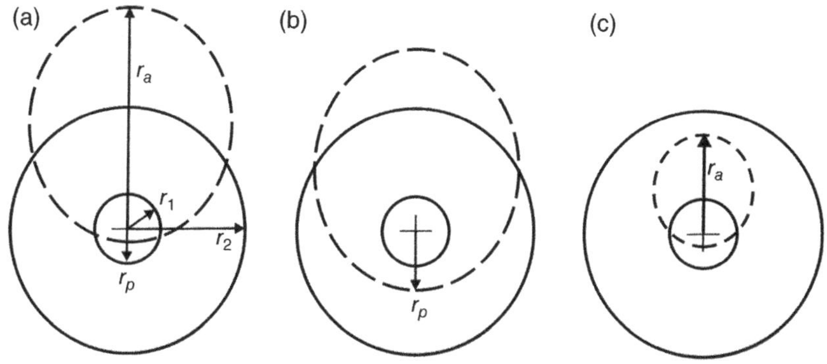

Figure 5.3 Transfer possibilities between two circular orbits. Cases (a) though (c) are explained in the preceding text.

Figure 5.4 Limiting ranges of transfer trajectories. H designates the Hohmann transfer.

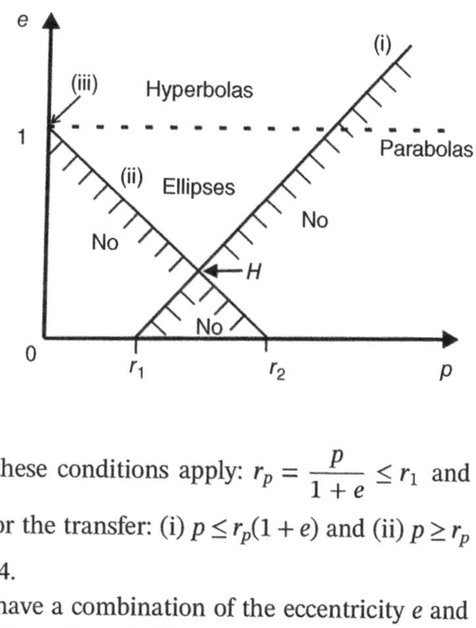

This indicates that for the transfer to be possible, these conditions apply: $r_p = \dfrac{p}{1+e} \le r_1$ and $r_a = \dfrac{p}{1-e} \ge r_2$. This dictates these two conditions for the transfer: (i) $p \le r_p(1+e)$ and (ii) $p \ge r_p(1-e)$. These two limiting lines are shown in Fig. 5.4.

In order to be possible, a transfer trajectory must have a combination of the eccentricity e and orbital parameter p that lies above both limiting lines (i) and (ii). Elliptic ($e < 1$), parabolic ($e = 1$), and hyperbolic ($e > 1$) transfer trajectories are all possible, depending on the value of the parameter p. The trajectory that is "just possible," that just satisfies the limiting conditions, is the Hohmann transfer. The special case labeled (iii) in Fig. 5.4, $p = 0$, $e = 1$, is a degenerate parabola, as discussed in Chapter 2.

5.4 Bi-Elliptic Hohmann Transfer

A clever variation on the basic Hohmann transfer is the *bi-elliptic* Hohmann transfer shown in Fig. 5.5. The bi-elliptic transfer is essentially the combination of two Hohmann transfers. The first transfer is to a "super apogee," r^*, that is at a larger distance than the radius of the target orbit, r_2. The second maneuver is another Hohmann transfer from the super apogee to the target orbit, as shown. The potential benefit of this combination of maneuvers is that the slow velocity at the point (*) can allow for substantial changes to the trajectory for a comparatively small change in velocity.

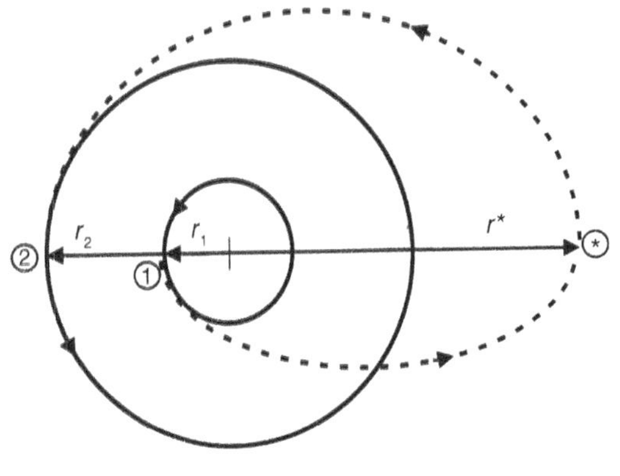

Figure 5.5 Bi-elliptic Hohmann transfer.

Since each maneuver in this transfer is a Hohmann transfer, the velocity at a given point is given by this variation on Eq. 2.42: $V = \sqrt{2\mu\left(\frac{1}{r} - \frac{1}{a}\right)}$, where at point ① $a = r_1 + r^*$ and $r = r_1$. The velocity boost needed at point ① is then

$$\Delta V_1 = \sqrt{2\mu\left(\frac{1}{r_1} - \frac{1}{r_1 + r^*}\right)} - \sqrt{\frac{\mu}{r_1}}. \tag{5.6}$$

Similarly, the velocity upon arrival at the point (*) is determined by setting $r = r^*$ still with $a = r_1 + r^*$. For the next trajectory, the transfer from (*) to ②, the semi-major axis becomes $a = r^* + r_2$, which gives the required velocity change at that point as:

$$\Delta V^* = \sqrt{2\mu\left(\frac{1}{r^*} - \frac{1}{r^* + r_2}\right)} - \sqrt{2\mu\left(\frac{1}{r^*} - \frac{1}{r_1 + r^*}\right)}. \tag{5.7}$$

This is again a boost maneuver as the transfer trajectory from (*) to ② has a higher energy than the transfer from ① to point (*). Finally, the maneuver for insertion into the target orbit ② requires the velocity change:

$$\Delta V_2 = \sqrt{2\mu\left(\frac{1}{r_2} - \frac{1}{r^* + r_2}\right)} - \sqrt{\frac{\mu}{r_2}}. \tag{5.8}$$

This is a braking maneuver needed to reduce the velocity to match that in the target orbit (expressed as a positive quantity here). The entire transfer then requires a total velocity change of $\Delta V_{total} = \Delta V_1 + \Delta V^* + \Delta V_2$.

The total velocity-change requirement for the bi-elliptic Hohmann transfer thus depends, for given radii r_1 and r_2, on the choice of the radius r^*. Example situations are shown below, where the total velocity change is normalized by the corresponding velocity change for the basic, two-impulse Hohmann transfer.

As can be seen in Fig. 5.6, not all bi-elliptic transfers offer an advantage in the required total velocity change compared with the standard Hohmann transfer. The advantage that does result is typically quite small (e.g. approximately 3.5% for $r_2/r_1 = 20$). There is, in fact, no total velocity advantage for r_2/r_1 less than approximately 12–15, depending on the value of r^*/r_1 (these values are farther out than the orbit of Saturn for an interplanetary mission from Earth). Determining the precise limiting values of r_2/r_1 is provided as an exercise.

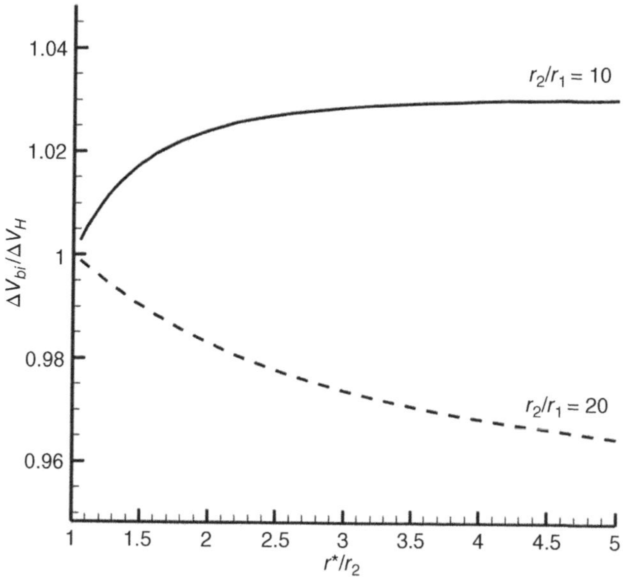

Figure 5.6 Total normalized velocity change for selected bi-elliptic Hohmann transfers.

Unsurprisingly, the transfer time for the bi-elliptic maneuver is always longer (and can be much longer) than that for the basic Hohmann transfer, as shown in Fig. 5.7. The total transfer time is the sum of the times on both half-elliptical segments of the transfer, or:

$$T_{bi} = \pi\sqrt{\frac{\left(\frac{r_1 + r^*}{2}\right)^3}{\mu}} + \pi\sqrt{\frac{\left(\frac{r^* + r_2}{2}\right)^3}{\mu}}. \tag{5.9}$$

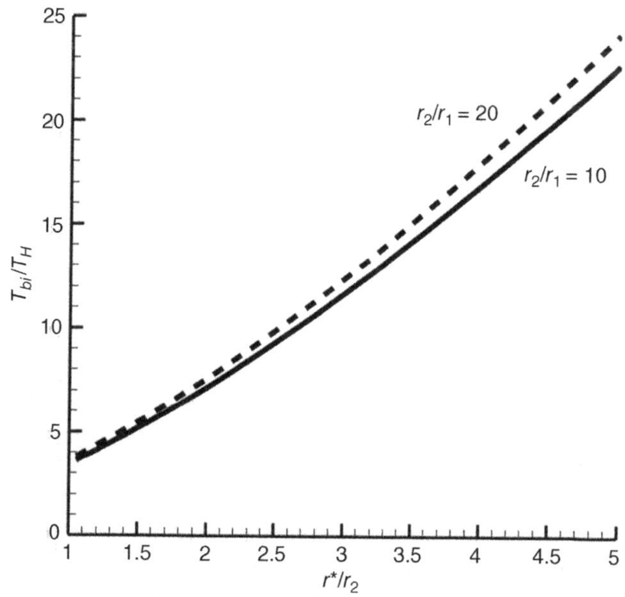

Figure 5.7 Normalized transfer time for selected bi-elliptic Hohmann transfers.

This can be compared to the transfer time for the corresponding basic Hohmann transfer of

$$T_H = \pi\sqrt{\frac{\left(\frac{r_1 + r_2}{2}\right)^3}{\mu}}.$$

It is apparent from Fig. 5.7 that the total time associated with a bi-elliptic transfer can considerably exceed that of a basic Hohmann transfer. This, combined with the small decrease in total velocity and in only some circumstances, results in the bi-elliptic maneuver rarely being used in practice.

5.5 Phasing Maneuver

Another variation on the Hohmann transfer is the *phasing maneuver* shown in Fig. 5.8. The phasing maneuver is essentially a Hohmann transfer from, and back to, the same orbit. This is useful for accomplishing a rendezvous with a target object in the same orbit, but somewhat ahead or behind the spacecraft.

The transfer orbit in a phasing maneuver can be either inside or outside of the target orbit, as shown in Fig. 5.8. Utilizing the inner transfer trajectory (identified as orbit ① in the figure) will result in arriving at the target more quickly; the outer trajectory, ②, will result in a delayed return to the target orbit. Thus, trajectory ① can be used to catch up with an object orbiting ahead of the spacecraft, and correspondingly trajectory ② would be useful in reaching an object orbiting behind.

The velocity decrease required to initiate transfer ① is given by a variation on Eq. 5.6:

$$\Delta V_1 = V_c - V_1 = \sqrt{\frac{\mu}{r_1}} - \sqrt{2\mu\left(\frac{1}{r_c} - \frac{1}{r_c + r_1}\right)}, \tag{5.10}$$

where V_c is the constant speed in the circular target orbit, and the velocity difference is taken to be positive. A boost maneuver of the identical magnitude will then be required for reinsertion into the target orbit. Thus, the total velocity change will be:

$$\Delta V_{total_①} = 2\Delta V_1 = 2\left[\sqrt{\frac{\mu}{r_c}} - \sqrt{2\mu\left(\frac{1}{r_c} - \frac{1}{r_c + r_1}\right)}\right]. \tag{5.11}$$

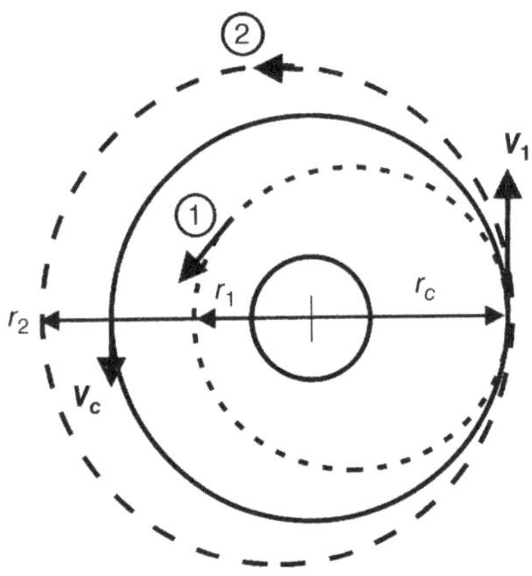

Figure 5.8 Phasing maneuver.

The corresponding total velocity change for transfer trajectory ② is:

$$\Delta V_{total_②} = 2\Delta V_2 = 2\left[\sqrt{2\mu\left(\frac{1}{r_c} - \frac{1}{r_c + r_2}\right)} - \sqrt{\frac{\mu}{r_c}}\right], \tag{5.12}$$

where in this case a boost in velocity is needed to start the transfer, followed by a corresponding decrease upon rearrival at the target orbit. The total transfer time in both cases is simply the period of the elliptical orbit, as given by Eq. 3.5.

Perhaps counter-intuitively, to reach an object ahead in the same orbit using a phasing maneuver first requires a velocity *decrease*, or braking maneuver, into a lower, faster orbit. Similarly, to reach an object behind in the same orbit is first accomplished by increasing velocity to boost into the higher, slower transfer orbit. It should be remarked that the phasing maneuver only suffices to bring a pursuing spacecraft into the vicinity of the target object – the actual close-in encounter (within a few kilometers) is more complex, as discussed in Chapter 9.

Example Problem 5b

The International Space Station (ISS) is in a circular orbit with an average altitude of 400 km (taking the Earth's radius to be 6378 km gives an orbital radius of $r_c = 6778$ km). Suppose a *Dragon* spacecraft is in the same orbit, but three minutes behind the ISS. What phasing maneuver would be required to reach the location of the ISS?

Solution:

The period of the orbit of the ISS is $T_{ISS} = 2\pi\sqrt{\frac{r_c{}^3}{\mu}} = 2\pi\sqrt{\frac{(6778)^3}{398{,}600}} = 5553\text{ s} = 92.56\text{ min}$. The period of the phasing transfer orbit needs to be three minutes shorter to catch up, so a braking maneuver into a lower orbit is required. The period of the *Dragon* on the transfer will be $T_{Dragon} = 2\pi\sqrt{\frac{\left(\frac{r_c + r_1}{2}\right)^3}{\mu}}$. Solving for the perigee radius of the transfer orbit taking into account the shorter period gives:

$$r_1 = 2\mu^{1/3}\left(\frac{T_{Dragon}}{2\pi}\right)^{2/3} - r_c = 2(398{,}600)^{1/3}\left(\frac{5553 - 180}{2\pi}\right)^{2/3} - 6778 = 6483\text{ km}.$$

The total velocity change needed by the *Dragon* spacecraft is then given by Eq. 5.11: $\Delta V_{total_1} = 2\left[\sqrt{\frac{\mu}{r_c}} - \sqrt{2\mu\left(\frac{1}{r_c} - \frac{1}{r_c + r_1}\right)}\right] = 2\left[\sqrt{\frac{398{,}600}{6778}} - \sqrt{2(398{,}600)\left(\frac{1}{6778} - \frac{1}{6778 + 6483}\right)}\right] = 0.172\frac{\text{km}}{\text{s}} = 172\text{ m/s}$, split between two identical impulses (first braking, then reboosting).

5.6 General Transfer

It is usually desirable to accomplish a transfer between two orbits more quickly than the Hohmann transfer. Such a general, more energetic transfer maneuver between two orbits (for example, between the Earth and a superior planet, such as Mars) is shown schematically in Fig. 5.9.

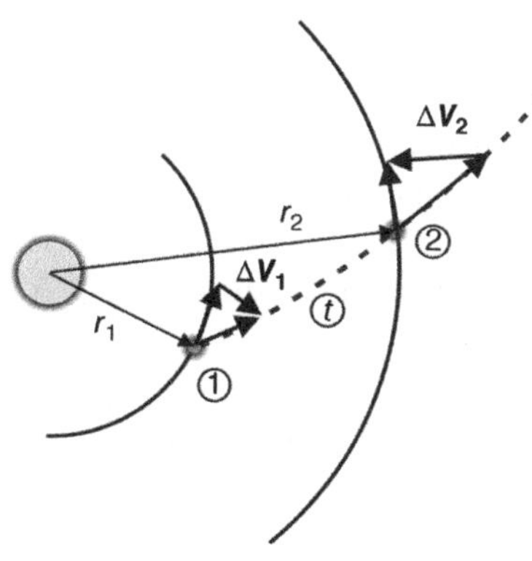

Figure 5.9 General transfer.

The transfer trajectory between the two orbits, designed ⓣ in Fig. 5.9, is generally an ellipse with an apoapsis greater than the target orbital radius r_2. Parabolic or hyperbolic transfer trajectories are theoretically possible, but involve unrealistically high velocity changes for interplanetary missions. In the Hohmann transfer, it was noted that all thrust maneuvers are aligned with the flight path. This is not the case for the more energetic, general

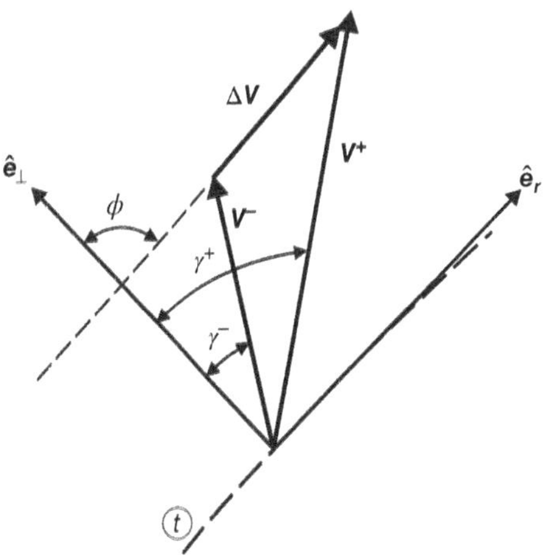

Figure 5.10 Geometry of a general-transfer maneuver.

transfer, as shown in the figure. In the example transfer shown, the orbital-insertion maneuver at point ② requires changing both the velocity and the flight path angle.

Details of general-transfer maneuver are shown in Fig. 5.10, which shows the vector velocities of the spacecraft both before (−) and after (+) an impulsive velocity change.

The vector velocities before and after the maneuver are related by: $\boldsymbol{V}^{+} = \boldsymbol{V}^{-} + \Delta\boldsymbol{V}$, where $\Delta\boldsymbol{V}$ is the vector velocity change, which can be in any direction. The corresponding change in flight path angle is $\Delta\gamma = \gamma^{+} + \gamma^{-}$. If the initial and postmaneuver velocity vectors are known, the necessary change in velocity can be determined by using either scalars or vectors. In the scalar approach, the velocity change can be determined from the law of cosines, which gives:

$$\Delta V^2 = V^{+^2} + V^{-^2} - 2V^{+}V^{-}\cos(\Delta\gamma). \tag{5.13}$$

Using the radial and tangential components of the vectors, the velocity change can also be expressed using the radial and tangential unit vectors. The initial and final velocity vectors are respectively:

$$\boldsymbol{V}^{-} = V_r^{-}\,\hat{\boldsymbol{e}}_r + V_{\perp}^{-}\,\hat{\boldsymbol{e}}_{\perp} = V^{-}\sin\gamma^{-}\,\hat{\boldsymbol{e}}_r + V^{-}\cos\gamma^{-}\,\hat{\boldsymbol{e}}_{\perp} \tag{5.14}$$

and

$$\boldsymbol{V}^{+} = V_r^{+}\,\hat{\boldsymbol{e}}_r + V_{\perp}^{+}\,\hat{\boldsymbol{e}}_{\perp} = V^{+}\sin\gamma^{+}\,\hat{\boldsymbol{e}}_r + V^{+}\cos\gamma^{+}\,\hat{\boldsymbol{e}}_{\perp}, \tag{5.15}$$

resulting in the vector velocity change:

$$\begin{aligned}\boldsymbol{V} = (V_r^{+} - V_r^{-})\hat{\boldsymbol{e}}_r + (V_{\perp}^{+} - V_{\perp}^{-})\hat{\boldsymbol{e}}_{\perp} &= (V^{+}\sin\gamma^{+} - V^{-}\sin\gamma^{-})\hat{\boldsymbol{e}}_r \\ &+ (V^{+}\cos\gamma^{+} - V^{-}\cos\gamma^{-})\hat{\boldsymbol{e}}_{\perp},\end{aligned} \tag{5.16}$$

Alternatively, Eqns.2.43 and 2.44 can be used to determine the velocity components, noting that the eccentricities, true anomalies, and angular momenta all change due to the impulsive maneuver. The radius r is the same in both cases. Lastly, the *thrust direction*, ϕ, relative to the horizontal/tangential direction as shown in Fig. 5.10 is given geometrically by:

$$\phi = \tan^{-1}\left(\frac{V_r^{+} - V_r^{-}}{V_{\perp}^{+} - V_{\perp}^{-}}\right). \tag{5.17}$$

Example Problem 5c

A satellite in LEO with an altitude of $r_1 = 6700$ km needs to be moved up to GEO (radius $r_2 = 42{,}164$ km). What are the required velocity changes and total transfer times for (i) a Hohmann transfer, and (ii) a faster, more energetic transfer on a trajectory with double the semi-major axis of the Hohmann transfer? The departure maneuver from LEO is tangential to the orbit in both cases.

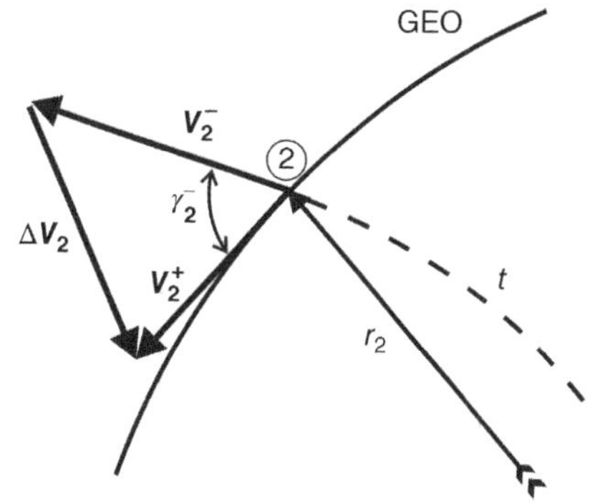

Figure 5c Arrival geometry at GEO.

Solution:

From the example case immediately after Example Problem 5a, the quantities for the Hohmann transfer are $\Delta V_1 = 2.42$ km/s and $\Delta V_2 = 1.47$ km/s, so the total velocity change needed is $\Delta V_{total} = \Delta V_1 + \Delta V_2 = 3.89$ km/s, with a transfer time of $T_t = 19{,}120$ s $= 5.28$ hr.

For the second case, (b), the departure velocity from LEO can be determined by a slight reformulating Eq. 2.41 to give:

$$V_1^+ = \sqrt{\frac{2\mu}{r} - \frac{\mu}{a}} = \sqrt{\mu\left(\frac{2}{r} - \frac{1}{(2)\frac{r_1 + r_2}{2}}\right)} = \sqrt{\mu\left(\frac{2}{r_1} - \frac{1}{r_1 + r_2}\right)}$$

$$= \sqrt{398{,}600\left(\frac{2}{6{,}900} - \frac{1}{6{,}900 + 42{,}164}\right)} = 10.36\frac{\text{km}}{\text{s}}.$$

This reflects that the semi-major axis on this transfer trajectory is doubled the value for the Hohmann transfer. Subtracting this velocity from the orbital velocity $V_{LEO} = 7.73$ km/s gives for the first velocity change $\Delta V_1 = V_1^+ - V_{LEO} = 2.64$ km/s. The velocity change maneuver needed at arrival is shown in Fig. 5c.

The velocity upon arrival at GEO, before insertion into that orbit, is determined similarly:

$$V_2^- = \sqrt{\mu\left(\frac{2}{r_2} - \frac{1}{r_1 + r_2}\right)} = \sqrt{398{,}600\left(\frac{2}{42{,}164} - \frac{1}{6900 + 42{,}164}\right)} = 3.28\frac{\text{km}}{\text{s}}.$$

The desired final velocity, V_2^+, is the circular velocity of the GEO orbit, $V_2^+ = \sqrt{\frac{\mu}{r_2}} = \sqrt{\frac{398{,}600}{42{,}164}} = 3.08\frac{\text{km}}{\text{s}}$, with a zero flight path angle ($\gamma_2^+ = 0$). The flight path angle before orbit insertion can be determined from the scalar angular momentum, $h = rV\cos\gamma$, to give:

$$\gamma_2^- = \cos^{-1}\left(\frac{h_t}{r_2 V_2^-}\right) = \cos^{-1}\left(\frac{r_1 V_1^+}{r_2 V_2^-}\right)$$

with the angular momentum on the transfer trajectory being $h_t = r_1 V_1^+$ since the first maneuver is tangent to the LEO orbit. This gives for the pre-orbit-insertion flight path angle $\gamma_2^- = \cos^{-1}\left(\frac{r_1 V_1^+}{r_2 V_2^-}\right) = \cos^{-1}\left[\frac{(6900)(10.36)}{(42{,}164)(3.28)}\right] = 58.88°$, so the needed change in flight path angle is $\Delta\gamma = \gamma_2^- - \gamma_2^{\mp} = 58.88° - 0 = 58.88°$.

The magnitude of the velocity change needed for orbit insertion is then from Eq. 5.13: $\Delta V_2 = \left(V_2^{+\,2} + V_2^{-\,2} - 2V_2^+ V_2^- \cos(\Delta\gamma)\right)^{1/2} = \left[(3.075)^2 + (3.284)^2 - 2(3.075)(3.284)\cos(58.88)\right]^{1/2} = 3.13\,\mathrm{km/s}$. Thus, the total velocity change for the more energetic transfer is $\Delta V_{total} = \Delta V_1 + \Delta V_2 = 2.64 + 3.13 = 5.77\frac{\mathrm{km}}{\mathrm{s}}$. This is 67% greater than the total velocity change of $\Delta V_{total} = 3.89\frac{\mathrm{km}}{\mathrm{s}}$ needed for the basic Hohmann transfer.

The transfer time for the more energetic maneuver can be determined as previously seen from Kepler's time equation, as presented in Eq. 3.11. The semi-major axis of the transfer trajectory is $a = (2)\frac{r_1 + r_2}{2} = r_1 + r_2 = 49{,}064\,\mathrm{km}$ with the eccentricity $e = 1 - \frac{r_1}{a} = 0.859$. The true anomaly at the target orbit determined from the *Orbit Equation* (Eq. 2.33) is $\cos\theta = \left[\frac{r_1(1+e)}{r_2} - 1\right]\frac{1}{e} = \left[\frac{6900(1+0.859)}{42{,}164} - 1\right]\frac{1}{0.859} = -0.810$, so $\theta = 144.1°$ (the true anomaly is zero at the point of departure from LEO since that is the perigee of the transfer trajectory). Inserting these values into Eqs. 3.12 and 3.14 gives the transfer time $T_t = 9966\,\mathrm{s} = 2.68\,\mathrm{hr}$, which is 49% shorter than the transfer time for the Hohmann transfer fof $T_t = 5.28\,\mathrm{hr}$.

5.7 Chase Maneuver

The general transfer discussed earlier suggests that it should be possible to achieve a rendezvous with a target in the same orbit faster than is possible for the phasing maneuver presented in Section 5.5. Such a *chase maneuver* is illustrated in Fig. 5.11. This maneuver is more direct than the phasing maneuver, in that it does not require a full orbit to accomplish, and is thus faster. Both the transfer initiation (point ①) and the reinsertion into orbit (point ②) are general transfer maneuvers and can be analyzed as in the previous section. A simple example is presented below.

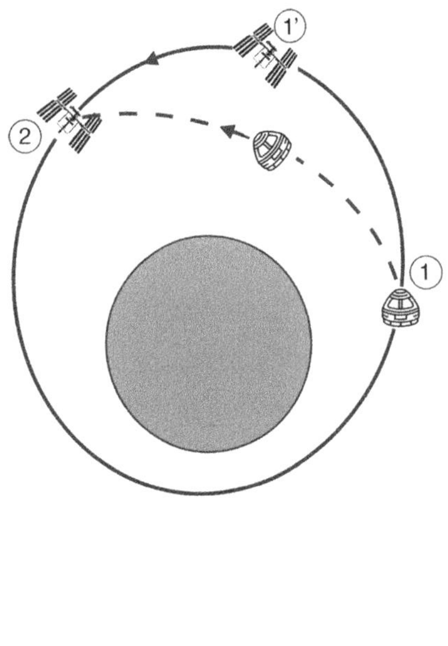

Figure 5.11 Chase maneuver.

Example Problem 5d

A space station is in a circular orbit with radius $r_c = 3000$ km around the Moon. Another spacecraft is in the same orbit, but behind the station, as shown in Fig. 5d. A chase maneuver is planned at point ① by increasing the velocity by a factor of 1.1 and changing the flight path angle to $\gamma_1 = -30°$. What will be the required velocity changes for this chase maneuver? The chase maneuver is calculated to require a time of $T_t = 1.26$ hr. What will be the position of the space station at the start of the maneuver (i.e. the angle ϕ shown in the figure)?

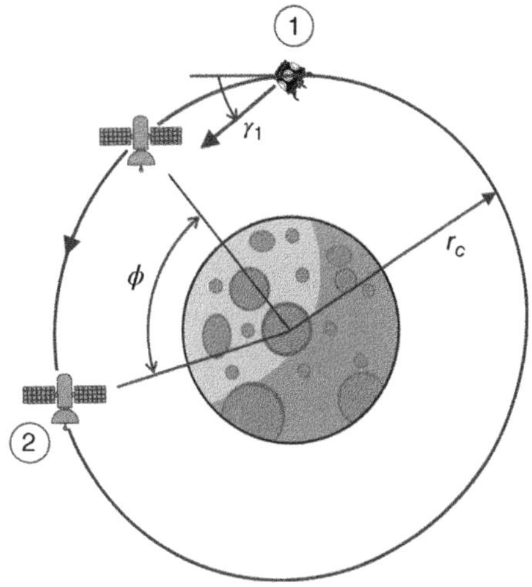

Figure 5d Lunar space station rendezvous via chase maneuver.

Solution:

The velocity in the circular orbit around the moon is

$$V_C = \sqrt{\frac{\mu_{moon}}{r}} = \sqrt{\frac{4903}{3000}} = 1.278\frac{\text{km}}{\text{s}}.$$

The magnitude of the first velocity change is from Eq. 5.12:

$$\begin{aligned}\Delta V_1 &= \left[V_c^2 + (1.1V_c)^2 - 2V_c(1.1V_c)\cos(\Delta\gamma)\right]^{1/2} \\ &= \left[(1.278)^2 + (1.406)^2 - 2(1.278)(1.406)\cos(-30)\right]^{1/2} \\ &= 0.706\ \text{km/s}.\end{aligned}$$

Since the target radius is equal to the initial radius, $r_2 = r_1 = r_c$, so the conservation of energy implies that the scalar velocity upon arrival at point ② is identical to the departure velocity at ①. The conservation of angular momentum then dictates that the cosines of the flight path angles at the two points are also equal given the identical scalar velocities. This symmetry in the trajectory implies that $V_2 = V_1 = 1.406$ km/s and $\gamma_2 = -\gamma_1 = +30°$, so the magnitude of the velocity change needed at point ② is the same as at point ①, $\Delta V_2 = 0.706$ km/s. The total velocity change requirement is $\Delta V_{total} = \Delta V_2 + \Delta V_2 = 1.412$ km/s. The period of the circular orbit is simply $T_c = 2\pi\frac{r_c}{V_c} = 2\pi\frac{3000}{1.278} = 14{,}749\,s = 4.10\,\text{hr}$. During the transfer maneuver, the station therefore has advanced $\frac{T_t}{T_c} = \frac{1.26}{4.10} = 0.308$ of an orbit, so the angle $\phi = (0.308)360° = 110.7°$.

5.8 Apse Line Rotation

In a few applications, it may be desired to change the location of the perigee or apogee of an orbit. This can be done with or without changing any of the other orbital parameters. In the simplest form, only the argument of periapsis is changed. Such rotation of the *apse line*, the line connecting the perigee and apogee through the focus, is shown in Fig. 5.12. The apse line is shown rotated counterclockwise by the amount ϕ.

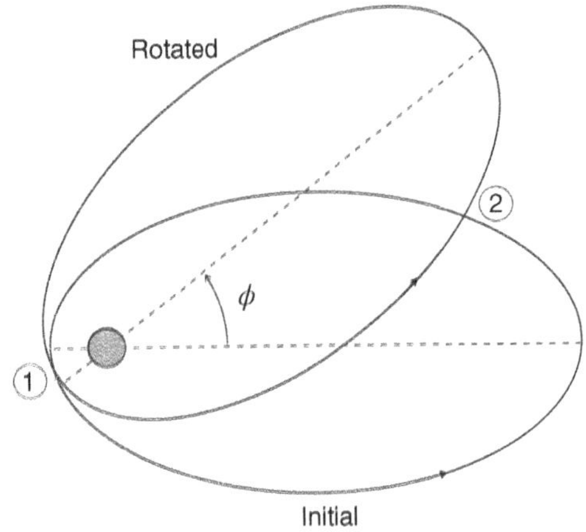

Figure 5.12 Apse line rotation.

As seen in the figure, the apse line rotation maneuver can be accomplished at either of the two points ① and ② that are common to both the original and rotated orbits. The required velocity changes for this maneuver can be derived using radial and perpendicular components. The problem is symmetric, since the two orbits are identical except for the apse line rotation. The true anomaly of point ① on the initial orbit is then $\theta^- = \phi/2$; by symmetry the corresponding true anomaly on the rotated orbit at that same point is $\theta^+ = -\phi/2$. The initial vector velocity at point ① using radial and perpendicular coordinates is (see Eqs. 2.43 and 2.44):

$$\boldsymbol{V}_{①}^- = \frac{\mu}{h} e \sin\theta^- \,\hat{\boldsymbol{e}}_{r_①} + \frac{h}{r_①}\hat{\boldsymbol{e}}_\perp = \frac{\mu}{h} e \sin\left(\frac{\phi}{2}\right)\hat{\boldsymbol{e}}_{r_①} + \frac{h}{r_①}\hat{\boldsymbol{e}}_\perp. \tag{5.18}$$

Similarly, the velocity components at point ① in the rotated orbit are:

$$\boldsymbol{V}_{①}^+ = \frac{\mu}{h} e \sin\theta^+ \,\hat{\boldsymbol{e}}_{r_①} + \frac{h}{r_①}\hat{\boldsymbol{e}}_\perp = \frac{\mu}{h} e \sin\left(-\frac{\phi}{2}\right)\hat{\boldsymbol{e}}_{r_①} + \frac{h}{r_①}\hat{\boldsymbol{e}}_\perp = -\frac{\mu}{h} e \sin\left(\frac{\phi}{2}\right)\hat{\boldsymbol{e}}_{r_①} + \frac{h}{r_①}\hat{\boldsymbol{e}}_\perp, \tag{5.19}$$

noting that the angular momentum of both orbits here is the same. The velocity change in this simple apse line rotation is then:

$$\Delta\boldsymbol{V}_① = \boldsymbol{V}_①^+ - \boldsymbol{V}_①^- = -\frac{\mu}{h} e \sin\left(\frac{\phi}{2}\right)\hat{\boldsymbol{e}}_{r_①} - \frac{\mu}{h} e \sin\left(\frac{\phi}{2}\right)\hat{\boldsymbol{e}}_{r_①} = -2\frac{\mu}{h} e \sin\left(\frac{\phi}{2}\right)\hat{\boldsymbol{e}}_{r_①}, \tag{5.20}$$

with the scalar magnitude

$$\Delta V_① = 2\frac{\mu}{h} e \sin\left(\frac{\phi}{2}\right). \tag{5.21}$$

The required change in flight path angle is:

$$\Delta\gamma_① = \gamma_①^+ - \gamma_①^- = \tan^{-1}\frac{V_{r_①}^+}{V_{\perp_①}^+} - \tan^{-1}\frac{V_{r_①}^-}{V_{\perp_①}^-} = \tan^{-1}\left[-\frac{\mu r_①}{h^2}\sin\left(\frac{\phi}{2}\right)\right] - \tan^{-1}\left[+\frac{\mu r_①}{h^2}\sin\left(\frac{\phi}{2}\right)\right]$$
$$= -2\tan^{-1}\left[+\frac{\mu r_①}{h^2}\sin\left(\frac{\phi}{2}\right)\right]. \tag{5.22}$$

It can be noted from Eqs. 5.18 and 5.19 that for this simple apse-line-change maneuver, the perpendicular component of the velocity is unchanged, and the radial components are equal in magnitude but opposite in direction.

Repeating the analysis at point ②, where $\theta^- = \phi/2 + 180°$ and $\theta^+ = 180° - \phi/2$ results in: $\Delta\boldsymbol{V}_② = \boldsymbol{V}_②^+ - \boldsymbol{V}_②^- = \frac{\mu}{h} e \sin\left(\frac{\phi}{2}\right)\hat{\boldsymbol{e}}_{r_②} - \left[-\frac{\mu}{h} e \sin\left(\frac{\phi}{2}\right)\hat{\boldsymbol{e}}_{r_②}\right] = +2\frac{\mu}{h} e \sin\left(\frac{\phi}{2}\right)\hat{\boldsymbol{e}}_{r_②}$, which has the magnitude $\Delta V_{②} = 2\frac{\mu}{h} e \sin\left(\frac{\phi}{2}\right)$, the exact same magnitude as for the maneuver being performed at point ①.

The corresponding change in flight path angle at that point is:

$$\Delta\gamma_{②} = \tan^{-1}\left[+\frac{\mu r_{②}}{h^2}\sin\left(\frac{\phi}{2}\right)\right] - \tan^{-1}\left[-\frac{\mu r_{②}}{h^2}\sin\left(\frac{\phi}{2}\right)\right] = +2\tan^{-1}\left[+\frac{\mu r_{②}}{h^2}\sin\left(\frac{\phi}{2}\right)\right] \quad (5.23)$$

It is apparent that for this basic apse line rotation, the magnitude of the velocity change is the same at both points ① and ②, but with a greater change in flight path angle at ② than at point ①

It should be noted that this analysis is for the case of a simple apse line change only; a change in apse line orientation can also be combined with simultaneous changes in other planar orbital elements (the semi-major axis and eccentricity).

5.9 Plane Changes

In many cases, it will be necessary to change the inclination of an orbit. Examples of such *plane change* maneuvers include moving a vehicle into a polar orbit for an Earth-surveillance mission, or making the inclination of an orbit to zero, or equatorial, for a satellite in geostationary orbit to remain exactly over the same spot on the Earth's surface (discussed in Chapter 7).

An example of a simple plane change maneuver would be an object in a circular orbit at a point exactly over the Earth's equator, as shown in Fig. 5.13.

In this case, an object in an orbit with the inclination δ undergoes an impulsive maneuver that reduces the inclination to zero (equatorial orbit). If the magnitudes and flight path angles of the orbit are unchanged, but only the inclination, the scalar velocity change can be determined from Eq. 5.13, with $V^+ = V^- = V$ and the change in angle δ to give:

$$\Delta V^2 = V^2 + V^2 - 2VV\cos(\delta) = 2V^2(1 - \cos\delta) \text{ so } \Delta V = \sqrt{2}V\sqrt{1 - \cos\delta}.$$

The use of the trig identity $\sqrt{\frac{1-\cos\delta}{2}} = \sin\left(\frac{\delta}{2}\right)$ then gives for the simple plane change:

$$\Delta V = 2V\sin\left(\frac{\delta}{2}\right). \quad (5.24)$$

As with many other orbital maneuvers, it is possible to change multiple orbital elements with a single impulsive maneuver. Thus, a maneuver to change the plane of the orbit might also include changes in the magnitudes of the radial and perpendicular velocity components. This more general situation is illustrated in Fig. 5.14 using a Cartesian coordinate system with the x and y axes aligned with the radial and perpendicular velocity components of the initial velocity vector, respectively.

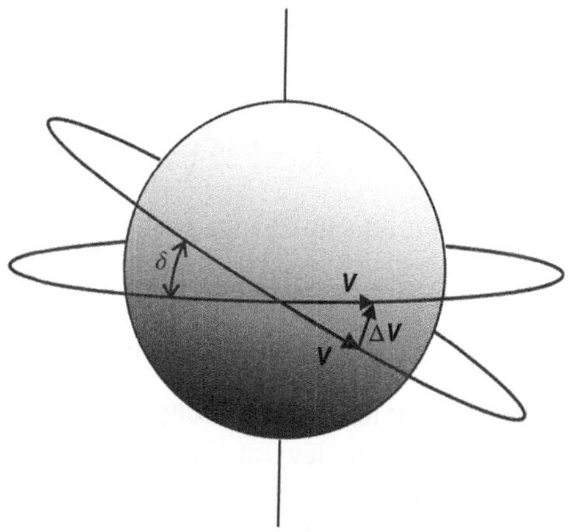

Figure 5.13 Simple plane change maneuver.

The initial velocity vector is $\boldsymbol{V}^- = V_x^-\hat{\boldsymbol{i}} + V_y^-\hat{\boldsymbol{j}}$, and the postmaneuver velocity can be written as

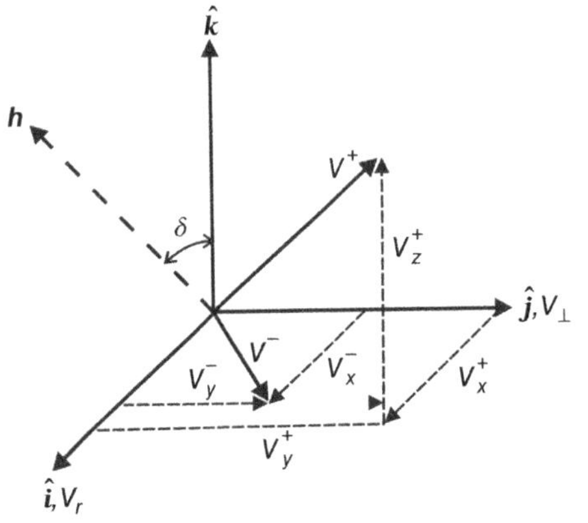

Figure 5.14 Velocity change in local Cartesian coordinates.

$\boldsymbol{V}^+ = V_x^+\,\hat{\boldsymbol{i}} + V_y^+\hat{\boldsymbol{j}} + V_z^+\,\hat{\boldsymbol{k}}$. The magnitude of the velocity difference due to this maneuver using these coordinates can then be determined using:

$$\Delta V^2 = \left(V_x^+ - V_x^-\right)^2 + \left(V_y^+ - V_y^-\right)^2 + \left(V_z^+ - V_z^-\right)^2. \tag{5.25}$$

The velocity components can be related to the radial and tangential components as follows. Since the axes shown are oriented to align with the components of the initial velocity vector, $V_x^- = V_r^-$, $V_y^- = V_\perp^-$, and $V_z^- = 0$. Similarly, $V_x^+ = V_r^+$. The postmaneuver perpendicular velocity is $V_\perp^+ = \left(V_y^{+\,2} + V_z^{+\,2}\right)^{1/2}$. This can be related to the change in inclination, δ, by the use of Eq. 4.1 to get $\cos\delta = \dfrac{\hat{\boldsymbol{k}} \cdot \boldsymbol{h}^+}{h^+} = \dfrac{\hat{\boldsymbol{k}} \cdot (\boldsymbol{r} \times \boldsymbol{V}^+)}{h^+} = \dfrac{\hat{\boldsymbol{k}} \cdot \left(r\hat{\boldsymbol{i}} \times \left(V_x^+\hat{\boldsymbol{i}} + V_y^+\hat{\boldsymbol{j}} + V_z^+\hat{\boldsymbol{k}}\right)\right)}{h^+} = \dfrac{V_y^+}{r\left(V_y^{+\,2} + V_z^{+\,2}\right)^{1/2}} = \dfrac{V_y^+}{V_\perp^+}$.

Thus, $V_y^+ = V_\perp^+ \cos\delta$ and from the geometry shown in Fig. 5.14, $V_z^+ = V_\perp^+ \sin\delta$. Inserting these components into Eq. 5.25 (and doing some algebra) then gives the velocity change for a general plane change maneuver as follows:

$$\Delta V = \left[\left(V_r^+ - V_r^-\right)^2 + V_\perp^{+\,2} + V_\perp^{-\,2} - 2V_\perp^+ V_\perp^- \cos\delta\right]^{1/2}. \tag{5.26}$$

For the case where there is no change in the radial velocity component, this reduces directly to the expression in Eq. 5.24 for the simple plane change maneuver.

Example Problem 5e

A satellite in LEO with an altitude of $r_1 = 6700$ km is launched due east from Cape Canaveral, which resulted in an initial orbital inclination of 28.4°. What velocity change will be needed to convert this into an equatorial orbit? A polar orbit?

Solution:

The velocity in LEO is $V = \sqrt{\dfrac{\mu}{r_1}} = \sqrt{\dfrac{398{,}600}{6700}} = 7.713\ \text{km/s}$. To transfer to the equatorial orbit requires the velocity change from Eq. 5.24 of $\Delta V = 2V\sin\left(\dfrac{\delta}{2}\right) = 2(7.713)\sin\left(\dfrac{28.4^\circ}{2}\right) = 3.784$ km/s, which is substantial. To change to the polar orbit requires an inclination change of $\delta = 90^\circ - 28.4^\circ = 61.6^\circ$, with a required velocity change of $\Delta V = 2(7.713)\sin\left(\dfrac{61.6^\circ}{2}\right) = 7.900$ km/s, comparable to the velocity increase required for launch. It is, thus, not surprising that there are comparatively few satellites in polar orbits. The velocity needed for transfer to an equatorial orbit is one motivation for the European Space Agency selecting a launch site in French Guiana (latitude 5.05°N), which would require a velocity change of only 0.680 km/s = 680 m/s.

5.10 Combined Maneuvers

As was pointed out in the previous sections, orbital maneuvers can be, and often are, combined to achieve multiple objectives with a single, impulsive maneuver. This might include circularizing a preliminary, elliptical orbit while also performing a plane change, rotating the apse line while changing the perigee of the orbit, and others. There is typically some flexibility in the choice of how the combined maneuvers can be applied, as shown in the following example.

Example Problem 5f

In Example Problem 5a, a GPS satellite in LEO was transferred to a 12-hour circular orbit. Suppose a GPS satellite is launched from Vandenberg Air Force base, latitude 34.74° N, and needs to be put into an equatorial 12-hour orbit, would it be more efficient to make the plane change in LEO, or upon arrival in the 12-hour GPS orbit? Assume that both plane change maneuvers occur exactly over the Earth's equator.

Solution:

From Example Problem 5a, the orbital velocity in LEO is V_{LEO} = 7.726 km/s, and in the GPS orbit, V_{GPS} = 3.870 km/s. The velocities needed for the transfer in LEO and the GPS orbit are, respectively, V_1 = 9.769 km/s and V_2 = 2.451 km/s. The required plane change is δ = 34.74°. There are no radial velocity components at either transfer point in this Hohmann transfer, so Eq. 5.26 simplifies to

$$\Delta V = \left[V_\perp^{+\,2} + V_\perp^{-\,2} - 2V_\perp^+ V_\perp^- \cos\delta \right]^{1/2}.$$

For the case where the plane change and boost are performed in LEO, $V_\perp^- = V_{LEO} = 7.726\frac{\text{km}}{\text{s}}$ and $V_\perp^+ = V_1 = 9.769\frac{\text{km}}{\text{s}}$, so the needed velocity change is $\Delta V_1 = [\,(9.769)^2 + (7.726)^2 - 2(9.769)(7.726)\cos(34.74°)\,]^{1/2}$ = 5.575 km/s. This needs to be combined with the velocity change needed to insert into the GPS orbit, ΔV_2= 1.419 km/s to give a total velocity change requirement of ΔV_{total} = 6.994 km/s. Alternatively, for the case where the plane change is performed upon arrival in the GPS orbit, $V_\perp^- = V_2 = 2.451\frac{\text{km}}{\text{s}}$ and $V_\perp^+ = V_{GPS} = 3.870\,\text{km/s}$. This gives the velocity change on arrival of $\Delta V_2 = [\,(2.451)^2 + (3.870)^2 - 2(2.451)(3.870)\cos(34.74°)\,]^{1/2}$ = 2.323 km/s. Adding this to the velocity change required to depart LEO of ΔV_1 = 2.043 km/s gives a total velocity change requirements of ΔV_{total} = 2.043 km/s. Thus, the "more efficient," that is, lower total velocity change situation, is the second scenario – combine the plane change with the arrival maneuver at the GPS orbit.

5.11 Problems

5.1 A Hohmann transfer is used to raise an Earth orbit with perigee and apogee radii of r_{p1} = 6930 km and r_{a1} = 29,100 km, respectively, to a geosynchronous orbit (period 23 hr 56 s) with an eccentricity of e_2 = 0.63. What are the required velocity changes, ΔV_1 and ΔV_2? Which velocity change is greater, and why?

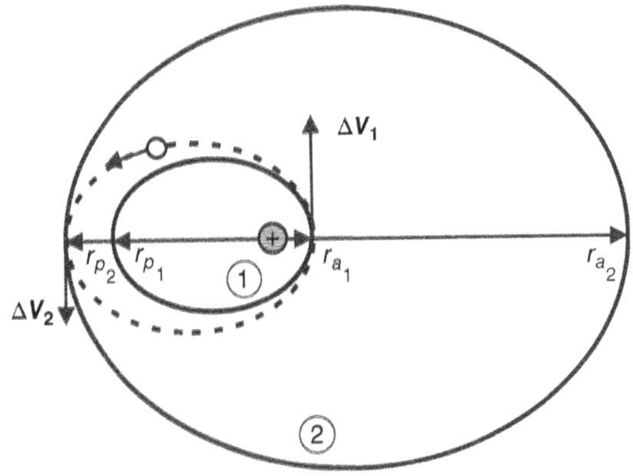

5.2 A spacecraft is in a circular orbit around Mars with a radius of 7000 km. A probe is to be launched from the spacecraft to enter Mars' atmosphere. It is desired to enter the atmosphere as "shallow" as possible. Given the thin atmosphere of Mars, the atmosphere can be neglected for the purposes of this problem. What would be the minimum velocity change required for the probe? The probe trajectory can be computed to just touch the surface of Mars.

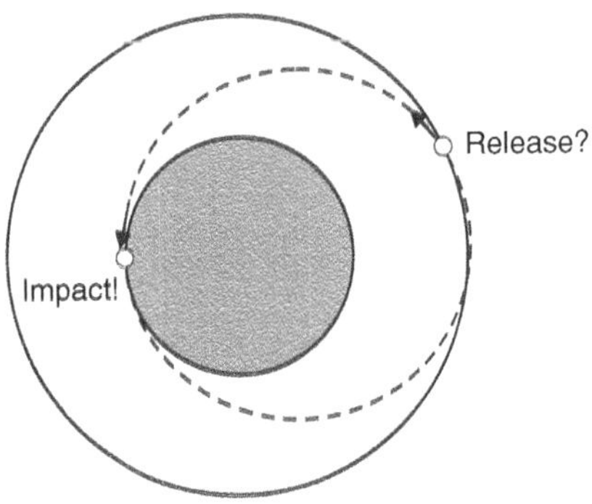

5.3 On launch, the Space Shuttle did not directly inject into its final orbit. Rather, main-engine shutdown occurs at the apogee of an elliptical orbit at 100 km altitude. The perigee of this orbit is only 30 km altitude, to ensure that the discarded external tank will safely burn up over open ocean. After the external tank is jettisoned, the orbiter used its Orbital Maneuvering System (OMS) to maneuver into another elliptical orbit with a perigee at 100 km altitude and an apogee at 250 km. This is called the OMS-1 burn. At the new apogee, the shuttle performed the OMS-2 burn to circularize the orbit at the mission altitude of 250 km.

What was the burnout speed at the apogee of the external tank disposal orbit?
What ΔV was required for the OMS-1 maneuver (km/s)? Does this represent a speed *increase* or *decrease*?
What was the ΔV required for the OMS-2 maneuver (km/s)? Was this a speed increase or decrease?

5.4 A bi-elliptic transfer is proposed to move an Earth satellite from LEO ($r_{LEO} = 7000$ km) to GEO ($r_{GEO} = 42{,}000$ km). The first bi-elliptic transfer segment has the eccentricity $e_2 = 0.8$. What will be the required velocity changes for this transfer, ΔV_1, ΔV_2, and ΔV_3, and the total velocity change? Compute also the total velocity change if the transfer were accomplished directly by a Hohmann transfer from orbit ① to orbit ④. Express all velocities in terms of $\sqrt{\dfrac{\mu}{r_{LEO}}}$.

5.5 Determine the optimum (smallest total velocity change) of r^*/r_2 for a bi-elliptic transfer with a given value of r_2/r_1. Make the plot of r^*/r_2 versus r_2/r_1 for $1 \le r_2/r_1 \le 20$.

5.6 Consider a modification of the *Orion* spacecraft/*LSAM* (Lunar Surface Access Module) rendezvous scenario from Problem 8 in Chapter 3. As before, the circular orbit of the Orion spacecraft has an altitude of 300 km. At one point in the test maneuver, the *LSAM* is at apogee of its orbit, while the *Orion* is exactly opposite the Earth (at the perigee of the Test orbit). What apogee of the LSAM orbit will allow for a rendezvous with *Orion* at its current position?

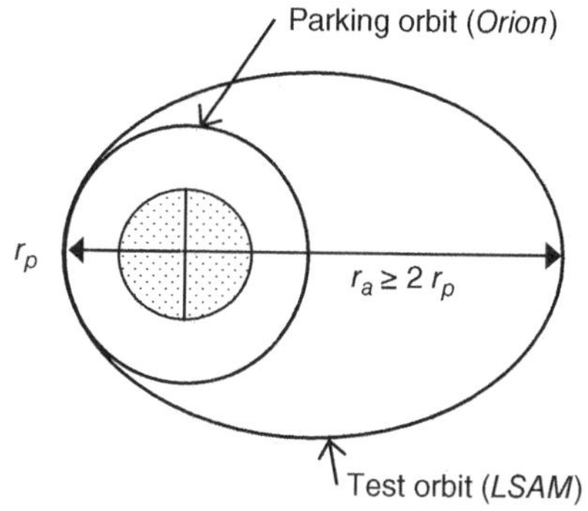

5.7 The ISS, a *Dragon* capsule, and a *Soyuz* spacecraft are all in the same LEO orbit with $h = 400$ km altitude. The *Dragon* is five minutes ahead of the ISS; the *Soyuz* is five minutes behind. Both spacecraft, at the same time, initiate a phasing maneuver to reach the ISS. Determine: (i) all required velocity changes and totals, and (ii) the duration of each spacecraft's maneuver. Which spacecraft will arrive at the ISS first?

5.8 A spacecraft in earth orbit ① has a velocity of $V_0 = 10$ km/s at a perigee radius of $r_p = 7000$ km. It is desired to insert the spacecraft into a circular orbit ② at the point orbit ① reaches a true anomaly of $\theta = 90°$, as shown. The period of the initial orbit is $T_1 = 13.45$ hr.

(a) What is the eccentricity of the initial trajectory, ①?
(b) What will be the magnitude of the required velocity change at circular orbit insertion, point $*$, ΔV (km/s)? What will be the change in flight path angle, $\Delta\gamma$ (degrees), at that point?

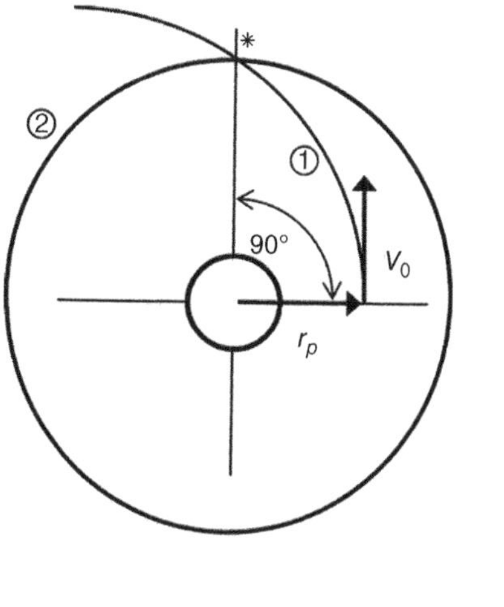

5.9 In the maneuver shown, the apse line is rotated exactly 180°, while all other elements of the orbit are unchanged. Plot the magnitude of the required velocity change versus the eccentricity of the orbit. For what value of the eccentricity does the required velocity change reach that of the initial orbit?

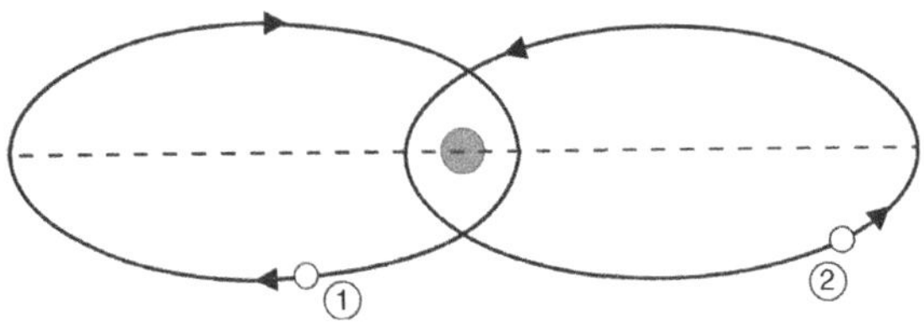

5.10 An earth satellite is launched into an orbit with the following orbital elements $i = 10°$, $p = 10{,}000$ km, $\omega = 315°$, $e = 0.5$.

What will be the *minimum* ΔV (km/s) needed to change this orbit into an equatorial orbit? This ΔV is accomplished by a single impulsive maneuver. It may be helpful to first draw a *careful* sketch of this orbit, showing the inclination, location of perigee, etc.

5.11 An earth satellite is in a circular, equatorial orbit with radius r_c and orbital speed V_c. The inclination of the orbit is to be increased to the value i. It is proposed to accomplish this by employing a bi-elliptic transfer. The apogee of the bi-elliptic trajectory (the point where the plane change occurs in this case) is exactly three times the initial orbital radius (i.e. $r_a = 3r_c$).

(a) What will be the total velocity, $\Delta V_{total}/V_c$, required for this bi-elliptic maneuver? (Treat *all* velocity changes as positive, whether boosting or braking.)
(b) How does this compare with the ΔV_{direct} required to make the inclination change directly on the circular orbit (i.e. what is the numerical value of $\Delta V_{total}/\Delta V_{direct}$)?

5.12 Consider a transfer from a 300 km altitude, inclination $i = 28.4°$ LEO to a GEO orbit with zero inclination. The plane change and the Hohmann transfer maneuvers are to be combined to reduce the total ΔV as much as possible.

(a) Develop an expression for the total ΔV for generalized maneuvers in which part of the plane change, δ_1, is made in LEO and the remainder, $\delta_2 = i - \delta_1$ is made in GEO.
(b) Generate a graph of ΔV_{total} as a function of δ_1 for reasonable values of δ_1.
(c) What are the optimum values of δ_1 and δ_2 (degrees) and ΔV_{total} (km/s)? How does that ΔV_{total} compare to that for the cases where the plane change is done entirely at departure ($\delta_1 = 28.5°$, $\delta_2 = 0°$) or at arrival ($\delta_1 = 0°$, $\delta_2 = 28.5°$)?

6

Interplanetary Trajectories and Maneuvers

Interplanetary missions typically require combinations of the trajectories and maneuvers presented in previous chapters. An added consideration is the need to evaluate the extent of the gravitational influences of the various primary bodies acting on a spacecraft during a mission, which would include the Earth, the Sun, and the target planet(s). The nature of the interaction of a spacecraft with a target planet depends on numerous factors and can have a substantial impact on the spacecraft's trajectory. Particularly important is the relative motion between a target planet and an incoming spacecraft. The positions and motions of the planets are also key factors in interplanetary missions. The chapter outline is as follows:

- Planetary phasing
- Patched-conic approximation
- Sphere of Influence concept
- Earth-departure trajectories
- Planetary Hohmann transfers
- General planetary approach
- Planetary encounters

6.1 Planetary Phasing

The relative position of a target planet relative to the Earth is crucial for the start of any interplanetary mission to ensure that a spacecraft arrives at the planet, not simply at the planet's orbit. This is the topic of planetary phasing (different from the orbital phasing maneuver described in the previous chapter).

An example of planetary phasing is shown in Fig. 6.1. In the case shown, the transfer trajectory ⓣ is from Earth to a superior planet. The location of the target planet at the time of the spacecraft's departure from the orbit of the Earth, t_1, is shown, along with the planet's position at time t_2 corresponding to the spacecraft's arrival. The angles θ_1 and θ_2 are, respectively, the corresponding true anomalies of the interplanetary transfer trajectory ($\theta_1 = 0$ for the case of a tangential departure from Earth's orbit). The angle γ is here the *phase angle*, the amount by which the target planet is ahead of the Earth at the time t_1.

The parameters of the transfer trajectory from Earth to the target planet are determined based on the Earth's orbit and the distance to the target planet at the desired arrival time. The radial distance

Introduction to Orbital Mechanics: A Concise and Practical Approach, First Edition. James C. Hermanson.

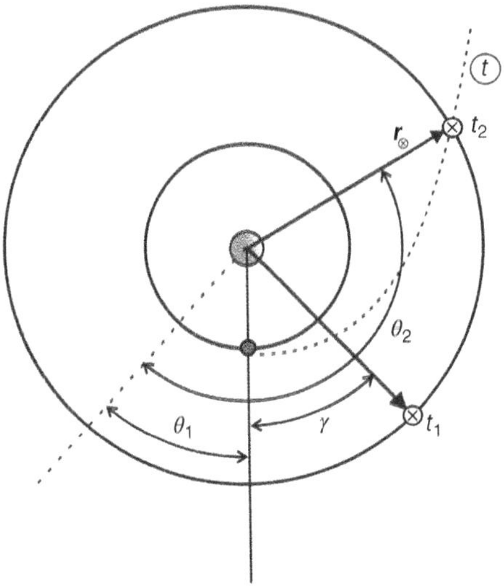

Figure 6.1 Planetary phasing.

from the sun to the Earth (at spacecraft departure) and the target planet (at spacecraft arrival) are given from these forms of the Orbit Equation (Eq. 2.30) as:

$$r_{\oplus} = \frac{h/\mu}{1 + e\cos\theta_1} \text{ and } r_{\otimes} = \frac{h/\mu}{1 + e\cos\theta_2}, \tag{6.1}$$

where h, a, and e are the parameters of the transfer trajectory, and the symbols ($\oplus$) and ($\otimes$) designate the Earth and the target planet, respectfully. These equations give the difference in the true anomalies of the transfer trajectory at arrival and departure, $\Delta\theta = \theta_2 - \theta_1$.

The time of the interplanetary transit, $\Delta t = t_2 - t_1$ can be determined from Kepler's equation (Eq. 3.13). Taking the orbits to be circular for simplicity, then during the transit time the target planet advances in its orbit by the amount $n_{\otimes}(t_2 - t_1)$, where the angular rate $n_{\otimes} = \sqrt{\frac{\mu}{a_{\otimes}^3}}$. The amount of advance can then be related to the angles shown in Fig. 6.1 as follows:

$$n_{\otimes}(t_2 - t_1) = (\theta_2 - \theta_1) - \gamma,$$

which gives for the required phase angle

$$\gamma = (\theta_2 - \theta_1) - n_{\otimes}(t_2 - t_1). \tag{6.2}$$

A key question is then does a particular geometric arrangement or phase angle repeat? This points to the issue of *launch windows* – when can a similar mission (similar trajectory, velocity-change requirements, and transit time) to a planet be conducted. This condition will occur when

$$\gamma + \tau_s(n_{\oplus} - n_{\otimes}) = \gamma + 2\pi,$$

where $\tau_s = \frac{2\pi}{|n_{\oplus} - n_{\otimes}|}$ is the *synodic period*. Taking the absolute value of the difference in angular rates allows this expression to apply to both the cases of transfer to a superior and to an inferior planet. The computed synodic periods of the major planets, relative to the Earth, are given in Table 6.1. The synodic periods between any other planetary combinations (for example, missions from Mars to Jupiter) can be calculated similarly by using the relevant mean angular rates.

Table 6.1 Synodic periods of the major planets.

Planet	Synodic period (years)
Mercury	0.32
Venus	1.60
Mars	2.13
Jupiter	1.09
Saturn	1.07
Uranus	1.01
Neptune	1.005

The tabulated synodic periods point to a major challenge in conducting repeated missions to the nearest planets to Earth. In particular, a given launch window to Mars will reoccur only after approximately two years, with substantial implications for resupply missions. Given the relatively slower orbital speeds of the outer planets, the situation is less severe and allows for repeat missions slightly more than one year apart.

6.2 Patched-Conic Approximation

An interplanetary mission can be regarded as consisting of several sequential segments, each with different characteristics. This is the essence of the *patched-conic approximation*, which is a practical and reasonable approach to designing and analyzing interplanetary trajectories. This approach is illustrated in Fig. 6.2.

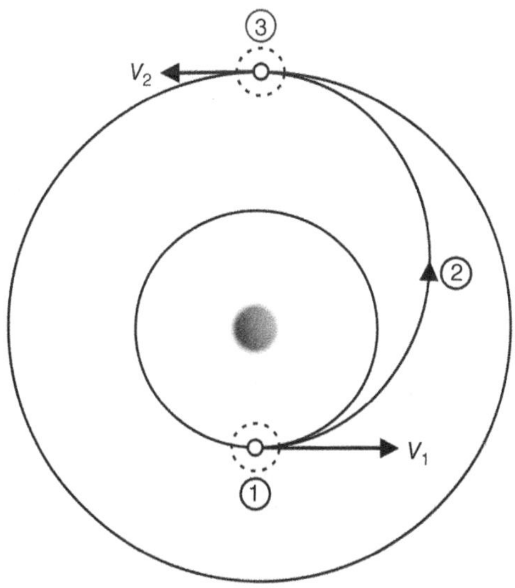

Figure 6.2 Patched-conic approximation (not to scale).

In the mission shown in Fig. 6.2, the trajectory is taken to consist of three segments:

1) ① Earth departure
2) ② Interplanetary transit
3) ③ Planetary arrival

The first segment is a hyperbolic departure from Earth orbit, as discussed in Chapter 2, Section 2.6.4. The departure must be hyperbolic relative to the Earth and leaves the Earth's vicinity with the excess velocity V_∞, which is typically aligned with the Earth's orbit to take advantage of the Earth's orbital velocity. Once sufficiently far from the Earth, the spacecraft is in an elliptical transfer trajectory, relative to the *Sun*, to the target planet. This trajectory results from the combination of the Earth's orbital velocity about the sun and the excess velocity resulting from the Earth-departure maneuver. This interplanetary trajectory departing Earth is always elliptical; a parabolic or hyperbolic trajectory about the Sun would require an excessive velocity change.[1] The final segment commences when the spacecraft arrives in the vicinity of the target planet. Relative to the target planet, the spacecraft will be seen to arrive on a hyperbolic trajectory resulting from the combination of the velocity on the transfer trajectory and the planet's orbital velocity.

Though in principle inexact, the patched-conic approach is usable, given that actual interplanetary trajectories typically require mid-course corrections during the transit to the target planet in any case. These corrections are needed to compensate for perturbations due to other planets (for example, Jupiter) and also for inaccuracies in the injection of the spacecraft onto the transfer trajectory, such as the result from small variations in direction and/or thrust duration. In addition, the distance of a spacecraft from the Earth during transfer injection, or from the planet upon arrival, is much smaller than the distances of the Earth and planet from the Sun, which are the bodies relevant to the transfer trajectory.

6.3 Sphere of Influence

Putting together the trajectory elements in the patched-conic approximation requires consideration of the appropriate points at which to perform the "patching." This leads to the concept of the *sphere of influence* of a planetary body, within which the spacecraft motion is dominated by the gravity of the planet. Correspondingly, outside of the sphere of influence, the spacecraft's trajectory is essentially governed by the gravitational field of the Sun alone, with the gravity of the planet being neglected. The sphere of influence concept is illustrated in Fig. 6.3, where r_{SOI} is the distance from the planet beyond which the sun's gravity is dominant.

A first approach to defining the location of the sphere of influence would be to find the location where the gravitational forces of the sun and the planet are in balance. This balance would suggest that

$$\frac{Gm_1 m}{r'^2} = \frac{Gm_2 m}{r^2}, \tag{6.3}$$

where the left side is the gravitational force on a spacecraft with mass m from the Sun; the right side is the force due to the planet. Expecting that the distance of the spacecraft to the Sun will be much greater than the distance from the planet, or $R \approx r'$, leads to

1 Hyperbolic trajectories relative to a *planet* can be achieved as a result of planetary flybys, as discussed in Section 6.6.

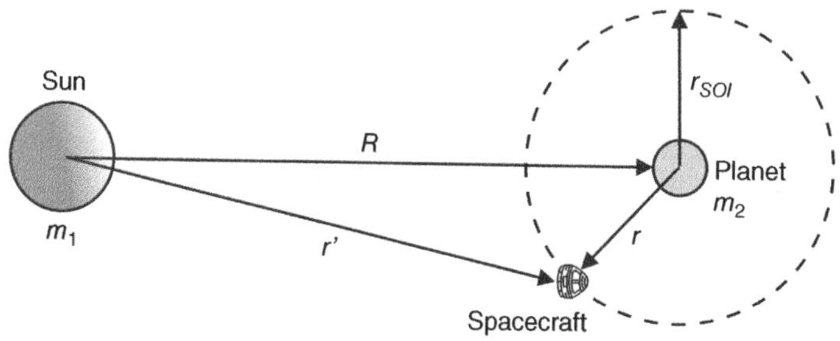

Figure 6.3 Sphere of influence.

$$\frac{m_1}{m_2} = \left(\frac{r'}{r}\right)^2 \approx \left(\frac{R}{r_{SOI}}\right)^2 \text{ suggesting that}$$

$$\frac{r_{SOI}}{R} \approx \left(\frac{m_2}{m_{21}}\right)^{1/2}. \tag{6.4}$$

A more exact result[2] is obtained by considering perturbations to the spacecraft motion. This leads to

$$\frac{r_{SOI}}{R} \approx \left(\frac{m_2}{m_1}\right)^{2/5} \tag{6.5}$$

which is similar to Eq. 6.4 but with a slightly different exponent. Tabulated values of the Spheres of Influence (SOI) as define by Eq. 6.5 are given in Table 6.2. The SOI values are compared with the semi-major axes of the orbits and the planet radii.

The sphere of influence of a planet is seen to be small compared with the size of the planet's orbit about the sun in all cases, consistent with the use of the patched-conic approximation. The sphere of influence is large compared with the planet's diameter in all cases and becomes larger with increasing distance from the Sun, as the Sun's gravitational influence becomes weaker.

Table 6.2 Spheres of influence for the major planets.

Planet	SOI (km)	a (AU)	SOI/a	SOI/R_{planet}
Mercury	1.13×10^5	0.387	1.9×10^{-3}	45
Venus	6.17×10^5	0.723	5.7×10^{-3}	100
Earth	9.24×10^5	1.000	6.2×10^{-3}	145
Mars	5.74×10^5	1.524	2.5×10^{-3}	170
Jupiter	4.83×10^7	5.203	6.2×10^{-2}	677
Saturn	5.45×10^7	9.539	3.8×10^{-2}	904
Uranus	5.17×10^7	19.191	1.8×10^{-2}	2023
Neptune	8.67×10^7	30.061	1.9×10^{-2}	3886

2 The details of the solution can be found in Curtiss, H.D, *Orbital Mechanics for Engingeering Students*, 4th Ed, Elsevier 2020.

6.4 Earth-Departure Trajectories

The first segment of a patched-conic interplanetary trajectory is the transfer from a parking Low Earth Orbit (LEO) out to the Earth's sphere of influence. The geometry of the departure trajectory for a mission to a superior planet is shown in Fig. 6.4 (this is similar to Fig. 2f).

The trajectory in Fig. 6.4 is only shown in a single plane. In fact, there are an infinite number of trajectories that depart Earth for given values of the orbital radius in LEO, the excess velocity V_∞, and offset Δ. This could include a departure from an equatorial parking orbit, a polar one, or any inclination in between. This leads to an envelope of possible hyperbolic departure trajectories, as shown in Fig. 6.5. Since the value of the offset will always be very small compared with the distance between the Earth and a target planet, the offset can be neglected when connecting the Earth departure hyperbola to the planetary transfer trajectory using the patched-conic approximation.

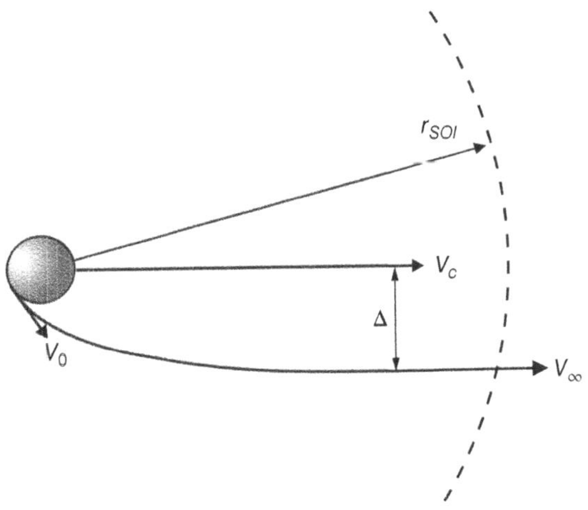

Figure 6.4 Earth departure trajectory to a superior planet.

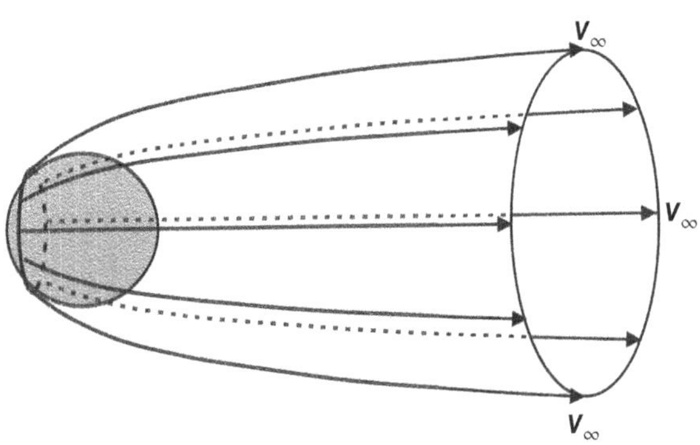

Figure 6.5 Envelope of departure hyperbolae.

The approach to the target planet will also involve hyperbolic trajectories, with an envelope of approach possibilities that will have significant implications for the nature of the planetary encounter.

Connecting the Earth-departure and interplanetary trajectories is accomplished by matching the hyperbolic excess velocity vector relative to the Earth, $\boldsymbol{V}_{\infty}$, to the required vector velocity on the transfer trajectory, $\boldsymbol{V}_1$. This matching is shown in the simple example below for the case of an interplanetary Hohmann transfer.

Example Problem 6a

A spacecraft in a parking orbit with an altitude of 300 km is to be launched on a Hohmann transfer to Mars. What will be the required velocity boost to depart the parking orbit, the excess velocity, the eccentricity of the departure hyperbola, and the offset distance?

Solution:

The velocity needed for the Hohmann transfer from Eq. 5.2 is:

$$V_1 = \sqrt{2\mu_{\odot}\left(\frac{1}{r_{\oplus}} - \frac{1}{r_{\oplus} + r_{\text{♂}}}\right)} = \sqrt{2(1.327\times 10^{11})\left(\frac{1}{149.6\times 10^{6}} - \frac{1}{149.6\times 10^{6} + 227.8\times 10^{6}}\right)}$$
$$= 32.72\,\text{km/s}.$$

This is the velocity on the transfer trajectory relative to the Sun. The excess velocity relative to the Earth is that velocity minus the orbit velocity of the Earth, $V_{\oplus} = \sqrt{\frac{\mu_{\odot}}{r_{\oplus}}} = \sqrt{\frac{1.327\times 10^{11}}{149.6\times 10^{6}}} = 29.78\frac{\text{km}}{\text{s}}$, so $V_{\infty} = V_1 - V_{\oplus} = 32.72 - 29.78 = 2.94\frac{\text{km}}{\text{s}}$. The velocity at the departure from LEO can be determined from Eq. 2.52:

$V_0 = \sqrt{\frac{2\mu_{\oplus}}{r_{LEO}} + V_{\infty}^2} = \sqrt{\frac{2(398{,}600)}{(300 + 6378)} + (2.94)^2} = 11.31\ \text{km/s}$. The needed velocity boost to leave Earth orbit and inject the spacecraft on the trajectory to Mars is then $\Delta V = V_0 - V_{LEO} = V_0 - \sqrt{\frac{\mu_{\oplus}}{r_{LEO}}} = 11.31 - 7.73 = 3.59\ \frac{\text{km}}{\text{s}}$.

The eccentricity of the departure trajectory from Earth can be determined from $r_{LEO} = a(e - 1)$ for a hyperbola, or $e = 1 + \frac{r_{LEO}}{a}$. Noting that $a = \frac{+\mu}{V_{\infty}^2}$ from Eq. 2.45 gives $e = 1 + \frac{r_{LEO}V_{\infty}^2}{\mu_{\oplus}} = 1 + \frac{(300 + 6378)(2.94)^2}{398{,}600} = 1.145$. The offset is then from Eq. 2.55 $\Delta = a\sqrt{e^2 - 1} = \frac{+\mu}{V_{\infty}^2}\sqrt{e^2 - 1} = \frac{398{,}600}{(2.94)^2}\sqrt{(1.145)^2 - 1} = 25{,}718\,\text{km}$, which is negligible compare to the Earth's orbit radius of $r_{\oplus} = 149.6\times 10^{6}$ km, as expected.

6.5 Planetary Hohmann Transfers

The Hohmann transfer is useful for illustrating some key features of interplanetary trajectories. The previous section discussed the first segment of a patched-conic interplanetary mission- the departure geometry and velocity relative to the Earth. The next segment in the mission is the interplanetary transfer trajectory, which being outside the Earth's sphere of influence describes the spacecraft motion relative to the Sun. This Sun-centered transfer trajectory, combined with the orbital velocity of the destination planet, then determines the velocity and geometry of the approach trajectory as seen by the planet. Interplanetary trajectories based on the Hohmann transfer are shown in Fig. 6.6 for transfers to both superior and inferior planets.

Given that for a Hohmann transfer the trajectory is collinear with the orbital velocity of the destination planet (taking the planet's orbit to be circular), the velocity relative to the planet for case (a) in Fig. 6.6 is $V_{\infty_2} = V_2 - V_{c_2}$, where V_{c_2} is the orbital velocity of the planet. The transfer to a superior planet is the same situation seen previously for a transfer from LEO to Geostationary Earth Orbit (GEO) – the velocity of the spacecraft on the transfer trajectory is less than that of the orbiting planet at arrival. Thus in this case, $V_{\infty_2} = V_2 - V_{c_2} < 0$, and the target planet overtakes the oncoming spacecraft, as shown in Fig. 6.7a. For transfer to an inferior planet (Fig. 6.6b), the situation is reversed, and the spacecraft velocity exceeds the orbital velocity of the planet, so in this case $V_{\infty_2} = V_2 - V_{c_2} > 0$, and the spacecraft overtakes the planet, as shown in Fig. 6.7b.

The circumstances of the first planetary-arrival scenario, at a superior planet on a Hohmann transfer, can be appreciated using a freeway-merging analogy, as shown in Fig. 6.8a. The "target planet" in this model is moving at a constant speed on the freeway, as seen from a remote location, the "Sun." The incoming spacecraft, the spacecraft "s/c," intercepts the path of the target planet by merging onto the freeway, on its Hohmann trajectory, via the "onramp" as shown. The s/c is moving more slowly than the planet, so it needs to pull onto the freeway ahead of the planet to enable

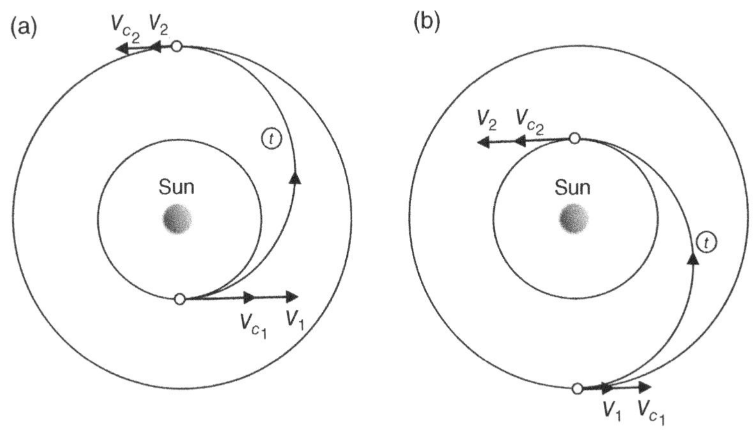

Figure 6.6 Interplanetary Hohmann transfers. (a) To a superior planet and (b) to an inferior planet. The velocities V_1 and V_2 and the transfer trajectories (t) are all relative to the sun.

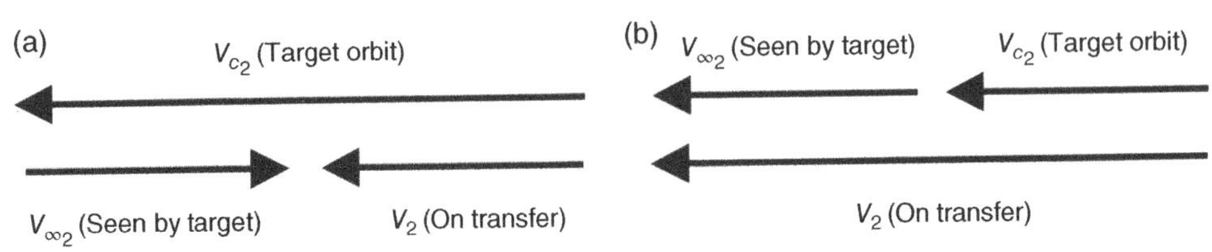

Figure 6.7 Arrival on a Hohmann transfer at (a) a superior planet and (b) an inferior planet.

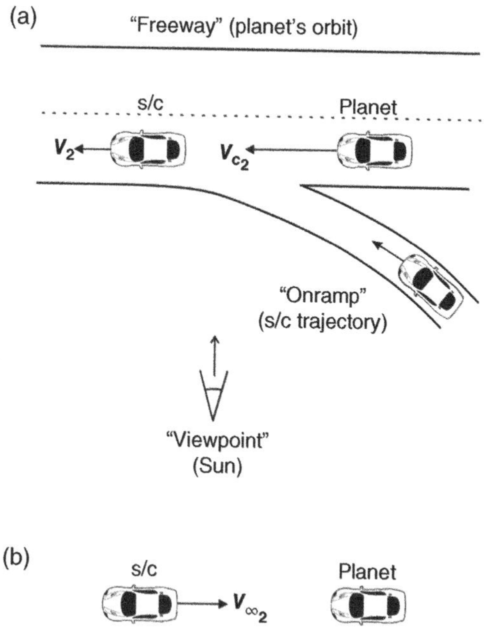

Figure 6.8 Freeway planetary arrival analogy. Vehicle motion as seen from the perspective of (a) the remote "Sun" and (b) the target vehicle "the planet."

the intercept. The "Sun" now sees both vehicles on the same path on the freeway, with the s/c ahead of the planet, but with a different speed.

From the vantage point of the planet, the s/c is seen upon arrival to be directly ahead, but since it is moving with a slower speed, it will appear to be directly oncoming, as shown in Fig. 6.8b.

Example Problem 6b

Determine the relative velocity (magnitude and direction) of spacecraft on a Hohmann transfer from Earth arriving at Jupiter and Venus.

Solution:

The velocities in the Hohmann transfers upon arrival at Jupiter (♃) and Venus (♀) are given correspondingly from Eq. 5.2:

$$V_{2_{♃}} = \sqrt{2\mu_{☼}\left(\frac{1}{r_{♃}} - \frac{1}{r_{\oplus} + r_{♃}}\right)} = \sqrt{2(1.327 \times 10^{11})\left(\frac{1}{778.3 \times 10^{6}} - \frac{1}{149.6 \times 10^{6} + 778.3 \times 10^{6}}\right)}$$
$$= 7.41\frac{\text{km}}{\text{s}}$$

and

$$V_{2_{♀}} = \sqrt{2\mu_{☼}\left(\frac{1}{r_{♀}} - \frac{1}{r_{\oplus} + r_{♀}}\right)} = \sqrt{2(1.327 \times 10^{11})\left(\frac{1}{108.2 \times 10^{6}} - \frac{1}{149.6 \times 10^{6} + 108.2}\right)}$$
$$= 37.72\frac{\text{km}}{\text{s}}.$$

The corresponding circular orbital speeds of the planets are:

$$V_{c_{♃}} = \sqrt{\frac{\mu_{☼}}{r_{♃}}} = \sqrt{\frac{1.327\times 10^{11}}{778.3\times 10^{6}}} = 13.1\,\frac{\text{km}}{\text{s}} \text{ and } V_{c_{♀}} = \sqrt{\frac{\mu_{☼}}{r_{♀}}} = \sqrt{\frac{1.327\times 10^{11}}{108.2\times 10^{6}}} = 35.02\,\frac{\text{km}}{\text{s}},$$

with velocities of the incoming probe relative to the planets of $V_{\infty_{♃}} = V_{2_{♃}} - V_{c_{♃}} = 7.41 - 13.06 = -5.65\ \text{km/s} < 0$ (planet overtakes spacecraft) and $V_{\infty_{♀}} = V_{2_{♀}} - V_{c_{♀}} = 37.72 - 35.02 = 2.71\ \text{km/s} > 0$ (spacecraft overtakes planet).

6.6 General Planetary Approach

Given the comparatively long transit times required for Hohmann transfers, interplanetary transfers are usually on faster, more energetic trajectories that do not arrive collinear with the planet's orbital velocity. This more general planetary approach is illustrated in vector form in Fig. 6.9.

In the figure, $\mathbf{V_2}$ is the velocity vector of the incoming spacecraft on its heliocentric transfer trajectory and $\mathbf{Vc_2}$ is the orbital velocity of the planet, both quantities relative to the Sun. Given the (presumably known) three-dimensional elements of the transfer trajectory (such as, a, e, E, and h), the incoming velocity vector $\mathbf{V_2}$ can be determined in heliocentric coordinates, for example, by using the perifocal coordinates in Eq. 4.7 or the radial and normal coordinates in 2.43 and 2.44 and transforming as discussed in Section 4.3.1 to account for any plane change, as needed.

The velocity vector of the incoming spacecraft relative to the planet is $\mathbf{V_3} = \mathbf{V_2} - \mathbf{Vc_2} = \mathbf{V_2} + (-\mathbf{Vc_2})$. This velocity relative to the planet can have a very different direction from the incoming velocity vector relative to the Sun, as suggested in Fig. 6.9. The magnitude of the velocity relative to the planet can be determined from the cosine law (see Eq. 5.13) as

$$V_3 = \sqrt{V_2^2 + (V_{c_2})^2 - 2V_2V_{c_2}\cos\gamma_2}. \tag{6.6}$$

This is the same scalar velocity as the hyperbolic excess velocity of the incoming spacecraft relative to the planet, V_∞. The magnitudes of the incoming and circular orbit velocities are from Eq. 2.41 as seen previously: $V_2 = \sqrt{2\mu_{☼}\left(\frac{1}{r_2} - \frac{1}{2a}\right)}$ and $V_{c_2} = \sqrt{\frac{\mu_{☼}}{r_2}}$, where r_2 is the planet's distance from the Sun at the time of spacecraft approach and a is the semi-major axis of the transfer trajectory. The flight path angle at approach (as seen by the Sun) is $\gamma_2 = \cos^{-1}\frac{h}{r_2V_2}$, where h is the angular momentum of the transfer trajectory ($h = r_{\oplus}V_1$ for the typical situation where the departure is tangential to the Earth's orbit). The angle that the incoming velocity vector makes with the direction of the orbit, as seen by the planet, ϕ, can be determined using the sine law in Fig. 6.9, which gives:

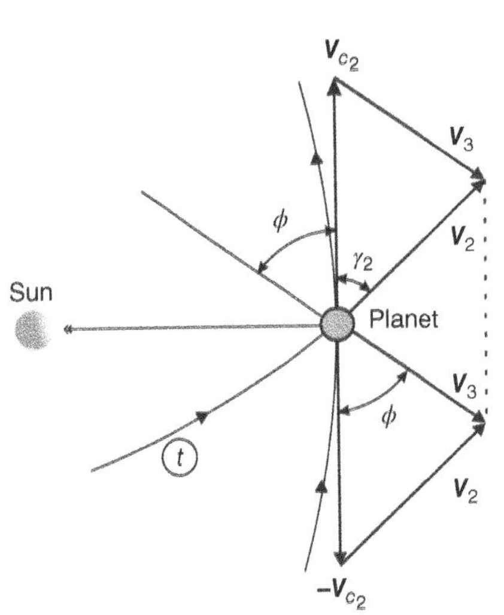

Figure 6.9 General planetary approach geometry.

$$\frac{\sin\phi}{V_2} = \frac{\sin\gamma_2}{V_3} \text{ or } \phi = \sin^{-1}\left(\frac{V_2}{V_3}\sin\gamma_2\right). \tag{6.7}$$

Example Problem 6c

The *Voyager 1* deep-space probe (launched September 5, 1977) headed to Jupiter on a transfer trajectory with the orbital elements[3] $a = 745{,}761{,}000$ km and $e = 0.7978$. What was the incoming velocity relative to Jupiter upon arriving at that planet? Determine both the magnitude of the velocity and the angle relative to the planet's orbital velocity, ϕ.

Solution:

The spacecraft's velocity upon arrival at Jupiter was $V_2 = \sqrt{2\mu_{☼}\left(\frac{1}{r_2} - \frac{1}{2a}\right)} = \sqrt{2(1.327\times10^{11})\left[\frac{1}{778.3\times10^6} - \frac{1}{2(745.8\times10^6)}\right]} = 12.77\,\text{km/s}$, and the velocity of Jupiter in its orbit is $V_{c_2} = \sqrt{\frac{\mu_{☼}}{r_2}} = \sqrt{\frac{1.327\times10^{11}}{778.3\times10^6}} = 13.06\,\text{km/s}$. The momentum of the transfer trajectory can be expressed as follows:

$h = \sqrt{a\mu_{☼}(1-e^2)} = \sqrt{(745{,}761{,}000)(1.327\times10^{11})\left[1-(0.7978)^2\right]} = 6.000\times10^9\,\text{km}^2/\text{s}$. The flight path angle, relative to the Sun, was then $\gamma_2 = \cos^{-1}\left(\frac{h}{r_2V_2}\right) = \cos^{-1}\left[\frac{6.000\times10^9}{(778.3\times10^6)(12.77)}\right] = 52.87°$. The incoming velocity relative to Jupiter was from Eq. 6.6: $V_3 = \sqrt{V_2^2 + (V_{c_2})^2 - 2V_2V_{c_2}\cos\gamma_2} = \sqrt{(12.77)^2 + (13.06)^2 - 2(12.77)(13.06)\cos(52.9°)} = 11.51$ km/s. The angle relative to Jupiter was then (Eq. 6.7) $\phi = \sin^{-1}\left(\frac{V_3}{V_2}\sin\gamma_2\right) = \sin^{-1}\left[\frac{11.51}{12.77}\sin(52.9°)\right] = 46.0°$. The approach geometry was similar to that shown in Fig. 6.9.

The preceding discussion outlined how to determine the velocity (magnitude and direction) of an incoming spacecraft upon planetary arrival. However, this does not address the *offset* – the distance between the line of the incoming velocity and the center of the target planet. The approach is exactly the reverse of the situation depicted in Fig. 6.5, in that there is an envelope of possible approach hyperbolae for a given value of offset, which is in turn related to the incoming excess hyperbolic velocity, the eccentricity of the hyperbola, and the periapsis radius. Since the incoming hyperbolic excess velocity, V_∞, is determined only by the transfer trajectory and the planet's orbital velocity, the semi-major axis of the approach hyperbola has the fixed value $a = \frac{\mu_{planet}}{V_\infty^2}$. The eccentricity can be determined from Eq. 2.55: $\Delta = a\sqrt{e^2-1}$, so $e = \sqrt{\left(\frac{\Delta}{a}\right)^2 + 1}$. The resulting periapis radius from the planet is then simply $r_p = a(e-1)$. In practice, the needed offset for the desired planetary approach (and thereby periapsis radius) can be easily adjusted, as needed, by small mid-course corrections during the transit to the planet.

Assuming there is no interaction with the planet's atmosphere, a spacecraft performing a flyby of a planet will exit the planet's sphere of influence with the same excess velocity that it had on approach, but with the trajectory turned by an amount $\delta = 2\sin^{-1}\left(\frac{1}{e}\right)$ depending on

3 https://voyager.jpl.nasa.gov/mission/science/hyperbolic-orbital-elements/ accessed 1/24.

the eccentricity of the hyperbola (Eq. 2.51). The amount of turning can thus be determined from the amount of the offset. The amount of turning, combined with the side of the planet over which a flyby encounter occurs, can result in significant changes in both velocity and inclination of the departure velocity from a planet relative to the Sun.

Insertion into orbit around a planet can be performed by providing a suitable velocity change (a braking maneuver) at periapsis to yield the desired planetary orbit.

Example Problem 6d

A spacecraft approaches Mars as shown in Fig. 6d. The velocity at Mars' SOI of $V_\infty = 2.67$ km/s and with an offset distance of $\Delta = 10{,}000$ km. The spacecraft is to be inserted into a circular orbit with radius $r = 5000$ km.

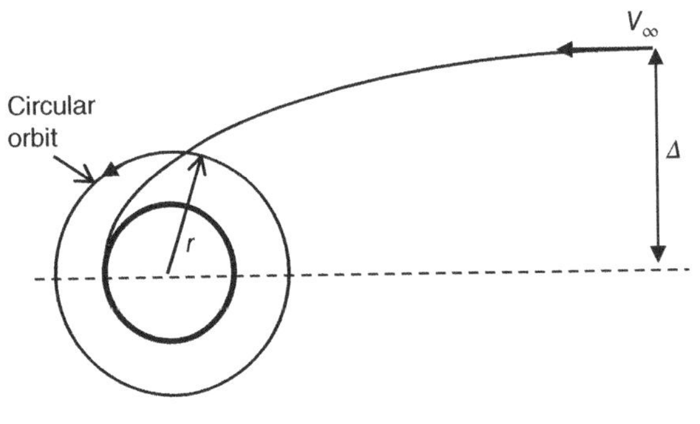

Figure 6d Mars approach geometry.

If the orbit insertion maneuver does not occur, will the spacecraft impact the surface of Mars? What will be the spacecraft's velocity, V (km/s), when it reaches the target radius of $r = 5000$ km? What will be the corresponding flight path angle, γ (degrees) at that point? Determine the magnitude of velocity change, ΔV (km/s), required for insertion into the required circular orbit.

Solution:

The periapsis radius in the absence of a maneuver can be determined from the semi-major axis and eccentricity of the approach hyperbola. The semi-major axis is $a = \dfrac{\mu_{♂}}{V_\infty^2} = \dfrac{4.3 \times 10^4}{(2.70)^2} = 5898.5$ km, so the eccentricity is $e = \sqrt{\left(\dfrac{\Delta}{a}\right)^2 + 1} = \sqrt{\left(\dfrac{8000}{5898.5}\right)^2 + 1} = 1.685$. The periapsis radius would then be $r_p = a(e-1) = 5898.5(1.685 - 1) = 4041$ km, which exceeds the planet's radius (3390 km), so the spacecraft will not hit the planet, but will fly past if the orbital-insertion maneuver does not occur. The velocity upon arrival at the target orbit is $V = \sqrt{2\dfrac{\mu_{♂}}{r} + (V_\infty)^2} = \sqrt{2\left(\dfrac{4.3\times10^4}{5000}\right) + (2.70)^2} =$ 4.95 km/s, with the flight path angle $\gamma = \cos^{-1}\left(\dfrac{V_\infty \Delta}{rV}\right) = \cos^{-1}\left[\dfrac{(2.70)(8000)}{(5000)(4.95)}\right] = -29.2°$, where the negative value is chosen since the trajectory is clearly inward at that point. Given the required velocity in the circular target orbit of $V_c = \sqrt{\dfrac{\mu_{♂}}{r}} = \sqrt{\dfrac{4.3\times10^4}{5000}} = 2.93\,\text{km/s}$, the magnitude of the velocity change needed for orbital insertion is (see Eq. 5.13) $\Delta V = \sqrt{V^2 + V_c^2 - 2VV_c\cos\gamma} = \sqrt{(4.95)^2 + (2.93)^2 - 2(4.95)(2.93)\cos(29.2°)} = 2.79\,\text{km/s}$.

6.7 Planetary Encounters

A trajectory associated with a complete planetary flyby is illustrated in Fig. 6.10. In this case, a Hohmann transfer has been performed to a superior planet, so the spacecraft appears to be incoming from ahead of the planet, with a hyperbolic excess velocity V_∞ that is collinear with the planet's orbital velocity, V_C. A sunlit-side approach is shown. A key point is that the motion of the spacecraft, relative to the planet, will be very different from its motion as seen by the Sun. Assuming that there is no interaction with the planet's atmosphere or other perturbations or maneuvers, the spacecraft will, relative to the planet, pass on a fixed, hyperbolic trajectory, as shown with a dashed line in each frame in Fig. 6.10. The spacecraft's position is indicated in each frame as it flies on its hyperbolic trajectory past the planet, with t_1 being the time at the start of the encounter; t_4, the end. Taking into account the orbital motion of the planet during the flyby maneuver and connecting the positions of the spacecraft results gives the spacecraft's flight path as seen by the sun, shown by the solid line in the figure. The trajectory of the spacecraft, both in terms of its velocity and flight path relative to the sun, has thus been changed by the encounter with the planet. The outward shift of the spacecraft trajectory shown in this case could be useful, for example, in reaching another superior planet beyond the planet just encountered.

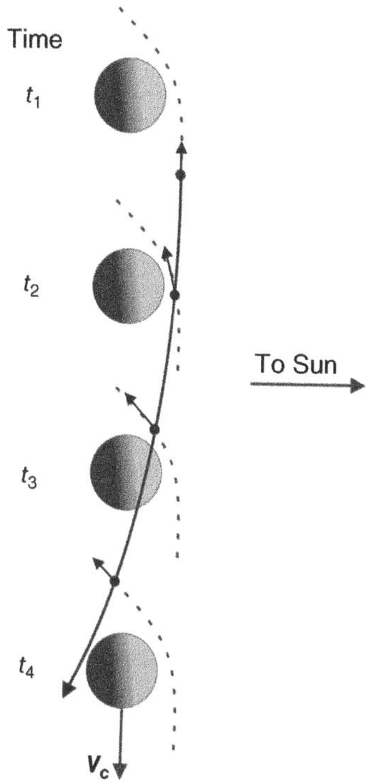

Figure 6.10 Superior planet flyby from solar and planetary perspectives.

In addition to selecting the desired offset for the planetary encounter as discussed earlier, a key consideration is the *side* of the planet on which the encounter is to occur. For example, the spacecraft might approach the planet on the planet's sunlit side, or the dark side, or alternatively, the "leading" or "trailing" side relative to the planet's orbital velocity. Other possibilities include a trajectory designed to pass directly over the north or south pole of the planet, or any inclination between polar and equatorial. A few such possibilities are shown in Fig. 6.11, all with the same incoming hyperbolic excess velocity, direction, offset, and periapsis distance. The hyperbolic flyby trajectories considered here are identical in shape, but different in orientation relative to the target planet. The orientation chosen will impact the departure geometry as seen by the Sun. This is the essence of the *gravity assist* maneuver, which can provide a velocity boost or braking useful in subsequent planetary encounters.

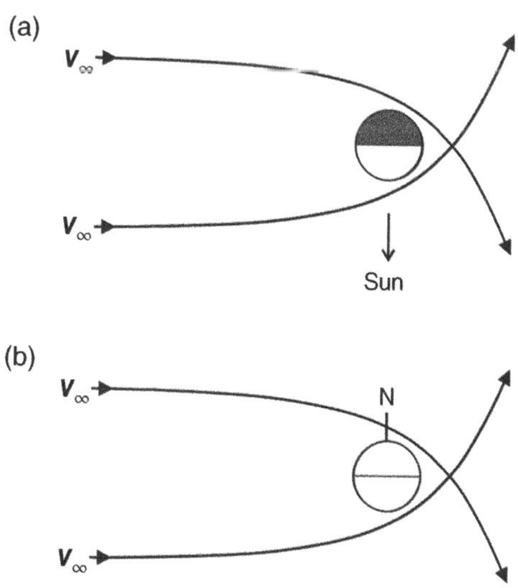

Figure 6.11 Some possible trajectories for a planetary encounter. (a) Dark side/sunlit side approaches; (b) polar approaches.

Example Problem 6e

A spacecraft arrives at Venus on a Hohmann transfer from Earth. The incoming spacecraft trajectory is directly over the north pole of Venus and turns by an angle of exactly $\delta = 90°$ during the flyby. Sketch the approach geometry relative to Venus using an appropriate Venus-centered axis system. After the flyby maneuver, determine the departing vector and scalar velocities *relative to Venus* and the radius at periapsis. Sketch the complete flyby in Venus-centered coordinates. Finally, determine the Venus-departure velocity *relative to the Sun* but in the same Venus-centered coordinates. What will be the inclination of the trajectory after the flyby, relative to the Sun (degrees)? Take the orbit of Venus to be circular, with the poles exactly normal to the ecliptic plane.

Solution:

In this case, the spacecraft overtakes the planet with the scalar velocity $V_{\infty ♀} = 2.71$ km/s (Example Problem 6b). The complete flyby maneuver, as seen by the planet, is shown in Fig. 6e-1. If there are no losses due to Venus' atmosphere or any thrust maneuver, the outgoing velocity has the same magnitude as the incoming value. Flying over Venus' north pole with a turning angle of $\delta = 90°$, as sketched in the figure, the incoming velocity vector is then $\boldsymbol{V}_{\infty ♀ in} = 2.71\ \hat{\boldsymbol{u}}_v$ km/s; the departing velocity vector is $\boldsymbol{V}_{\infty ♀ out} = -2.71\ \hat{\boldsymbol{u}}_n$ km/s.

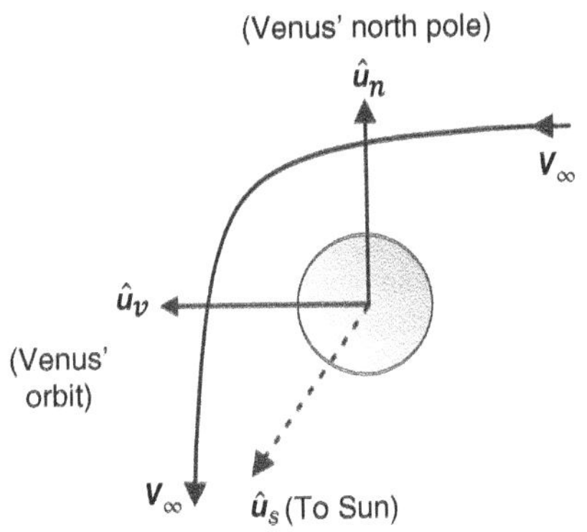

Figure 6e-1 Venus fly-by trajectory (relative to the planet).

The eccentricity of the flyby hyperbola can be determined from Eq. 2.51 to give $e = \dfrac{1}{\sin\left(\dfrac{\delta}{2}\right)} = \dfrac{1}{\sin\left(\dfrac{90°}{2}\right)} = \sqrt{2} = 1.414$. The periapsis radius can be determined from Eq. 2.31 and noting that $a = \dfrac{V_\infty^2}{\mu_♀}$ to get $e = 1 + \dfrac{r_p V_\infty^2}{\mu_♀}$. Thus $r_p = \dfrac{\mu_♀(e-1)}{V_\infty^2} = \dfrac{\mu_♀(e-1)}{V_\infty^2} = \dfrac{3.25 \times 10^5(1.414-1)}{(2.71)^2} = 18{,}330$ km. This is roughly three times the radius of Venus, $r_♀ = 6052$ km.

The flyby trajectory, as seen by the planet, is shown in Fig. 6e-1. From the Sun, the velocity departing Venus is obtained by combining the departing velocity vector, as seen by Venus, with the planet's orbital velocity vector. Thus, relative to the Sun, the departure velocity is (using the same coordinates) $\boldsymbol{V}^+ = \boldsymbol{V}_{c_♀} + \boldsymbol{V}_{\infty ♀ out} = +35.02\ \hat{\boldsymbol{u}}_v - 2.71\ \hat{\boldsymbol{u}}_n$. The incoming velocity, again relative to the sun, would be $\boldsymbol{V}^- = \boldsymbol{V}_{c_♀} + \boldsymbol{V}_{\infty ♀ in} = +35.02\ \hat{\boldsymbol{u}}_v + 2.71\ \hat{\boldsymbol{u}}_v$. The inclination of the post-flyby trajectory relative to the Sun will be $\delta_{sun} = \tan^{-1}(-2.71/35.20) = -4.43°$. This is shown in Fig. 6e-2. In this case, there is no change in the magnitude of the velocity due to the flyby encounter.

Figure 6e-2 Departure geometry (relative to the Sun).

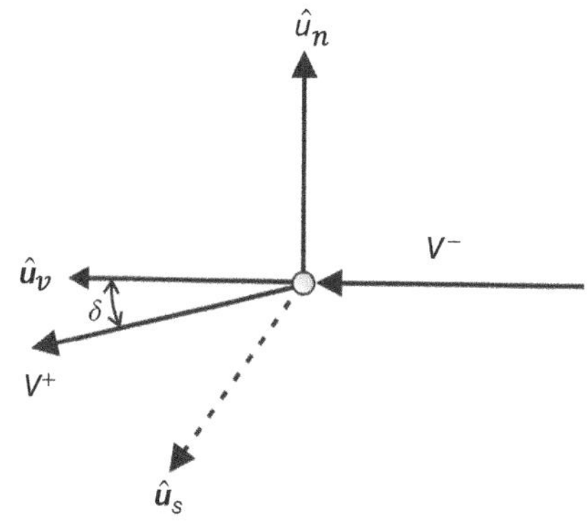

In general, a spacecraft's approach to a planet will not be aligned with the planet's orbital velocity, as was the case for the Hohmann transfer between circular orbits considered in Example Problem 6e. The geometry of a general planetary approach was discussed in Section 6.6 and was illustrated in Fig. 6.9. That geometry, while providing the velocity and orientation of the incoming spacecraft relative to the target planet, does however not specify the *side* of the planet over which the spacecraft approaches. As discussed earlier, the choice of on which side of the planet the encounter occurs, along with the selected periapsis radius, has a profound effect on the post-flyby trajectory. Two encounter possibilities are shown in Fig. 6.12. For simplicity, assume that both encounters are in the plane of the planet's orbit.

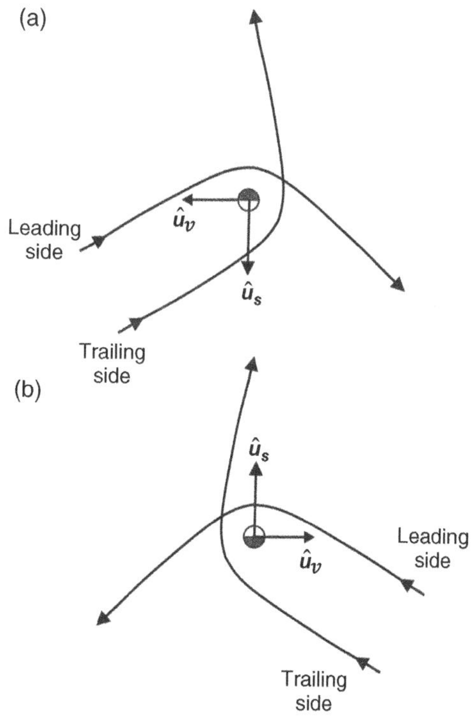

Figure 6.12 Leading-side versus trailing-side planetary encounters. Flyby maneuvers for encounters with (a) a superior planet and (b) an inferior planet.

Assuming no losses due to a planet's atmosphere or any braking maneuver, the spacecraft will depart the planet with the same excess velocity, V_∞, that it had on approach. However, the departure angle from the planet will be different for a dark-side versus a sunlit-side approach, as shown in Fig. 6.12. The amount of turning for the hyperbolic trajectory will depend on the incoming excess velocity and the periapsis radius (or the chosen offset). The geometry shown in Fig. 6.12a is that of a spacecraft encountering a superior planet. In that case, a sunlit-side approach results in a deflection of the spacecraft outward, away from the Sun. The corresponding dark-side approach causes an inward deflection, towards the Sun. Similar effects are seen for an encounter with an inferior planet, as shown in Fig. 6.12b. The change in the spacecraft's trajectory relative to the Sun will be a combination of the turning about the planet and the planet's orbital velocity, as shown in the following example.

Example Problem 6f

The famous *Voyager 1* deep-space probe was launched on September 5, 1977, and reached Jupiter on March 5, 1979. The periapsis radius at Jupiter was $r_p = 349{,}000$ km.[4] What was the eccentricity of the flyby trajectory and the turning angle? What was the incoming angle relative to Jupiter?

Next, consider both a leading side and trailing side flyby of Jupiter. Take all maneuvers to be in-plane for simplicity (the actual inclinations and inclination changes were only a few degrees), and assume circular orbits for the Earth and Jupiter. For both flyby possibilities, determine the outbound velocity components, relative to both Jupiter and to the Sun. Sketch both trajectories. Which of those two flyby scenarios (leading side, trailing side) would result in the *Voyager* to move out beyond the orbit of Jupiter? Verify that a resulting trajectory would be sufficient to reach Saturn (it was!).

Solution:

Voyager 1 arrived at Jupiter on the transfer trajectory from Earth with the velocity $V_2 = 12.77$ km/s and flight path angle $\gamma_2 = 52.87°$ relative to the sun, and the hyperbolic excess velocity relative to Jupiter of V_∞=11.51 km/s (see Problem 6c). This gives for the eccentricity of the flyby trajectory $e = 1 + \dfrac{r_p V_\infty^2}{\mu_{♃}} = 1 + \dfrac{349{,}000(11.51)^2}{1.268 \times 10^8} = 1.364$, with the turning angle $\delta = \sin^{-1}\left(\dfrac{1}{1.3643}\right) = 94.27°$.

The incoming angle relative to Jupiter would be from Eq. 6.7 $\phi_1 = \sin^{-1}\left(\dfrac{V_2}{V_\infty}\sin\gamma_2\right) = \sin^{-1}\left(\dfrac{12.77}{11.51}\right)\sin(52.87°) = 62.21°$.

The two in-plane flyby trajectories are shown in Fig. 6fa and b. Both trajectories necessarily have the same incoming angle of $\phi_1 = 62.2°$, but the outgoing angle, ϕ_2, will be different depending on which side of Jupiter is approached. For the leading-side approach, the outgoing angle will be $\phi_2 = \phi_1 - \delta$, and for a trailing-side approach, it will be $\phi_2 = \phi_1 + \delta$, where the turning angle is the same for both trajectories. This gives for the leading-side and trailing-side approaches the outgoing angles of $\phi_2 = 62.21° - 94.27° = -32.06°$ and $\phi_2 = 62.21° + 94.27° = 156.48°$, respectively.

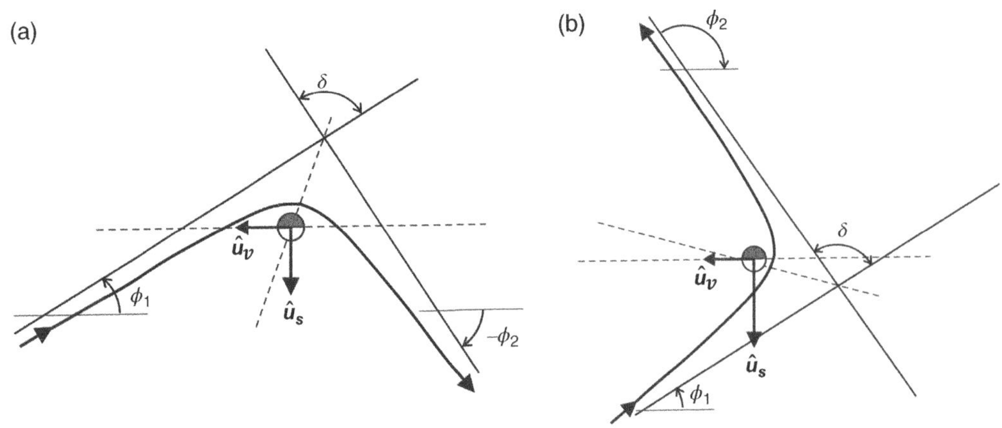

Figure 6f Jupiter flyby geometries. (a) Leading-side approach and (b) trailing-side approach.

4 https://voyager.jpl.nasa.gov/mission/science/hyperbolic-orbital-elements/ accessed 1/24.

It is apparent from the figure that the trailing-side flyby (Fig. 6fb) gives the outward velocity boost needed to reach a planet in a higher orbit. The post-flyby departing velocity vector in that case is given by $\mathbf{V}_{♃}^{+} = -V_{\infty} \cos \phi_2 \hat{\mathbf{u}}_v - V_{\infty} \sin\phi_2 \hat{\mathbf{u}}_s = -11.51[\cos(156.48°)\hat{\mathbf{u}}_v + \sin(156.48°)\hat{\mathbf{u}}_s] = 10.55\ \hat{\mathbf{u}}_v - 4.59\hat{\mathbf{u}}_s$ km/s relative to Jupiter (the corresponding departing velocity vector for the leading-side approach would be $\mathbf{V}_{♃}^{+} = -11.51[\cos(-32.06°)\hat{\mathbf{u}}_v + \sin(-32.06)\hat{\mathbf{u}}_s] = -9.75\hat{\mathbf{u}}_v + 6.11\hat{\mathbf{u}}_s$ km/s). Relative to the Sun, the post-flyby velocity is determined by adding Jupiter's orbital velocity to get $\mathbf{V}_{sun}^{+} = (10.55 + 13.06)\hat{\mathbf{u}}_v - 4.59\hat{\mathbf{u}}_s = 23.61\hat{\mathbf{u}}_v - 4.59\hat{\mathbf{u}}_s$ km/s for the trailing-side flyby. The magnitude of the Sun-centered departure velocity is $V_{sun}^{+} = \sqrt{(23.61)^2 + (4.59)^2} = 24.05$ km/s. This velocity exceeds the escape velocity from the solar system at the distance of Jupiter's orbit, which would be $V_{esc} = \sqrt{2}V_{♃} = \sqrt{2}(13.06) = 18.46\frac{\text{km}}{\text{s}}$. Thus, the departure trajectory from Jupiter will clearly be able to reach the orbit of Saturn (which it did on November 12, 1980).

The preceding example is a classic case of the gravity assist that can be achieved by a planetary flyby. The energy is conserved in the maneuver, so the change in the spacecraft's velocity and energy is accompanied by (infinitesimally small!) changes in the velocity and energy of the planet. Such gravity-assist maneuvers were used repeatedly in the subsequent *Voyager 2* mission (launched August 20, 1977) to take advantage of the unique alignment of the outer planets to be able to do flybys in turn of Jupiter, Saturn, Uranus, and Neptune. *Voyager 2* remains the only space probe to date to have visited either Uranus or Neptune.

The changes in velocity relative to the sun of the *Voyager 2* spacecraft are shown in Fig. 6.13, showing the velocity increases resulting from each planetary encounter. After Jupiter, an escape velocity from the solar system was achieved, which enabled both *Voyager 1* and *Voyager 2* to eventually reach the termination shock, effectively leaving the solar system for interstellar space. Gravity assists involving combinations of flybys past superior planets, inferior planets, and even the Earth have been frequently used in planetary missions. A notable example of this is the *Cassini-Huygens* mission (launched October 15, 1997) which performed, in turn, a flyby of Venus, the Earth, then Venus again, and finally Jupiter before arriving at Saturn on July 1, 2004. The mission profile is rendered in Fig. 6.14. This highly efficient (but long-duration) approach to planetary exploration is viable for un-crewed probes; crewed missions will require more direct transfers.

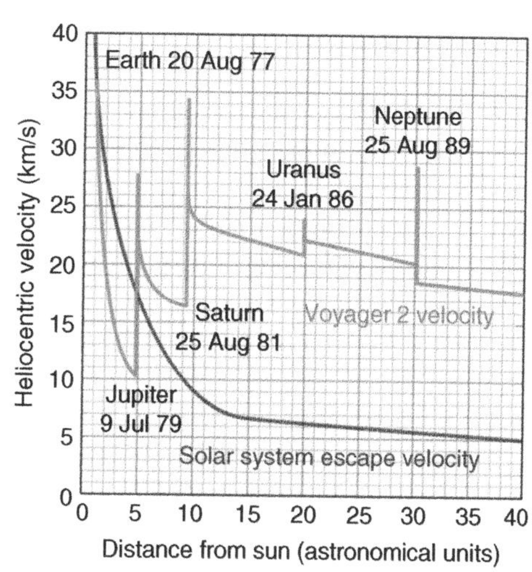

Figure 6.13 Heliocentric velocity of the *Voyager 2* spacecraft. *Source:* https://physics.stackexchange.com/questions/54979/when-did-voyager-1-achieve-solar-system-escape-velocity, accessed 2/24.

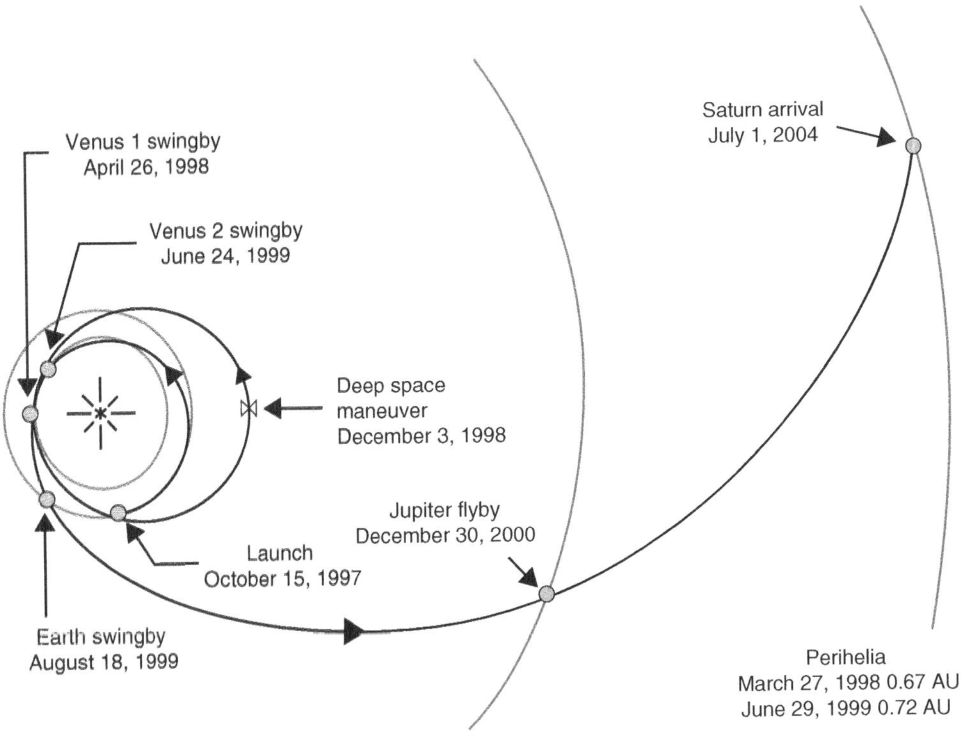

Figure 6.14 Interplanetary trajectory of the *Cassini-Huygens* spacecraft. *Source:* https://www.intechopen.com/chapters/64528, accessed 2/24.

6.8 Problems

6.1 Consider a mission from the Earth to Jupiter and Saturn designed to use only Hohmann transfers for all trajectories. What would be the transfer times from Earth to Jupiter and then from Jupiter to Saturn if Hohmann transfers were performed in both cases? Compare those times with the actual times for the *Voyager 1* given in Example problem 6f. Compare the needed velocity changes for the Hohmann-transfer mission with the actual values for *Voyager 1*.

6.2 Verify the synodic periods of Mars and Jupiter relative to Earth given in Table 6.1. What are the implications of these periods for space missions to those planets?

6.3 For the case of a Hohmann transfer to Jupiter from the Earth, plot the variation in the amount of turning during a flyby encounter for periapsis radii ranging from 75,000 km (just outside Jupiter's atmosphere) to 400,000 km. Plot also the turning angle for hyperbolic excess velocities of 1.5 and 2.0 times that of the Hohmann transfer.

6.4 A spacecraft in LEO with an altitude of 300 km altitude circular earth orbit is to be launched into an elliptical orbit around the sun with a perihelion distance of 100×10^6 km. Calculate the ΔV required in LEO for this maneuver.

6.5 Make a plot of a spacecraft's velocity departing Earth from LEO out to a distance of 1.5×10^6 km. Plot the velocities for a parabolic departure trajectory and for hyperbolic excess velocities of 2, 4, 6, 8, and 10 km/s. Identify the location of the Earth's Sphere of Influence, and comment on the utility of the Sphere of Influence concept for the different cases.

6.6 Spacecraft reentry involves a narrow "entry corridor." For the Apollo lunar missions, the width of the corridor was approximately 30 km – if the entry is too shallow, the spacecraft would skip off the atmosphere and enter solar orbit; if too steep, there would be excessive heating and g-loading of the spacecraft. Define the entry corridor as starting at the radius r_p and with the thickness $r_p + \Delta r_p$. For Earth reentry, take the bottom of the corridor to be at $r_p = 100$ km. What would be the approximate variation in the offset distance on approach, Δy, where y is the offset corresponding to the bottom of the reentry corridor? Express the result in terms of the hyperbolic excess velocity, V_∞, and the gravitational parameter, μ.

6.7 A space probe to Mars departs Earth with a hyperbolic excess velocity of 5 km/s, on a trajectory exactly aligned with the Earth's orbit (which can be assumed to be circular). Given that the orbit of Mars is highly elliptical, calculate the probe's velocity upon arrival at Mars (relative to the sun), the hyperbolic excess velocity, and the incoming angle, ϕ_1, at Mars (see Fig. 6.9). Perform the calculations for an arrival at (i) Mars' perihelion, (ii) Mars' aphelion, and (iii) the parameter of Mars' orbit ($\theta = 90°$). The orbit of Mars can be taken to be coplanar to with that of the Earth.

6.8 For the third case in Problem 6.7 (part iii) calculate both the vector velocities before and after the flyby, for both leading- and trailing-side encounters. How much did the scalar velocity relative to the sun change by the flybys?

6.9 Suppose that in Problem 6.7 the inclination of the Earth parking orbit is 28°, relative to the Earth, what will be the needed inclination of the Mars-transfer trajectory, relative to the ecliptic? Does achieving that orbit require a plane change at trans-Mars injection, and if so, how much of the required injection velocity change, ΔV_1, is due to said plane change?

6.10 A spacecraft is to perform a flyby of Venus. The spacecraft departs parallel to the Earth's orbit and reaches Venus at a true anomaly of −45° on the transfer trajectory. The design periapsis distance at Venus is 6500 km (designed to remain completely outside of Venus' atmosphere). Taking both the orbits of Earth and Venus to be circular and coplanar and all maneuvers to be in-plane, determine both the vector velocities before and after the flyby, for both leading- and trailing-side encounters. How much did the scalar velocity relative to the sun change by the flybys, and would either maneuver allow the spacecraft to then reach the orbit of Mercury?

6.11 For Problem 6.10, what velocity change will be needed to insert the spacecraft into a circular orbit around Venus? How does this compare with the velocity boost needed to depart Earth orbit?

6.12 An earth-launched probe is to impact Ceres. It is required for the probe to impact Ceres with velocity V_{sfc}, on a trajectory *exactly* perpendicular to the surface and directly in line with the sun, as sketched.

Use these properties for Ceres	
Radius of asteroid:	$r_c = 487$ km
Gravitational constant:	$\mu_c = 63$ km^3/s^2
Orbital radius around sun:	$R_c = 3.0$ AU

For the sun, the gravitational constant is $\mu_{☼} = 1.33 \times 10^{11}$ km^3/s^2.

(a) What is the analytical expression for the velocity of the probe as it enters Ceres' SOI, relative to Ceres, V_∞? If $V_{sfc} = 2.0$ km/s, what is the numerical value of V_∞ (km/s)?
(b) What is the approach velocity at Ceres' SOI relative to the sun, V_2, and the flight path angle, γ_2? Give both the analytical expressions and the numerical values. Sketch all velocities and the flight path angle on the diagram provided.
(c) Develop the expressions and get the numerical values for the semi-major axis, a (km), and the eccentricity, e, of the transfer orbit to Ceres.
(d) What single-burn transfer will accomplish this mission from the orbit of the earth? Explain your answer.

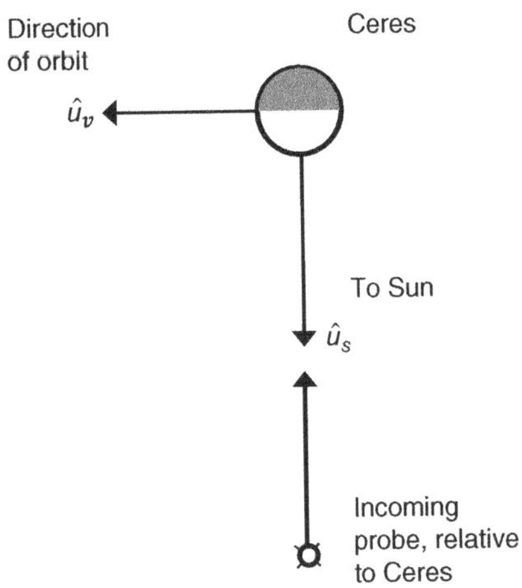

6.13 A solar probe is to use a Mercury flyby to put it into an orbit with a perihelion inside the orbit of Mercury. The probe arrives at Mercury via a Hohmann transfer from Earth. The radius of Mercury is 2487 km, and note that the planet has no atmosphere.

(a) Given that the probe is to be deflected toward the sun, on which side of Mercury must the flyby occur (the sunlit side, or the dark side)?
(b) What is the velocity relative to Mercury at the probe's entry into Mercury's SOI (V_∞, km/s)? Note: The perihelion velocity on an inward-bound Hohmann transfer can be quite high – so a rather energetic flyby of Mercury might be expected.
(c) What is the maximum possible deflection angle during the Mercury flyby, δ (degree)? What would be the periapsis radius of such a flyby? Sketch the approach and departure velocities using Mercury-centered coordinates.
(d) By what angle is the probe deflected relative to the Sun?

7

Earth Satellites and Ground Tracks

The orbital mechanics associated with Earth satellites is vitally important, given the very large number of satellites in Earth orbit (approaching 10,000 in 2024).[1] These satellites serve many important functions, such as Earth surveillance, resources monitoring, communications, navigation, intelligence, weather monitoring and prediction, microgravity research, and more. In many cases, the position of a satellite relative to the Earth's surface needs to be precisely known. This problem requires extending the basic formulation of 2-body orbits developed in Chapters 2, 3, and 4 to include the effects of a rotating, nonspherical Earth. That the Earth is rotating generally causes the orbit of a satellite to pass over different portions of the Earth with each orbit. In addition, the Earth is not a perfect sphere, and this lack of sphericity can cause perturbations to the orbit not considered in the point-mass, 2-body approach discussed thus far. The chapter outline is as follows:

- Satellite ground tracks
- Earth rotation effects
- Geosynchronous/geostationary orbits
- Orbital perturbations
  - Effects of gravity
  - Effects of atmospheric drag

7.1 Satellite Ground Tracks

The motion of a spacecraft in Earth orbit is typically displayed in Mission Control, as seen in Fig. 7.1. In the example shown, the large, center screen displays a map of the Earth, over which the spacecraft's motion is overlaid for easy visual reference and position tracking.

One issue in displaying a ground track is a problem that has challenged mapmakers for centuries-how to best portray the surface of a three-dimensional Earth on a two-dimensional page or surface. Many such projections have been developed, but all introduce distortions, in one form or another, unless only a very small area is being mapped. A map display commonly used the *Mercator Projection*, which was developed by Gerardus Mercator in 1569. An example of an Earth map based on this projection is shown in Fig. 7.2a. The Mercator Projection is widely used for navigation and has the advantages that all lines of latitude and longitude are straight and mutually

1 https://nanoavionics.com/blog/how-many-satellites-are-in-space/

Introduction to Orbital Mechanics: A Concise and Practical Approach, First Edition. James C. Hermanson.

Figure 7.1 Mission control (NASA)[2]. *Source:* Joel Kowsky / NASA/ Wikimedia Commons / Public domain.

perpendicular, resulting in a very simple display. One way to visualize the Mercator Projection is to imagine a point light source at the center of a transparent, spherical Earth, as shown in Fig. 7.2b. Projecting the resulting image onto a cylinder outside of the globe, then unrolling the cylinder flat produces the Mercator Projection. A drawback is that this projection substantially exaggerates the land areas at higher latitudes and, in fact, becomes singular as the poles are approached, precluding displaying the entirety of the polar regions. The classic example is to compare the projected sizes of Greenland and South America. Those two landmasses appear similar in area in the Mercator Projection, when in fact South America is more than eight times the size of Greenland. As another example, to avoid creating a 1% distortion in the North–South direction, a map can cover an extent no larger than approximately 67 km near the equator. This decreases to about 45 km at 40° latitude and down to around 28 km at 60°. Given that many satellites (and essentially all crewed missions to date) have moderate orbital inclinations, the extreme polar distortion is usually not a problem for ground tracking.

7.2 Earth Rotation Effects

A typical display of ground tracks of an orbiting spacecraft is shown in Fig. 7.3 (this is from the famous *Freedom 7* mission, the United States' first crewed orbital mission in 1963). It can be seen in the figure that the ground track of each orbit appears shifted to the west from that of the previous orbit. This is primarily due to the rotation of the Earth toward the east.

2 https://en.wikipedia.org/wiki/Mission_control_center#/media/File:Expedition_55_Soyuz_Docking_(NHQ201803230003).jpg

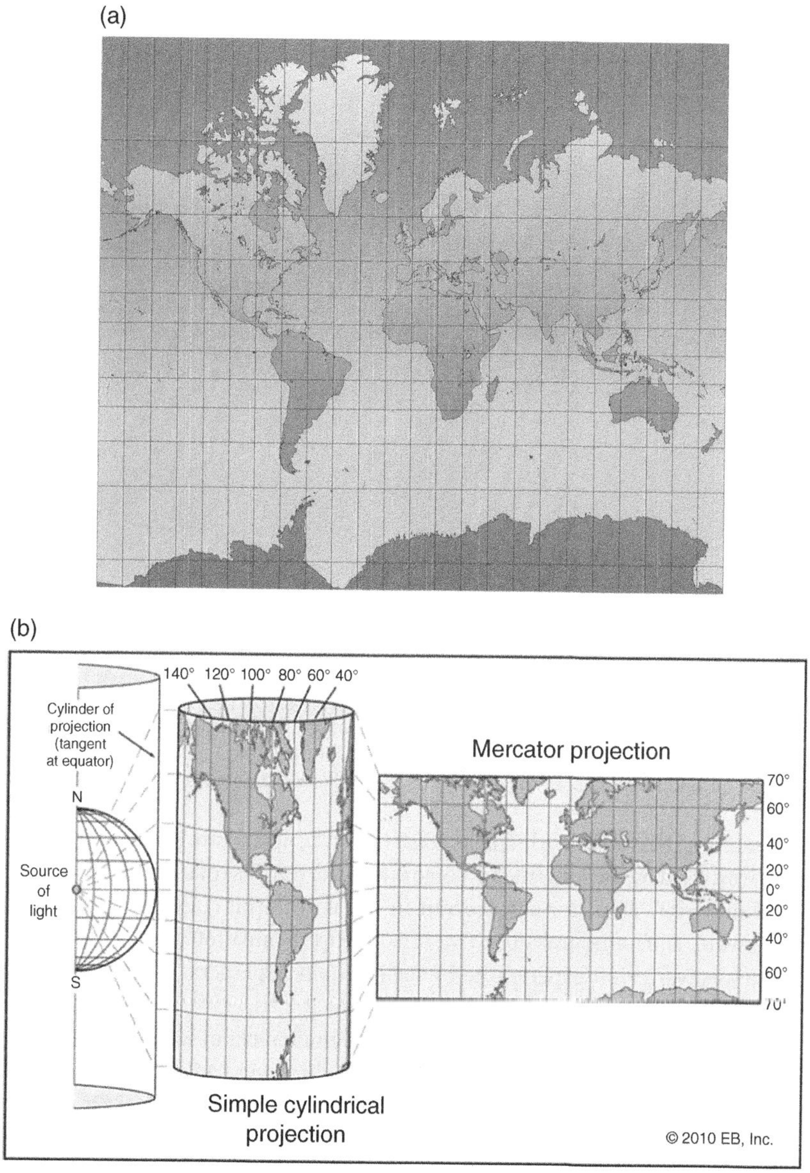

Figure 7.2 (a) Map of the Earth based on the Mercator projection. (b) The Mercator projection visualized versus the spherical Earth.[3]

The Earth rotates with the angular velocity $\omega_{\oplus} = 1$ rotation/(23 hr 56 min) = 360°/23.933 hr = 15.042°/hr. The speed of the Earth's surface at the equator is then $V_{\oplus_{eq}} = \omega_{\oplus} r_{\oplus} = \frac{2\pi}{23.933\text{hr}}(6378\text{ km}) = 1674\frac{\text{km}}{\text{s}}$. This high surface velocity points to the advantage of launching toward the east at a site as close to the equator as possible, as the surface velocity is added to that

3 https://www.britannica.com/science/Mercator-projection

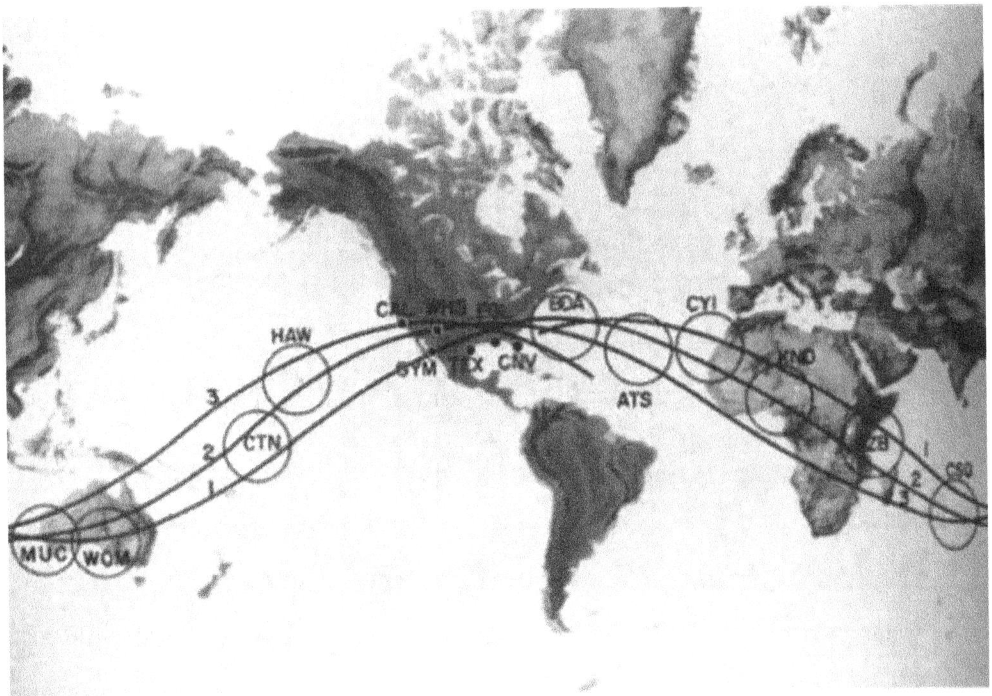

Figure 7.3 Ground track of the *Freedom 7* (Mercury-Atlas 6) orbital mission in 1963. The ground tracks of each of the three orbits are shown, from launch (start of orbit 1) to splashdown (end of orbit 3). The circles represent the coverage areas of various ground tracking stations[4]. *Source:* NASA/ Wikimedia Commons / Public domain.

of the launch vehicle.[5] Assuming that all the orbital elements (a, e, i, Ω, and ω) are known and fixed, each orbit will be shifted westward by the longitude change due to the Earth's rotation of $\Delta L = \omega_{\oplus} T$, where T is the period of the orbit. Thus, for a spacecraft in low earth orbit (LEO) with a period of $T = 95$ min, the westward shift amounts to $\Delta L = (15.04°/\text{hr})(95\ \text{min})/(60\ \text{min/hr}) = 23.8°$. This is approximately the westward shift experienced by the *Freedom 7* spacecraft after each complete orbit, as seen in Fig. 7.3.

There are software packages that are very useful for simulating ground tracks of satellites. One such program is the *Systems Tool Kit* (STK) available from Analytical Graphics, Inc. (AGI).[6] The simulated tracks of a satellite in LEO with an orbital inclination corresponding to Cape Canaveral (28.4°) are shown using STK in Fig. 7.4. Each subsequent orbit is seen to be shifted to the west, producing ground tracks similar to those shown in Fig. 7.3 for the *Freedom 7* mission. STK uses a somewhat different projection than the Mercator Projection shown in Fig. 7.3 that compresses the polar regions somewhat, allowing for better presentation of the Earth's surface at extreme latitudes.

The Earth rotation effect is illustrated again in Fig. 7.5, which shows both the ground tracks and 3D rendering of two satellites in LEO: one with an orbital inclination of $i = 28.4°$, and the other with $i = 67°$. Each ground track shows the position over ground at the start of a particular orbit, and the position exactly one orbital period later, which reflects the westward shift due to the Earth's

4 https://space.stackexchange.com/questions/19402/where-to-look-for-historical-or-reconstructed-orbit-data-for-early-nasa-missions

5 As an aside, this also explains why total solar eclipses have a maximum duration near the equation, where the Earth best "keeps up" with the speed of the Moon's shadow over ground.

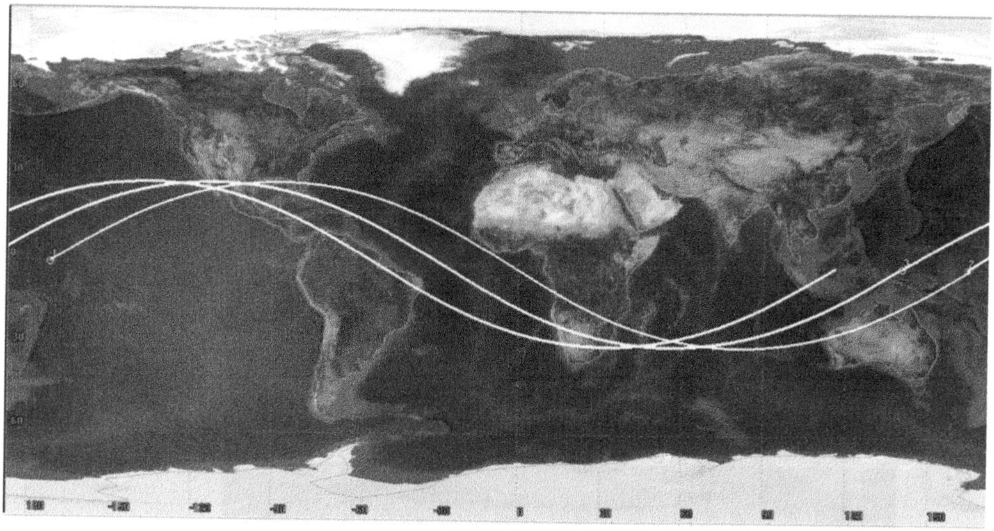

Figure 7.4 Simulated ground tracks of a satellite in LEO using STK. *Source:* with permission of ANSYS, Inc.

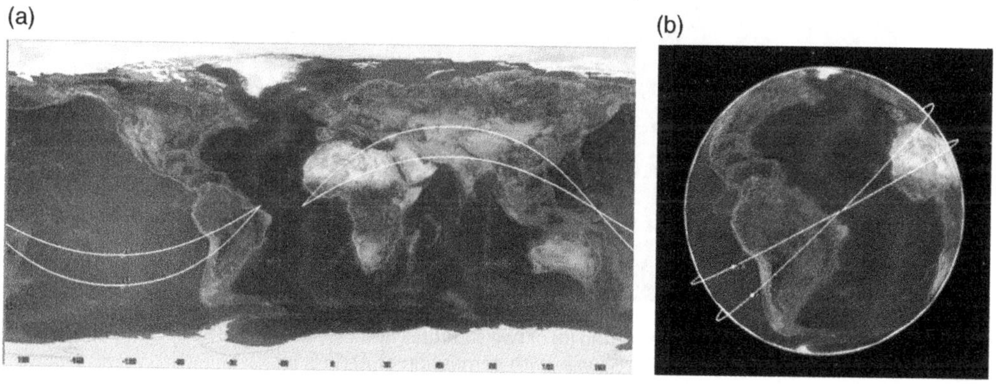

Figure 7.5 Ground tracks (a) and 3D rendering (b) of two satellites in LEO.[6] Both satellites have the same circular orbit altitude of 500 km. *Source:* with permission of ANSYS, Inc.

rotation. The westward shift in longitude of 23.8° is as calculated earlier. The amount of westward shift depends only on the period, which is here the same for both orbits, and is not affected by the inclination of the orbit.

As a different example, the ground tracks of two satellites in six-hour orbits are shown in Fig. 7.6. In this case, the orbits have the same period, but different eccentricities (circular and $e = 0.5$), which is apparent in the 3D rendering shown in Fig. 7.5b. However, given that the periods are identical, the westward shift is the same for both orbits, amounting to $\Delta L =$ (15.04°/hr)(360 min)/(60 min/hr) = 90.2°. The nonsinusoidal appearance of the ground track of the eccentric orbit is due to the fact that the orbital speed, and hence the speed over ground, is not constant in that case.

6 STK from AGI is available in both trial and purchased versions at https://www.ansys.com/products/missions/ansys-stk.

(a)

(b)

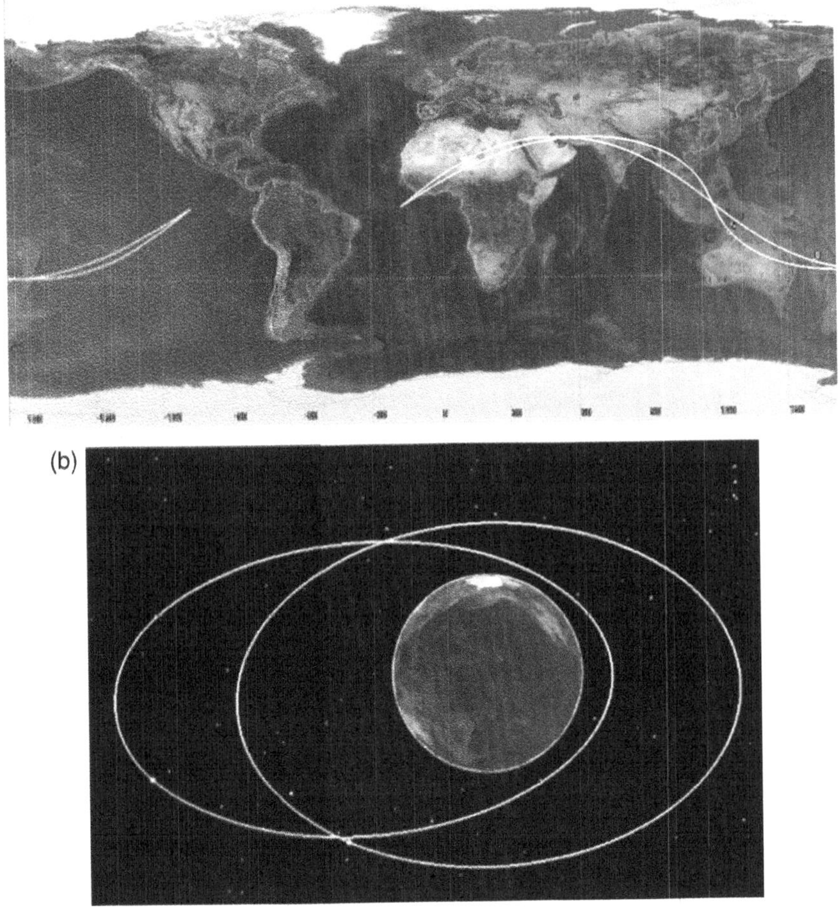

Figure 7.6 Ground tracks (a) and 3D rendering (b) of two satellites in six-hour orbits.[6] The eccentricities of the two orbits are $e = 0$ (circular) and $e = 0.5$. *Source:* with permission of ANSYS, Inc.

7.3 Geosynchronous/Geostationary Orbits

An important class of Earth-orbiting satellites is the *geosynchronous* orbit, where the period of the orbit is designed to exactly match the rotational period of the Earth (23 hr 56 min). In this case, the ground track remains essentially unchanged from orbit to orbit. A special case of the geosynchronous orbit is the *geostationary* orbit, in which the satellite remains precisely over the same spot on the Earth at all times. This is of great benefit, for example, in weather forecasting and communications. There were, in 2023, approximately 580 active satellites in geostationary orbits.[7]

There are different approaches to designing a geosynchronous orbit. Several different geosynchronous orbit possibilities are shown in Fig. 7.7. As an example, suppose it is desired to have a satellite that passes directly overhead the University of Washington in Seattle every day at the same exact time. The minimum inclination of such an orbit would be $i = 47.7°$. The resulting ground track for this orbit is shown in Fig. 7.7a as Case I. The figure-8 shape of that ground track reflects the fact

7 https://en.wikipedia.org/wiki/List_of_satellites_in_geosynchronous_orbit

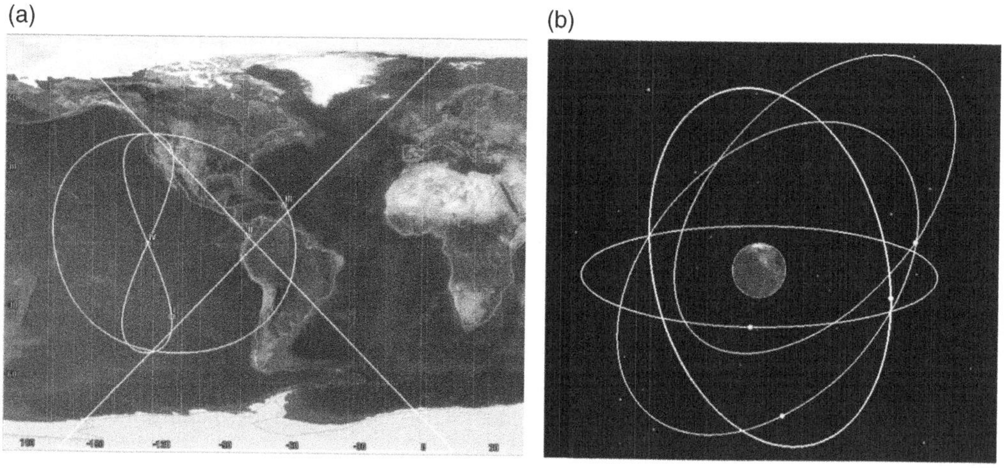

Figure 7.7 Geosynchronous satellite Ground tracks (a) and 3D rendering (b) of several satellites in geosynchronous orbits.[6] The special case of a geostationary satellite appears as a fixed point over the Earth's surface. *Source:* with permission of ANSYS, Inc.

that a satellite in an inclined orbit necessarily flies over both the northern and southern hemispheres. An alternative approach might be to use a geosynchronous polar orbit (designated Case II), which also satisfies the requirement of passing directly overhead during each orbit. The extreme motion of the ground track near the polar regions shown in Fig. 7.7a is due to the limitations of the map projection as discussed earlier. Another concept might be to have the satellite linger over Seattle for as long as possible. This can be accomplished by an eccentric orbit, designed so that the apogee is over the desired target. This situation (designated Case III) is also shown in Fig. 7.7a.

A geostationary orbit must necessarily be exactly over the equator. This is shown in Fig. 7.7a as Case IV as a fixed point on the Earth's surface. A 3D representation of the four cases is shown in Fig. 7.7b. All orbits are geosynchronous, with periods exactly equal to the Earth's rotation, but the orbital parameters (other than the semi-major axis, which governs the period) are quite different.

Example Problem 7a

Determine the effective ground speed of satellites in circular, equatorial orbits with periods of 6, 12, and 36 hours. The ground speed would be the speed of the spot directly beneath the satellite at the Earth's surface.

Solution:

Knowing that angular rate of a circular orbit is $\omega = \frac{2\pi}{T}$, the ground speed of a satellite over a non-rotating Earth would be given by $V_s = \omega R_e = \frac{2\pi}{T} R_e$. Similarly, the eastward velocity of the Earth's surface at the equator is $V_{eq} = \frac{2\pi}{T_s} R_e$, where $T_s = 23.933$ hr is the sidereal rotation time of the Earth. The ground speed of the satellite over the rotating Earth in a prograde orbit is then $V_{gnd} = V_s - V_{eq} = \frac{2\pi}{T} R_e - \frac{2\pi}{T_s} R_e = \frac{2\pi}{T_s} R_e \left(\frac{T_s}{T} - 1\right)$. For a six-hour circular orbit, this gives

$$V_{gnd} = \frac{2\pi}{(23.933\text{hr})\left(3600\frac{\text{s}}{\text{hr}}\right)}(6378\text{ km})\left(\frac{T_s}{T} - 1\right) = \left(0.4651\frac{\text{km}}{\text{s}}\right)\left(\frac{23.933\text{ hr}}{6\text{ hr}} - 1\right) = 1.390\text{ km/s}$$

due east. For a geostationary orbit, $T_s = T$, resulting in a fixed point (zero ground speed), as expected. Repeating the calculation for the 12-hour orbit gives $V_{gnd} = 0.463\dfrac{\text{km}}{\text{s}}$. For the 36-hour orbit, $V_{gnd} = -0.156\,\dfrac{\text{km}}{\text{s}}$, which is motion to the *west* over ground. This illusion of a retrograde orbit as seen from the ground is due to the orbital period of the prograde orbit exceeding the rotation time of the Earth. A satellite in an equatorial LEO at a constant altitude of 500 km has an orbital period of 1.6 hours, which gives a ground speed of 6.594 km/s, compared to the orbital speed of 7.613 km/s.

Example Problem 7b

Calculate the radius of circular, synchronous orbits around Mars and compare the orbit with those of the Martian moons Phobos and Deimos. Repeat the calculation for the Moon, and compare the resulting radius with the Moon's Sphere of Influence (SOI). Comment on the viability of both orbits.

Solution:

The radius of a circular synchronous orbit can be determined by rewriting Eq. 3.1 to give $r = \left[\mu\left(\dfrac{T}{2\pi}\right)^2\right]^{1/3}$, where T is the rotation period of the planet (or the Moon). The rotation period of Mars is $T = 24.617$ hr $= 88{,}621$ s, so using that and the gravitational parameter of Mars gives $r = \left[\mu\left(\dfrac{T}{2\pi}\right)^2\right]^{1/3} = \left[42{,}828\text{km}^3/\text{s}^2\left(\dfrac{88{,}621\text{s}}{2\pi}\right)^2\right]^{1/3} = 20{,}424$ km. The orbits of Phobos and Deimos are nearly circular, with orbital periods of 7.65 and 30.35 hours, respectively. The corresponding circular orbit radii are 9377 km for Phobos and 23,460 km for Deimos. The calculated synchronous orbit for Mars is practical, lying between the orbits of the two moons. For the Moon, the sidereal orbital period is $27.322\text{ d} = 2.360 \times 10^6$ s, so repeating the calculation gives for the Moon $r = \left[4903\left(\dfrac{2.360 \times 10^6}{2\pi}\right)^2\right]^{1/3} = 88{,}439$ km. The Moon's SOI can be estimated using Eq. 6.5 to give

$$r_{SOI_{☽}} \approx R_{\oplus-☽}\left(\frac{m_{☽}}{m_{\oplus}}\right)^{2/5} = 384{,}400\text{ km}\left(\frac{7.348 \times 10^{22}\text{kg}}{5.974 \times 10^{24}\text{kg}}\right)^{2/5} = 66{,}183\text{ km}.$$

This distance is smaller than the calculated orbital radius, indicating that such an orbit would be strongly perturbed by the presence of the Earth and would not be viable (see Chapter 9 for more on the Earth-Moon system).

7.4 Orbital Perturbations – Gravitational Effects

Up to this point, all orbits have been considered on a 2-body, point-mass basis. This model is completely satisfactory for many situations in orbital mechanics, such as planetary departure and arrival, interplanetary transit, etc. The Earth however is *not* a perfect sphere, which results in perturbations to the orbit of spacecraft in Earth orbit, which in turn impact the corresponding ground tracks. This can have important implications for situations where precise ground tracking is required, such as for Earth surveillance.

The dominant perturbation on an orbit due to the nonspherical Earth is due to the *oblateness* caused by the Earth's rotation. This rotation, combined with the deformable properties of the entire Earth, results in an equatorial bulge, so that the Earth's radius at the equator slightly exceeds that at

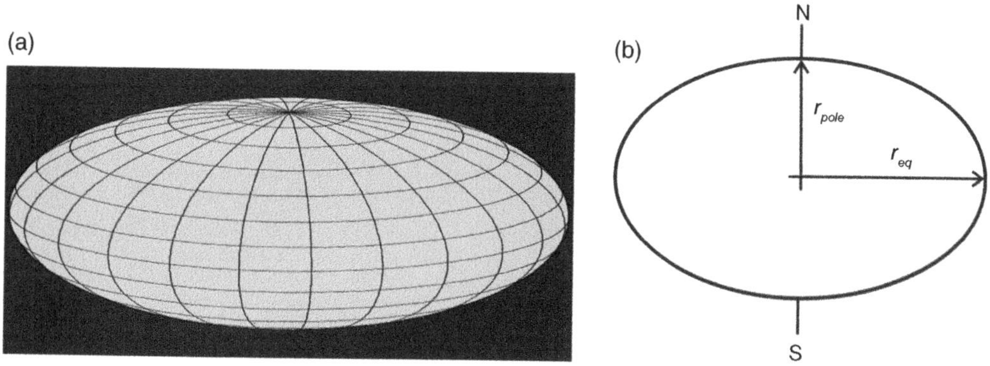

Figure 7.8 Oblateness illustrated. (a) Oblate spheroid and (b) Cross-section.

Table 7.1 Planetary oblateness.

Object	$\frac{r_{eq} - r_{pole}}{r_{eq}}$
Mercury	≈0
Venus	≈0
Earth	0.0034
Mars	0.0065
Jupiter	0.065
Saturn	0.098
Uranus	0.023
Neptune	0.017

the poles. The resulting shape is an oblate spheroid that is symmetric about the polar axis, as shown in Fig. 7.8a; the equatorial bulge concept is shown in cross-section in Fig. 7.8b.

One way to characterize the amount of oblateness is to compare the polar and equatorial radii, to give the nondimensional parameter $(r_{eq} - r_{pole})/r_{eq}$. This degree of oblateness for the various planets is tabulated in Table 7.1.

The oblatenesses of Mercury and Venus are negligible as their rotation rates are very low. By contrast, the outer planets are much more oblate, given their high rotation rates and gas/liquid composition (the oblateness of Saturn is readily apparent visually).

The oblateness of the Earth results in an out-of-plane force on an orbiting object, as shown in Fig. 7.9. This additional force applies a torque to the orbit, which changes the direction of the angular momentum vector, causing the orbit to precess and not remain in a fixed plane in space.

The perturbation due to the oblateness of the Earth results in a modification of the original equation of motion (Eq. 2.11) to give:

$$\ddot{\boldsymbol{r}} = -\frac{\mu}{r^3}\boldsymbol{r} + \boldsymbol{p}, \tag{7.1}$$

where the vector $\boldsymbol{p}$ is the additional acceleration due to the perturbation. This perturbation can be broken down into components relative to the orbital trajectory in the radial, tangential, and normal directions:

$$\boldsymbol{p} = p_r\hat{\boldsymbol{u}}_r + p_\perp\hat{\boldsymbol{u}}_\perp + p_n\hat{\boldsymbol{u}}_n. \tag{7.2}$$

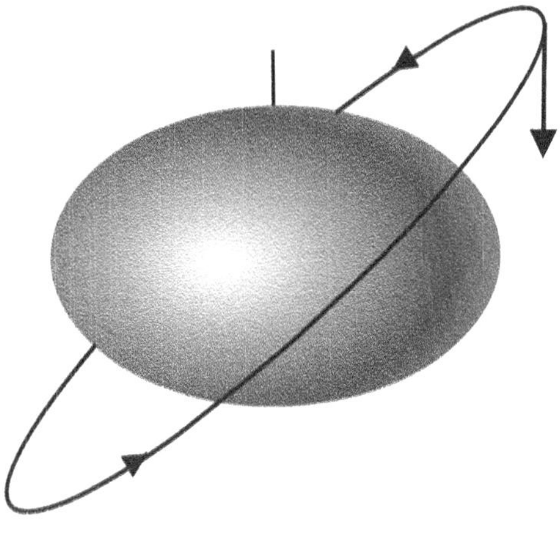

Figure 7.9 Perturbing effect of the Earth's oblateness.

The perturbation forces, like all forces, can be written as the gradient of a suitable potential function. Considering the perturbation due to the oblateness only gives the potential function[8]:

$$U_p = -\frac{\mu}{r}\left(\frac{R}{r}\right)^2 J_2\left[\frac{1}{2}\left(2-3\sin^2\phi\right)\right], \tag{7.3}$$

where ϕ is the co-latitude (the latitude subtracted from 90°) and R is the radius of the central body. The constant J_2 is the zonal harmonic constant that reflects the degree of oblateness.[9] The complete gravitational potential in this case is then:

$$U = \frac{\mu}{r}\left\{1 - J_2\left(\frac{R}{r}\right)^2\left[\frac{1}{2}\left(2-3\sin^2\phi\right)\right]\right\}, \tag{7.4}$$

where the first term is simply the central-force gravitational potential. Taking the gradient of the perturbation potential (Eq. 7.3) and transforming coordinates gives the components in Eq. 7.2 as follows:

$$p_r = -\frac{\mu}{r^2}\frac{3}{2}\left(\frac{R}{r}\right)^2 J_2\left[1-3\sin^2 i\,\sin^2(\omega+\theta)\right],$$

$$p_\perp = -\frac{\mu}{r^2}\frac{3}{2}\left(\frac{R}{r}\right)^2 J_2\sin^2 i\,\sin[2(\omega+\theta)],$$

$$p_n = -\frac{\mu}{r^2}\frac{3}{2}\left(\frac{R}{r}\right)^2 J_2\sin 2i\,\sin(\omega+\theta). \tag{7.5}$$

The perturbations due to oblateness can vary throughout an orbit as the true anomaly varies and also depend on the inclination and argument of perigee. Noting that all perturbations have the same prefactor points to several important effects. For one, the perturbation effects will be stronger for higher values of the gravitational parameter, the radius of the central body, and the constant J_2

8 A detailed mathematical description of the perturbation effects on an orbit can be found in Prussing, J.E. and Conway, B.A., *Orbital Mechanics*, Oxford University Press, 1993.
9 There are higher zonal harmonics, J_3, J_4, etc., but these are much smaller than J_2 and have much less effect on an orbit.

(all of which are different for different planets). In addition, for a given planet, the perturbations become smaller with increasing distance. Completing the analysis[8] and integrating over an entire orbit gives these formulas for the rates of change of the longitude of the ascending node and the argument of periapsis:

$$\dot{\Omega} = \left[-\frac{3}{2} \frac{\sqrt{\mu} J_2 R^2}{(1-e^2)^2 a^{\frac{7}{2}}} \right] \cos i \tag{7.6}$$

$$\dot{\omega} = \left[\frac{3}{2} \frac{\sqrt{\mu} J_2 R^2}{(1-e^2)^2 a^{\frac{7}{2}}} \right] \left(\frac{5}{2} \sin^2 i - 2 \right). \tag{7.7}$$

The daily change in the longitude of the ascending node indicated by Eq. 7.6 is shown in Fig. 7.10 for the case of circular orbits ($e = 0$). The nodal regression is toward the west (negative) for orbital inclinations up to 90° (prograde orbits); for inclinations greater than 90° (retrograde orbits), the nodal displacement is eastward. As expected from the perturbation forces (Eq. 7.5), the effect of the perturbation on the nodal regression weakens with increasing altitude of the orbit. The regression rate is also strongly dependent on the orbit inclination and goes to zero for the case of an exactly polar orbit, where there is no perturbation due to the planet's oblateness. Orbits with higher eccentricities will, for a given orbital period, have a comparatively higher nodal regression rate due to the closer proximity of perigee than that of the corresponding circular orbit.

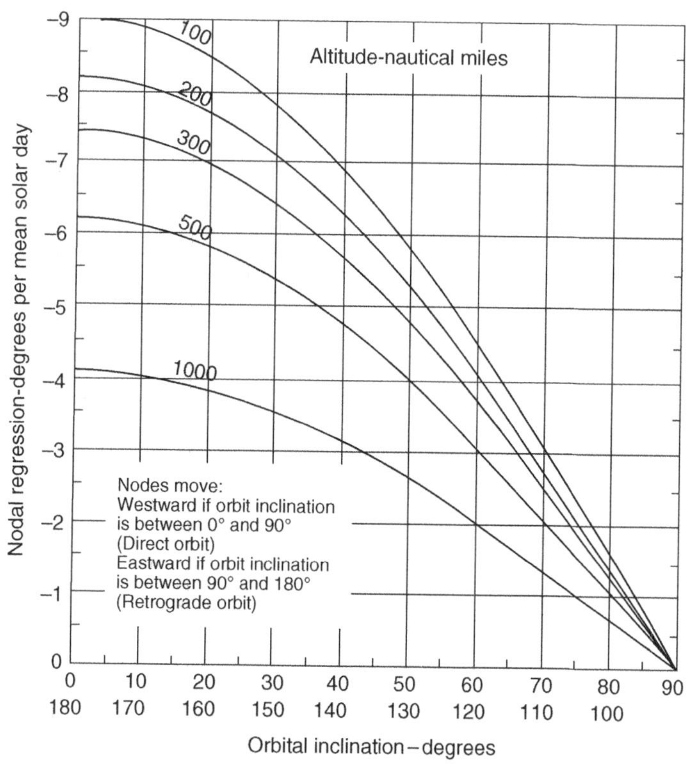

Figure 7.10 Daily change in the location of nodes. *Source:* BMWS.

Example Problem 7c

The International Space Station (ISS) is in a nominal orbit with an altitude of 400 km and an inclination of 51.3°. What is the spacing, in kilometers, between successive ground tracks at the equator, including the effect of Earth's oblateness?

Solution:

The period of the orbit is $T = 2\pi\sqrt{\frac{r^3}{\mu}} = 2\pi\sqrt{\frac{(400 + 6378)^3}{398{,}600}} = 5553.5\ \text{s} = 1.543\ \text{hr}$. The westward shift in longitude of each orbit due to the Earth's rotation is then $\Delta L = \omega_{\oplus} T = (15.04°/\text{hr})(1.543\ \text{hr}) = 23.21°$. The additional shift due to the perturbation to the orbit can be determined from Eq. 7.6: $\dot{\Omega} = \left[-\frac{3}{2}\frac{\sqrt{\mu} J_2 R^2}{(1-e^2)^2 a^{\frac{7}{2}}}\right]\cos i = \left[-\frac{3}{2}\frac{\sqrt{398{,}600}\,(0.001082)(6378)^2}{(400+6378)^{\frac{7}{2}}}\right]\cos(51.3°) = 1.017 \times 10^{-6}$ rad/s $= 5.827 \times 10^{-5}$ deg/s. The total perturbation over a single orbit would be $(5.827 \times 10^{-5}$ deg/s)(5553.5 s) $= 0.324°$. Thus, the total westward shift for each orbit is $23.21 + 0.324 = 23.53°$. As an aside, the total perturbation over a 24-hour period would be a westward shift of $(5.827 \times 10^{-5}$ deg/s)(24 hr)(3600 s/hr) $= 5.035°$, which matches the data in Fig. 7.10 for an orbital inclination of 51.3° and an altitude of 400 km = 248.5 mi.

The corresponding effect of the perturbation on the argument of perigee for circular orbits is illustrated in Fig. 7.11. As with the nodal regression shown in Fig. 7.10, the effect diminishes with increasing altitude. The rotation is positive (increasing argument of perigee) for prograde orbits, up to a value of $i = 63.435°$, which is the point where the term $\frac{5}{2}\sin^2 i - 2$ in Eq. 7.7 is exactly zero.

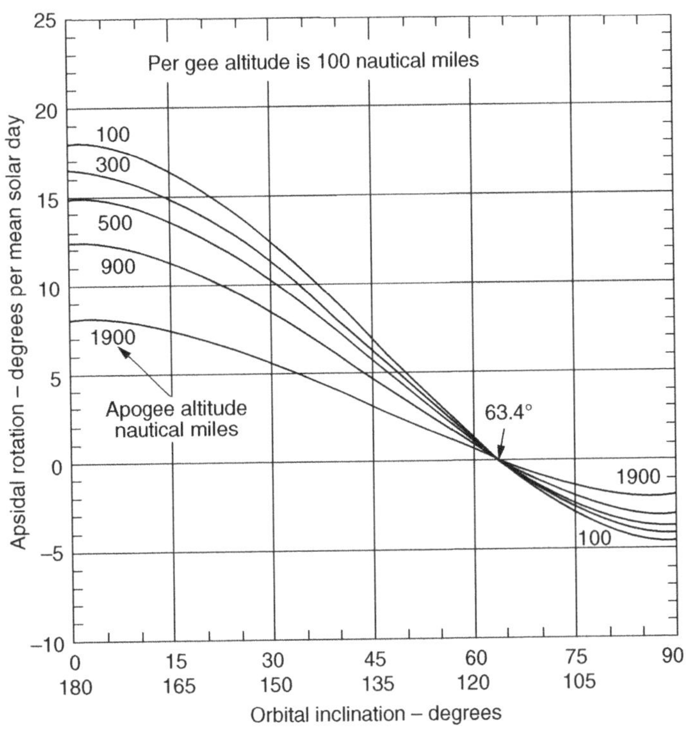

Figure 7.11 Daily change in the argument of perigee. *Source:* BMWS

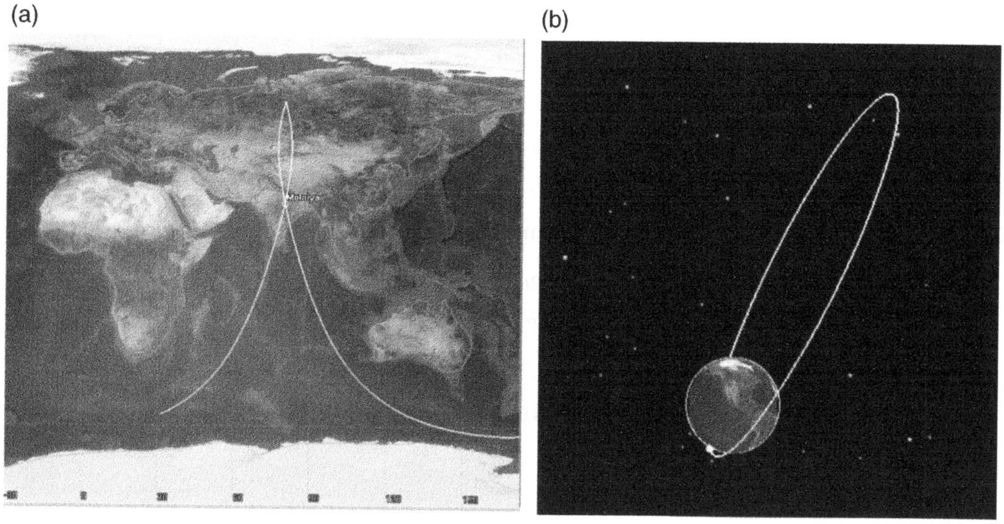

Figure 7.12 Ground track (a) and 3D rendering (b) of a *Molniya orbit*.[6] *Source:* with permission of ANSYS, Inc.

That there is a value of inclination of $i = 63.435°$ for which the argument of perigee remains constant was exploited by engineers in the Soviet Union to deal with high orbital inclinations dictated by the high latitude of the launch site in Kazakhstan of nearly 46°. This was the basis for the *Molniya orbit*, which was used for communications.[10] The unchanging argument of perigee resulted in the apogee of the orbit always being at the same longitude. This was combined with a suitably high value of eccentricity to allow a Molniya satellite to effectively "dwell" over the desired high-latitude locations ideal for Russia. In addition, a comparatively small velocity change was required to achieve an orbital inclination of $i = 63.4°$ than would have been needed for insertion into an equatorial orbit. The velocity-change requirement was further reduced by inserting the satellites into 12-hour, rather than geosynchronous, orbits. A Molniya orbit is shown in Fig. 7.12. The orbital parameters for the orbit shown are $a = 26{,}600$ km, $e = 0.74$, $i = 63.4°$, and $\omega = 270°$. This is for a single Molniya satellite; a constellation of these satellites then allowed switching from one to another to ensure effective, continuous communications.

Example Problem 7d

For the Molniya orbit described earlier, calculate the time that a satellite will be overhead for latitudes exceeding 41° N (the southernmost latitude in Russia).

Solution:

The Molniya orbit first needs to be transformed from perifocal (*pqw*) to geocentric (*IJK*) coordinates. In perifocal coordinates, the orbit is given by Eq. 4.5: $\boldsymbol{r} = r\cos\theta\,\hat{p} + r\sin\theta\,\hat{q}$. Taking for simplicity the longitude of the ascending node to be $\Omega = 0$, and using the inverse transformation matrix in Eq. 4.19 then results in the transformation (see Example Problem 4d):

$$\begin{bmatrix} r_I \\ r_J \\ r_K \end{bmatrix} = \begin{bmatrix} 0 & 1 & 0 \\ -0.4472 & 0 & -0.8944 \\ -0.8944 & 0 & 0.4472 \end{bmatrix} \begin{bmatrix} r\cos\theta \\ r\sin\theta \\ 0 \end{bmatrix} = r \begin{bmatrix} \sin\theta \\ -0.4472\cos\theta \\ -0.8944\cos\theta \end{bmatrix},$$

10 https://earthobservatory.nasa.gov/features/OrbitsCatalog/page2.php

so $\boldsymbol{r}_{IJK} = r\sin\theta\,\hat{I} - 0.4472r\cos\theta\,\hat{J} - 0.8944r\cos\theta\,\hat{K}$ km. When the satellite is exactly over 41° N latitude, the $\hat{K}$-component of the orbital radius is $r\sin(41°) = r\,(0.6561) = -0.8944r\cos\theta$. The true anomaly in the orbit is then given by $\cos\theta = \dfrac{0.6561}{-0.8944} = -0.7335$, so $\theta = 137.18°$. The time the satellite will be above this latitude is $T-2(t-t_0)$, where $T = 12$ hr is the orbital period and $t-t_0$ is the time from perigee to $\theta = 137.18°$. The latter quantity is determined combining Eqs. 3.12 and 3.13 to give: $t - t_0 = \dfrac{T}{2\pi}(E - e\sin E)$. The eccentric anomaly can be extracted from Eq. 3.16: $\tan\left(\dfrac{E}{2}\right) = \left(\dfrac{1-e}{1+e}\right)\tan\left(\dfrac{\theta}{2}\right) = \left(\dfrac{0.26}{1.74}\right)^{\frac{1}{2}}\tan\left(\dfrac{137.18}{2}\right) = 0.9859$, giving $E = 89.18° = 1.557$ rad. Thus, $t - t_0 = \dfrac{(12)(3600)}{2\pi}(1.557 - 0.74\sin(1.557)) = 5617.8$ s $= 1.56$ hr. The total time of the Molniya satellite above the latitude 41°N is then $T_{41+} = 12-(2)1.56 = 8.88$ hr, nearly three-fourths of the entire orbital period.

7.5 Orbital Perturbations – Atmospheric Drag Effects

The Earth's atmosphere is extremely rarified at orbital altitudes, but is not a perfect vacuum. For example, the density of the atmosphere at an altitude of 400 km (the nominal altitude of the ISS) is[11] approximately 4×10^{-12} kg/m^3 for average solar activity, which is roughly 12 orders of magnitude less than the typical density at sea level. At an altitude of only 100 km the density is substantially higher, but still nearly seven orders of magnitude less than that at sea level. Although the gas molecules are sparse in the outer atmosphere, they impact a spacecraft with orbital velocities which can result in substantial aerodynamic drag on a spacecraft, affecting its orbit. A representative time trace of the altitude of the ISS shown in Fig. 7.13 shows the effect of atmospheric drag and the impulsive maneuvers enabled by onboard thrusters on the *Zvezda* Service Module to reboost the altitude to the desired value.

The rate of orbital decay depends on numerous factors, such as the size, mass, and geometry of the spacecraft, as well as the initial orbital elements (altitude, inclination, etc.). In addition, the density of the upper atmosphere is not constant, but can change even daily depending on the level of solar activity. The result of a sample calculation is shown in Fig. 7.14, showing the decay in orbital elements for a satellite in a polar orbit with an initial perigee altitude of 245 km and an eccentricity of 0.12. For much of the satellite's orbital lifetime (100 days in this example), the perigee altitude changes only slightly as the eccentricity continually decreases and the orbit becomes more circular. There is also a small change in the inclination of the orbit. At the final altitude of 100 km shown, a satellite will normally enter the lower atmosphere in a matter of hours.

The effects of orbital decay are illustrated (not to scale) in Fig. 7.15. The first case (a) depicts an eccentric orbit; case (b), a nearly circular one. The eccentric orbit shows a continual drop in apogee altitude, while the altitude at perigee is relatively unaffected (see Fig. 7.14). A nearly circular orbit experiences a gradual decline in altitude throughout the spacecraft's lifetime, as sketched in Fig. 7.15b.

The expected orbital lifetime of a spacecraft in Earth orbit increases significantly with increasing altitude, as the density of gas molecules becomes less and less. The result of detailed calculations for a range of orbital perigee altitudes and eccentricities is shown in Fig. 7.16. The reduced lifetime

11 http://www.braeunig.us/space/atmos.htm

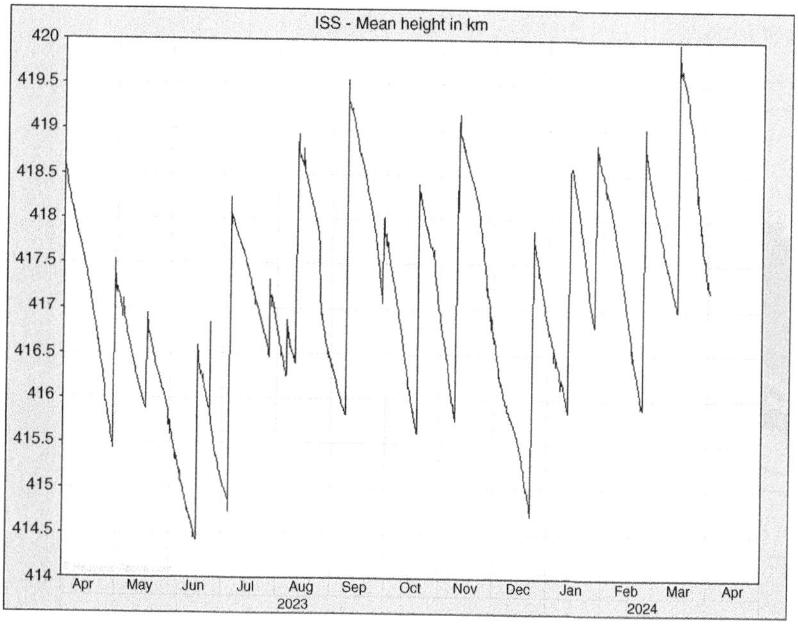

Figure 7.13 Altitude change in the orbit of the International Space Station due to atmospheric drag.[12]

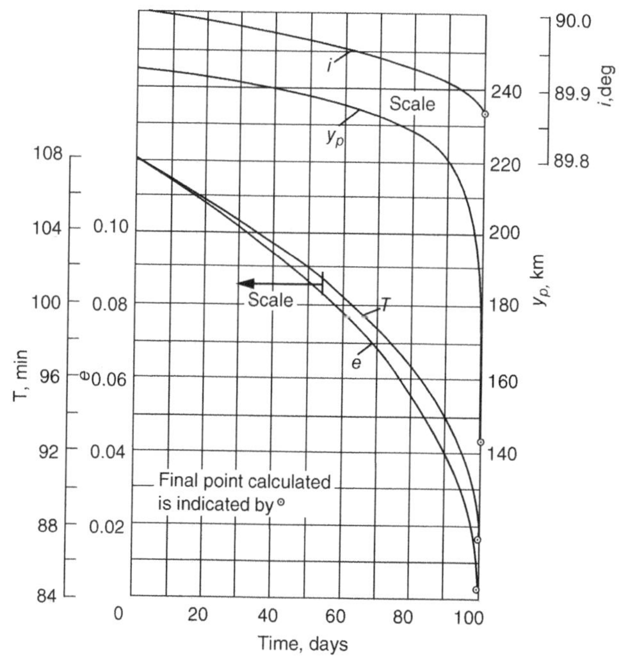

Figure 7.14 Example change in the orbital elements of satellite experiencing atmospheric drag.[13] *Source:* King-Hele, D., Theory of Satellite Orbits in an Atmosphere, Butterworths Mathematical Texts, London, 1964.

---

12 https://www.heavens-above.com/IssHeight.aspx?lat=0&lng=0&loc=Unspecified&alt=0&tz=UCT

13 King-Hele, D.G. (1978). "Methods for Predicting Satellite Orbital Lifetimes". *Journal of the British Interplanetary Society* **31**: 181–196.

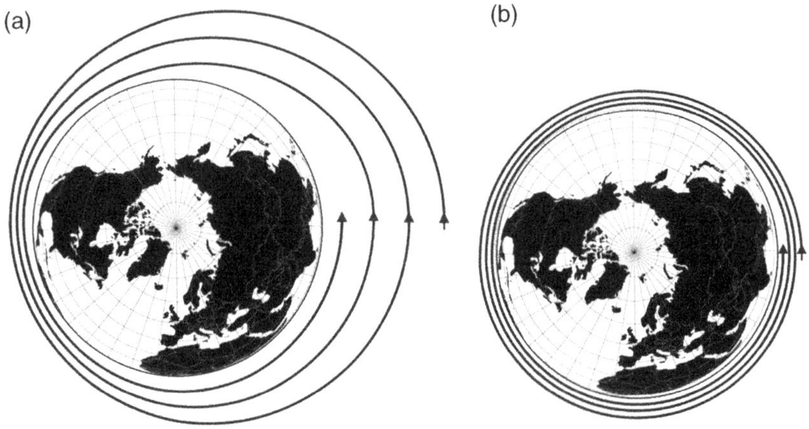

Figure 7.15 Illustration of orbital decay due to atmospheric drag for (a) an eccentric orbit, and (b) a nearly-circular orbit.

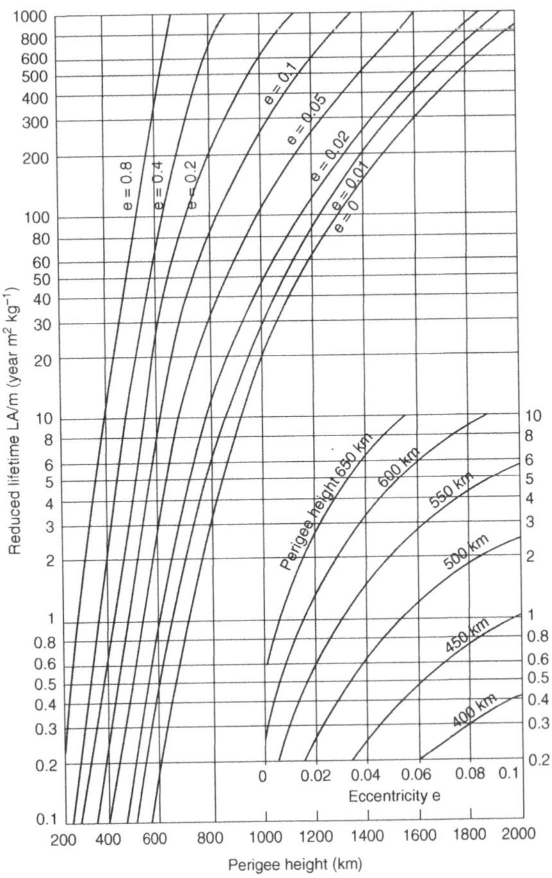

Figure 7.16 Variation in orbital lifetime with perigee altitude and eccentricity.[13]

shown is the calculated lifetime scaled by the frontal area of the spacecraft and the spacecraft mass. The significant increase in lifetime with increasing altitude is evident in the figure. For a satellite in geostationary orbit, the orbital lifetime essentially limitless, far beyond the operational lifetime of any satellite. Given that these orbits will not decay and enter the Earth's atmosphere on any practical time scale, satellites at the end of their service life are normally boosted into a somewhat higher, "graveyard" orbit, normally 200–300 higher than the geostationary orbit. Insertion into a graveyard orbit requires less velocity change that would be needed for transfer to a lower orbit to enable atmospheric entry.

7.6 Problems

7.1 Considering only the zonal harmonic J_2, which of the following are correct regarding perturbations to a satellite's orbit?

(a) The perturbations decrease with increasing altitude.
(b) For retrograde orbits, the line of nodes moves eastward.
(c) Polar, circular orbits are not perturbed.
(d) The value of J_2 increases with the mass of the planet.

7.2 A satellite in a sun-synchronous orbit is continually in sunlight, never in the Earth's shadow. This orbit can be achieved by exploiting the precession of the longitude of the ascending node due to the orbital perturbation. Calculate the orbital inclination required to place an Earth satellite in a 500 km by 1000 km sun-synchronous orbit.

7.3 The Hubble Space Telescope is in a nominal orbit with an altitude of 540 km and an inclination of 28.47°. What is the spacing, in kilometers, between successive ground tracks at the equator, including the effect of Earth's oblateness? Consider both the Earth's rotation as well as the perturbation to the orbit.

7.4 Determine the orbital parameters of a Jupiter-synchronous circular orbit designed to pass exactly over the famous Red Spot during each orbit. Take into account both the rotation of Jupiter as well as the perturbation to the orbit due to the planet's oblateness. The Red Spot is situated at a Jupiter-based latitude of 22° S.

7.5 Use STK to simulate the orbital motion of four satellites in circular equatorial orbits. These include a satellite in LEO, a 12-hour orbit, a geostationary orbit, and a 36-hour orbit. Show both the ground tracks and the orbits in space. Discuss the motion of the satellites over ground – are the individual cases prograde or retrograde?

7.6 What fraction of the Earth's surface can be seen (line of sight) for the satellites in the 12-hour, geostationary, and 36-hour orbits in Problem 5?

7.7 A satellite is in elliptical orbit around Mars with an inclination of 67°. The argument of periapsis advances 4° per Martian day. At what rate does the longitude of the ascending node progress and in which direction (east/west?)

7.8 If the period of the orbit in Problem 7 is two hours, what is the eccentricity of the orbit and the periapsis and apoapsis altitudes?

7.9 An Earth satellite is launched into an orbit with the following orbital elements:

$a = 9000$ km, $e = 0.3$, $i = 25°$, $\Omega = 30°$, $\omega = 315°$
For Earth, $\mu = 398{,}600\ \text{km}^3/\text{s}^2$, $R = 6378$ km, and $J_2 = 1.083 \times 10^{-3}$

(a) What is the period of the orbit (hours)?
(b) How much does the resulting ground track shift in longitude (ΔL) at the equator with each complete orbit? Is the shift westward or eastward? Neglect perturbations to the orbit.
(c) How much *additional* shift in the ground track, $\Delta Lpert$, will occur at the equator due to perturbation of the orbit?

7.10 Use STK to design a geosynchronous orbit that will linger directly overhead at the latitude of the reader for as long as possible.

7.11 It is desired to have a satellite in a circular orbit exactly return to, and retrace, a given ground track after a number of orbits. Assuming the satellite has an inclination of 30°, what will be the radius of the orbit and the orbital period that will so retrace an orbit in the shortest time (minimum number of orbits)?

7.12 Using the results shown in Fig. 7.15, estimate the orbital lifetime of a spacecraft in a circular orbit with an altitude of 600 km. The spacecraft mass is 10,000 kg, and it has a frontal surface area of 50 m^2. Suppose the orbit has the same perigee altitude but an eccentricity of $e = 0.1$. What will be the orbital lifetime in that case? Comment on why the orbital lifetime will be different from that of the circular orbit.

8

Methods for Orbit Determination

In previous chapters, the elements of an orbit have been either specified (e.g. Chapter 2), determined from a combination of known velocity and position information (Chapter 4), or from mission requirements, such as the orbital maneuvers discussed in Chapter 5 or the interplanetary maneuvers presented in Chapter 6. In some cases, however, those sets of information are not known in advance, and the elements of an orbit have to be determined by observation. This chapter discusses methods commonly used for orbit determination based on ground observations. Such observations can include radar tracking and ranging and astronomical measurements of positions in the sky.

The outline of this chapter is as follows:

- Simultaneous velocity and radius vectors known
- Gibbs' method (three position vectors known)
- Lambert's method (two vector positions plus times known)
- Topo-based systems
  - The "radar" case – range, azimuth, and elevation (and their rates) known
  - The "astronomical" case – azimuths and elevations known at multiple times

8.1 Simultaneous Velocity and Radius Vectors Known

The simplest (though highly idealized) case of determining orbital elements is, if at a moment in time, both the radius vector, $\boldsymbol{r}$, and velocity vector, $\boldsymbol{V}$, of an object are known. In this case, the angular momentum and energy of the object can be directly determined (see Fig. 2.6 and Eq. 2.21), which in turn allow determination of all orbital elements (Section 4.1).

8.2 Three Position Vectors Known (Gibbs'/Gauss' Method)

It is possible to determine the elements of an orbit from three known position vectors that might be determined, for example, from several radar observations. This situation is shown in Fig. 8.1

Introduction to Orbital Mechanics: A Concise and Practical Approach, First Edition. James C. Hermanson.

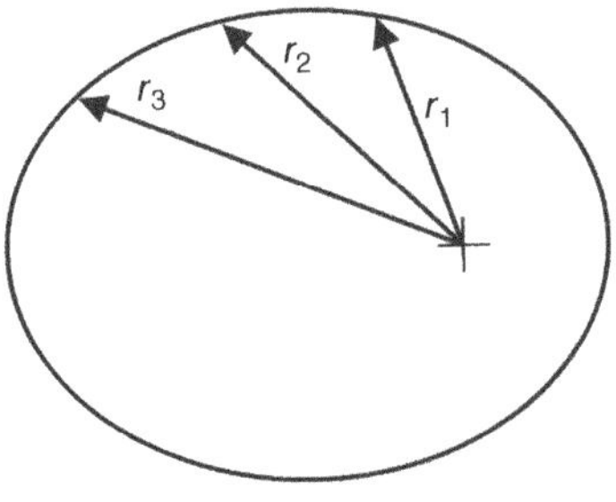

Figure 8.1 Orbit determination from three position vectors.

The analytical method employed in determining the orbit has been alternately attributed to J.W. Gibbs[1] or Karl Friedrich Gauss,[2] and it has correspondingly been termed Gibb's or Gauss' method. The detailed mathematical formulation of the method is provided in several reference works[3] and is summarized here.

It is assumed that the three position vectors $\boldsymbol{r}_1$, $\boldsymbol{r}_2$, and $\boldsymbol{r}_3$ are in a single plane (the plane of the orbit). These vectors are combined with the corresponding scalar magnitudes r_1, r_2, and r_3 to define three new vectors:

$$\boldsymbol{N} = r_1(\boldsymbol{r}_2 \times \boldsymbol{r}_3) + r_2(\boldsymbol{r}_3 \times \boldsymbol{r}_1) + r_3(\boldsymbol{r}_1 \times \boldsymbol{r}_2), \tag{8.1}$$

$$\boldsymbol{D} = \boldsymbol{r}_1 \times \boldsymbol{r}_2 + \boldsymbol{r}_2 \times \boldsymbol{r}_3 + \boldsymbol{r}_3 \times \boldsymbol{r}_1, \tag{8.2}$$

and

$$\boldsymbol{S} = \boldsymbol{r}_1(r_2 - r_3) + \boldsymbol{r}_2(r_3 - r_1) + \boldsymbol{r}_3(r_1 - r_2). \tag{8.3}$$

The vectors $\boldsymbol{N}$ and $\boldsymbol{D}$ are both normal to the plane of the orbit and can be shown to be related to the angular momentum as follows:

$$\boldsymbol{N} = \frac{h^2}{\mu}\boldsymbol{D} \text{ or in scalar form } N = \frac{h^2}{\mu}D. \tag{8.4}$$

The eccentricity of the orbit is then simply determined from the scalar values of the vectors $\boldsymbol{S}$ and $\boldsymbol{D}$:

$$e = \frac{S}{D}, \tag{8.5}$$

and the parameter of the orbit is given by

$$p = \frac{\boldsymbol{N}}{\boldsymbol{D}} = \frac{N}{D}. \tag{8.6}$$

The three unit vectors in the perifocal plane (Section 4.2.3) are

$$\hat{\boldsymbol{w}} = \frac{\boldsymbol{N}}{N}, \hat{\boldsymbol{q}} = \frac{\boldsymbol{S}}{S} \text{ and } \hat{\boldsymbol{p}} = \hat{\boldsymbol{q}} \times \hat{\boldsymbol{w}} = \frac{\boldsymbol{S} \times \boldsymbol{N}}{SN}. \tag{8.7}$$

The perifocal coordinates, combined with the eccentricity and parameter given respectively by Eqs. 8.5 and 8.6, then allow the determination of all orbital parameters (including the inclination, longitude of the ascending node, and argument of periapsis) in three dimensions.

The velocity corresponding to each of the three known radii is given by

$$\boldsymbol{V} = \sqrt{\frac{\mu}{ND}}\left(\frac{\boldsymbol{D} \times \boldsymbol{r}}{r} + \boldsymbol{S}\right), \tag{8.8}$$

1 Josiah Willard Gibbs (1839–1903), American scientist working in physics, mathematics and chemistry.
2 Carl Friedrich Guass (1777–1855), German mathematician, astronomer and physicist.
3 Curtis, H.D. *Orbital Mechanics for Engineering Students*, 4th edition, Elsevier, 2019, or Bate, R.R., Mueller, D.D., White, J.E., and Saylor, W.W. *Fundamentals of Astrodynamics*, 2nd edition, Dover, 2020.

where $\boldsymbol{r}$ is any of the three known vectors. Knowing the simultaneous velocity and radius vectors then allows for the determination of all of the orbital elements, as discussed in Section 8.1.

Alternatively, the Gibbs/Gauss problem can be solved using the three known scalar distances and the angles between them. The angle between the first and second vectors in Fig. 8.1 is $\Delta\theta_2 = \cos^{-1}\left(\frac{\boldsymbol{r}_1 \cdot \boldsymbol{r}_2}{r_1 r_2}\right)$; similarly $\Delta\theta_3 = \cos^{-1}\left(\frac{\boldsymbol{r}_1 \cdot \boldsymbol{r}_3}{r_1 r_3}\right)$, taking the smaller, positive two angles given by the inverse cosine. The Orbit Equation (Eq. 2.30) applies for each of the three radii, so that

$$r_1 = \frac{p}{1 + e\cos\theta_1}, \tag{8.9}$$

$$r_2 = \frac{p}{1 + e\cos\theta_2} = \frac{p}{1 + e\cos(\theta_1 + \Delta\theta_2)}, \tag{8.10}$$

and

$$r_3 = \frac{p}{1 + e\cos\theta_2} = \frac{p}{1 + e\cos(\theta_1 + \Delta\theta_3)}. \tag{8.11}$$

Equations 8.9–8.11 are three equations with the three unknowns p, e, and the true anomaly of the first radius, θ_1. Dividing Eq. 8.9 by Eq. 8.10 gives:

$$\xi_2 \equiv \frac{r_1}{r_2} = \frac{1 + e\cos(\theta_1 + \Delta\theta_2)}{1 + e\cos\theta_1}, \tag{8.12}$$

and

$$\xi_3 \equiv \frac{r_1}{r_3} = \frac{1 + e\cos(\theta_1 + \Delta\theta_3)}{1 + e\cos\theta_1}, \tag{8.13}$$

yielding two equations with the two unknowns, e and θ_1. Combining these two equations to eliminate the eccentricity gives a single, implicit equation for the true anomaly at the first radius:

$$\left(\frac{1-\xi_2}{1-\xi_3}\right) = \frac{\xi_2\cos\theta_1 - \cos(\theta_1 + \Delta\theta_2)}{\xi_3\cos\theta_1 - \cos(\theta_1 + \Delta\theta_3)}. \tag{8.14}$$

Given the value of θ_1 from solving Eq. 8.14 numerically, the eccentricity can be determined from

$$e = \frac{1-\xi_2}{\xi_2\cos\theta_1 - \cos(\theta_1 + \Delta\theta_2)}, \tag{8.15}$$

with the parameter and semi-major axis directly determined from the Orbit Equation:

$$p = a\left(1 - e^2\right) = r_1(1 + e\cos\theta_1). \tag{8.16}$$

This method works for both elliptical orbits and hyperbolic trajectories (in the latter case, the semi-major axis a would be taken to be negative). For the exceptional case of a parabolic trajectory, the parameter p is provided by Eq. 8.16, but the semi-major axis, a, is infinite. It should be noted that this technique only defines the planar orbital elements (e.g. a, e, p); to get the remaining elements in three dimensions (i, Ω, and ω), it is necessary to know at least two of the vector radii $\boldsymbol{r}_1$, $\boldsymbol{r}_2$, and $\boldsymbol{r}_3$ (since the three vectors are in principle coplanar, knowing two allows the determination of the third). The remaining orbital elements can be determined by transforming $\hat{\boldsymbol{w}}$, $\hat{\boldsymbol{q}}$, and $\hat{\boldsymbol{p}}$ from Eq. 8.7 into geocentric *IJK* coordinates.

Example Problem 8a

Three orbital radii are determined in geocentric coordinates to be $\boldsymbol{r_1} = -5471\hat{\boldsymbol{I}} - 3621\hat{\boldsymbol{J}} + 3712\hat{\boldsymbol{K}}$ km, $\boldsymbol{r_2} = -7513\hat{\boldsymbol{I}} - 119\hat{\boldsymbol{J}} - 3116\hat{\boldsymbol{K}}$ km, and $\boldsymbol{r_3} = 8354\hat{\boldsymbol{I}} - 3548\hat{\boldsymbol{J}} - 1963\hat{\mathbf{K}}$ km. Determine the orbital elements a and e and the true anomaly corresponding to $\boldsymbol{r_1}$.

Solution:

The three scalar radii are $r_1 = 7538$ km, $r_2 = 8134$ km, and $r_3 = 9287$ km. The angles between them are $\Delta\theta_2 = \cos^{-1}\left(\frac{\boldsymbol{r_1}\cdot\boldsymbol{r_2}}{r_1 r_2}\right) = \cos^{-1}(0.866) = 30°$ and $\Delta\theta_3 = \cos^{-1}\left(\frac{\boldsymbol{r_1}\cdot\boldsymbol{r_3}}{r_1 r_3}\right) = \cos^{-1}(0.574) = 55°$. The true anomaly at the first radius can then be determined using Eq. 8.14, where $\xi_2 = \frac{r_1}{r_2} = \frac{7538}{8134}$ $= 0.9267$, and $\xi_3 = \frac{r_1}{r_3} = \frac{7538}{9287} = 0.8116$. Inserting values into Eq. 8.14 gives: $\left(\frac{1-0.9267}{1-0.8117}\right) =$ $0.3893 = \frac{(0.9267)\cos\theta_1 - \cos(\theta_1 + 35°)}{(0.8117)\cos\theta 1 - \cos(\theta_1 + 55°)}$. Solving numerically gives the true anomaly of the first radius as $\theta_1 = 10.0°$. The eccentricity then follows from Eq. 8.15 as $e = \frac{1-\xi_2}{\xi_2 cos\theta_1 - \cos(\theta_1 + \Delta\theta_2)_1} = \frac{1-0.9267}{(0.9267)\cos(10°) - \cos(10° + 30°)} = 0.500$, which gives the parameter from Eq. 8.16 as $p = r_1(1 + e\cos\theta_1) = (7538)(1 + 0.500\cos(10.0°)) = 11{,}250$ km, so the semi-major axis is then $a = \frac{p}{1-e^2} = \frac{11{,}250}{1-(0.500)^2} = 15{,}000$ km. The perigee radius is $r_p =$ $\frac{p}{1+e} = \frac{11{,}250}{1+0.500} = 7500$ km.

Example Problem 8b

Determine all elements of the orbit from the radii vectors given in Example Problem 8a, using the Gibbs/Gauss method.

Solution:

First, the needed radius vector cross products need to be calculated. These are $\boldsymbol{r_1}\times\boldsymbol{r_2} = 1.084\times10^7\hat{\boldsymbol{I}} - 1.084\times10^7\hat{\boldsymbol{J}} + 2.655\times10^7\hat{\boldsymbol{K}}$ km^2, $\boldsymbol{r_2}\times\boldsymbol{r_3} = 1.129\times10^7\hat{\boldsymbol{I}} - 1.129\times10^7\hat{\boldsymbol{J}} + 2.764\times10^7\hat{\boldsymbol{K}}$ km^2, and $\boldsymbol{r_3}\times\boldsymbol{r_1} = -2.027\times10^7\hat{\boldsymbol{I}} + 2.028\times10^7\hat{\boldsymbol{J}} - 4.966\times10^7\hat{\boldsymbol{K}}$ km^2. Combining these with the scalar radii $r_1 = 7538$ km, $r_2 = 8134$ km, and $r_3 = 9287$ km then allows the determination of the quantities in Eqs. 8.1, 8.2 and 8.3 as follows:

$\boldsymbol{N} = (7538)\left(1.129\times10^7\hat{\boldsymbol{I}} - 1.129\times10^7\hat{\boldsymbol{J}} + 2.764\times10^7\hat{\boldsymbol{K}}\right) + (8134)\left(-2.027\times10^7\hat{\boldsymbol{I}} + 2.027\times10^7\hat{\boldsymbol{J}} - 4.966\times10^7\hat{\boldsymbol{K}}\right) + (9287)\left(1.084\times10^7\hat{\boldsymbol{I}} - 1.084\times10^7\hat{\boldsymbol{J}} + 2.655\times10^7\hat{\boldsymbol{K}}\right) = 2.090\times10^{10}\hat{\boldsymbol{I}} - 2.089\times10^{10}\hat{\boldsymbol{J}} + 5.099\times10^{10}\hat{\boldsymbol{K}}$ km^2 with scalar magnitude $N = 5.893\times10^{10}$ km^3,

$\boldsymbol{D} = \left(1.084\times10^7\hat{\boldsymbol{I}} - 1.084\times10^7\hat{\boldsymbol{J}} + 2.655\times10^7\hat{\boldsymbol{K}} + \left(1.129\times10^7\hat{\boldsymbol{I}} - 1.129\times10^7\hat{\boldsymbol{J}} + 2.764\times10^7\hat{\boldsymbol{K}}\right) + \left(-2.027\times10^7\hat{\boldsymbol{I}} + 2.028\times10^7\hat{\boldsymbol{J}} - 4.966\times10^7\hat{\boldsymbol{K}}\right) = 1.860\times10^6\hat{\boldsymbol{I}} - 1.850\times10^6\hat{\boldsymbol{J}} + 4.530\times10^6\hat{\boldsymbol{K}}$ km with scalar magnitude $D = 5.235\times10^6$ km, and

$\boldsymbol{S} = \left(-5471\hat{\boldsymbol{I}} + 3621\hat{\boldsymbol{J}} + 3712\hat{\boldsymbol{K}}\right)(8134-9287) + \left(-7513\hat{\boldsymbol{I}} + 119\hat{\boldsymbol{J}} + 3116\hat{\boldsymbol{K}}\right)(9287-7538) + \left(-8356\hat{\boldsymbol{I}} - 3547\hat{\boldsymbol{J}} + 1963\hat{\boldsymbol{K}}\right)(7538-8134) = -1.852\times10^6\hat{\boldsymbol{I}} - 1.853\times10^6\hat{\boldsymbol{J}} - 0.1165\times10^6\hat{\boldsymbol{K}}\ km^2$ with scalar magnitude $S = 2.622\times10^6$km^2.

The eccentricity of the orbit is determined from Eq. 8.5, giving $e = \frac{S}{D} = \frac{2.622 \times 10^6}{5.235 \times 10^6} = 0.501$. The semi-major axis can be determined by rearranging Eq. 8.4 to give $a = \frac{h^2}{\mu(1-e^2)} = \frac{N}{D(1-e^2)} = \frac{5.893 \times 10^{10}}{(5.235 \times 10^6)(1-(0.501)^2)} = 15{,}029\ km$. These results are in good agreement with those of Example Problem 8a, as expected. The unit vectors and the normal and periapsis directions (in geocentric coordinates) are then from Eq. 8.7:

$$\hat{\boldsymbol{w}} = \frac{\boldsymbol{N}}{N} = \frac{2.0902\times10^{10}\hat{\boldsymbol{I}} - 2.089\times10^{10}\hat{\boldsymbol{J}} + 5.099\times10^{10}\hat{\boldsymbol{K}}}{5.893\times10^{10}} = 0.3547\hat{\boldsymbol{I}} - 0.3545\hat{\boldsymbol{J}} + 0.8653\hat{\boldsymbol{K}},\ \text{and}\ \hat{\boldsymbol{p}} = \frac{\boldsymbol{S}\times\boldsymbol{N}}{SN}$$

$$= \frac{\left(-1.852\times10^6\hat{\boldsymbol{I}} - 1.853\times10^6\hat{\boldsymbol{J}} - 0.1165\times10^6\hat{\boldsymbol{K}}\right)\times\left(2.090\times10^{10}\hat{\boldsymbol{I}} - 2.089\times10^{10}\hat{\boldsymbol{J}} + 5.099\times10^{10}\hat{\boldsymbol{K}}\right)}{(2.622\times10^6)(5.893\times10^{10})} =$$

$-0.6273\hat{\boldsymbol{I}} + 0.5953\hat{\boldsymbol{J}} + 0.5010\hat{\boldsymbol{K}}$. The remaining orbital elements are determined following the steps in Section 4.1. The inclination (from Eq. 4.1) is $i = \cos^{-1}(\hat{\boldsymbol{K}}\cdot\hat{\boldsymbol{w}}) = \cos^{-1}(0.8653) = 30.1°$.

The longitude of the ascending node is then (from Eq. 4.2) $\Omega = \cos^{-1}\left[\hat{\boldsymbol{I}}\cdot\left(\frac{\hat{\boldsymbol{K}}\times\hat{\boldsymbol{w}}}{|\hat{\boldsymbol{K}}\times\hat{\boldsymbol{w}}|}\right)\right] = \cos^{-1}$

$$\left[\hat{\boldsymbol{I}}\cdot\left(\frac{\hat{\boldsymbol{K}}\times(0.3547\hat{\boldsymbol{I}} - 0.5173\hat{\boldsymbol{J}} + 0.8653\hat{\boldsymbol{K}})}{|0.3545\hat{\boldsymbol{I}} + 0.3544\hat{\boldsymbol{J}}|}\right)\right] = \cos^{-1}\left[\hat{\boldsymbol{I}}\cdot\left(\frac{0.3545\hat{\boldsymbol{I}} + 0.3547\hat{\boldsymbol{J}}}{0.5015}\right)\right] = \cos^{-1}[0.7062]$$

$= 45.1°$. In a similar fashion, the argument of periapsis (from Eq. 4.3) is $\omega = \cos^{-1}\left[\left(\frac{\hat{\boldsymbol{K}}\times\hat{\boldsymbol{w}}}{|\hat{\boldsymbol{K}}\times\hat{\boldsymbol{w}}|}\right)\cdot\hat{\boldsymbol{p}}\right] =$

$$\cos^{-1}\left[\left(\frac{0.3545\hat{\boldsymbol{I}} + 0.3547\hat{\boldsymbol{J}}}{0.5015}\right)\cdot\left(-0.6273\hat{\boldsymbol{I}} + 0.5936\hat{\boldsymbol{J}} + 0.5010\hat{\boldsymbol{K}}\right)\right] = \cos^{-1}[-0.0236] = 91.4°$$

8.3 Lambert's Problem (Two Vector Positions Plus Times Known)

The next class of orbital-determination methods is the case where two position vectors and the elapsed time between them are known. This situation, commonly termed Lambert's problem,[4] has several known solution methods,[5] one of which is summarized below.[6] The first step in that approach is to use the two scalar radii, r_1 and r_2, and the difference in true anomaly to define a constant A:

$$A = \frac{\sqrt{r_1 r_2}\sin(\Delta\theta)}{\sqrt{1-\cos(\Delta\theta)}}. \tag{8.17}$$

The value of $\Delta\theta$ in Eq. 8.17 is not unique, so it is necessary to know if the orbit is prograde or retrograde. These two possibilities are shown in Fig. 8.2.

Two auxiliary functions are then defined as follows:

$$C(z) = \frac{1-\cos\sqrt{z}}{z} = \frac{1-\cosh\sqrt{-z}}{z}\ \text{and}\ S(z) = \frac{\sqrt{z}-\sin\sqrt{z}}{\sqrt{z^3}} = \frac{\sqrt{-z}-\sinh\sqrt{-z}}{\sqrt{(-z)^3}}. \tag{8.18}$$

4 Johann Heinrich Lambert (1728–1777), a Swiss/French mathematician, physicist, and astronomer.
5 Vallado, D.A. *Fundamentals of Astrodynamics and Applications*, 4th edition, Space Technology Library, 2007.
6 Bate, R.R., Mueller, D.D., White, J.E., and Saylor, W.W. *Fundamentals of Astrodynamics*, 2nd edition, Dover, 2020.

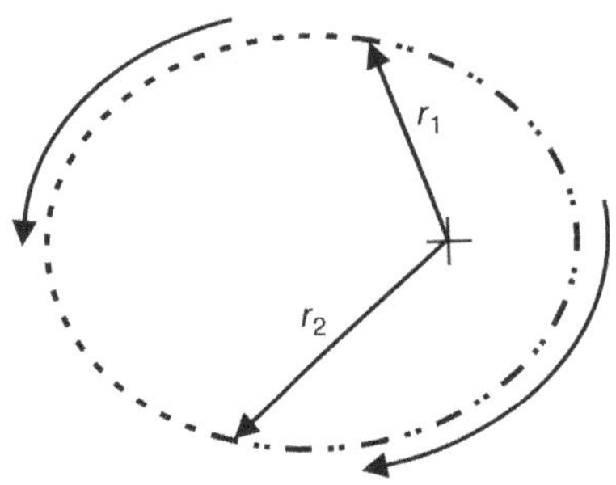

Figure 8.2 Lambert's problem showing two possible solutions.

For elliptical orbits, the circular trigonometric functions are used, with $z = \Delta E^2$; for hyperbolic trajectories, the hyperbolic functions, with $z = -\Delta F^2$, where E and F are, respectively, the eccentric and hyperbolic anomalies (Sections 3.2 and 3.4). The functions in Eq. 8.18 are then used to define two other variables, namely:

$$y = r_1 + r_2 - A\frac{(1 - zS(z))}{C(z)} \text{ and } x = \sqrt{\frac{y}{C(z)}}. \quad (8.19)$$

These relations can then be combined with the time elapsed in passing between the two radii, Δt:

$$\Delta t = \frac{S(z)x^3}{\sqrt{\mu}} + A\sqrt{\frac{y}{\mu}}. \quad (8.20)$$

Implicit in Eq. 8.20 is the unknown anomaly E (or F). Lambert's problem generally requires an iterative solution. This is done in this case by iterating on the variable z until Eq. 8.20 is satisfied. Having done so, the velocities at the locations of the two radii are then given by:

$$\boldsymbol{V}_1 = \frac{\boldsymbol{r}_2 - f\boldsymbol{r}_1}{g} \text{ and } \boldsymbol{V}_2 = \frac{\tilde{g}\boldsymbol{r}_2 - \boldsymbol{r}_1}{g}, \quad (8.21)$$

where the functions f, g, and $\tilde{g}$ are given by:

$$f = 1 - \frac{y}{r_1}, g = A\sqrt{\frac{y}{\mu}}, \text{ and } \tilde{g} = 1 - \frac{y}{r_2}. \quad (8.22)$$

The radius and velocity vectors at either point ($\boldsymbol{r}_1$, $\boldsymbol{V}_1$ or $\boldsymbol{r}_2$, $\boldsymbol{V}_2$) then allow the determination of all of the orbital elements, as discussed previously.

Alternatively, the planar orbital elements can be determined using two scalar distances, the angles between them, and the elapsed time. For the case of two known radii and the angle between them, $\Delta\theta$, Eq. 8.13 applies and can be rewritten as follows:

$$\frac{r_1}{r_2} = \frac{1 + e\cos(\theta_1 + \Delta\theta)}{1 + e\cos\theta_1}. \quad (8.23)$$

Assuming a value of the eccentricity, e, Eq. 8.23 can be solved numerically to give the corresponding value of the true anomaly at the first radius, θ_1. The semi-major axis is then readily determined from Eq. 8.16. The corresponding elapsed time for the assumed value of the eccentricity is given by Kepler's equations (Eqs. 3.13–3.14 for ellipses; Eqs. 3.28–3.30 for hyperbolae). Iterating on the eccentricity in Eq. 8.23 until the correct elapsed time is achieved then gives the solution for the planar elements e, p, and a using Eq. 8.16. If the type of trajectory is known, appropriate initial values of the eccentricity would be used. If not known, one approach would be to assume an elliptical orbit, and then if a solution is not found for the eccentricity approaching unity, a hyperbolic solution would be sought. In either case, the remaining three-dimensional orbital elements can be determined using Section 4.1, and the steps presented in Example Problem 8b.

Example Problem 8c

A spacecraft orbiting the Earth has two measured radii of $r_1 = 10{,}900$ km and $r_2 = 17{,}480$ km, with an angle between them of $\Delta\theta = 100°$. The elapsed time between the two positions is 5400 s = 1.4 hr. What are the planar orbital elements a and e, and the periapsis radius, r_p?

Solution:

The solution begins by inserting an assumed value of the eccentricity and solving Eq. 8.23 numerically with $r_1/r_2 = 10{,}900/17{,}480 = 0.6236$ and $\Delta\theta = 100°$. If the value of the assumed eccentricity is too low, there will not be an orbit that fits both r_1 and r_2. A first guess for the eccentricity can be made by taking the first radius to be at periapsis (i.e. $\theta_1 = 0$). In this case, Eq. 8.15 gives $e = \frac{\frac{r_1}{r_2} - 1}{\cos\Delta\theta - \frac{r_1}{r_2}} = \frac{-0.3764}{-0.1736 - 0.3746} = 0.684$. Setting $e = 0.68$ and solving Eq. 8.23 then gives a true anomaly of the first radius to be $\theta_1 = 115.27°$. The true anomaly of the second radius would be $\theta_2 = \theta_1 + \Delta\theta = 215.27°$. The corresponding semi-major axis can be computed from Eq. 8.16, which can be rewritten to give $a = \frac{r_1(1 + e\cos\theta_1)}{(1 - e^2)} = 10{,}900\frac{[1 + (0.684)\cos(115.27°)]}{[1 - (0.684)^2]} = 14{,}502$ km. The reported elapsed time is shorter than the complete period of such an orbit (Eq. 3.1) of $= 2\pi\sqrt{\frac{a^3}{\mu}} = 2\pi\sqrt{\frac{(14{,}502)^3}{398{,}600}} = 17{,}380$ s, so the spacecraft has only traversed a portion of a single orbit. In this case, Kepler's time equation can be employed as in Section 3.2, with the eccentric anomalies of $E_1 = 2\tan^{-1}\left[\left(\frac{1-e}{1+e}\right)\tan\left(\frac{\theta_1}{2}\right)\right] = 2\tan^{-1}\left[\left(\frac{1-0.684}{1+0.684}\right)\tan\left(\frac{115.27°}{2}\right)\right] = 68.706° = 1.199$ rad, $E_2 = 2\tan^{-1}\left[\left(\frac{1-e}{1+e}\right)\tan\left(\frac{\theta_2}{2}\right)\right] = 2\tan^{-1}\left[\left(\frac{1-0.684}{1+0.684}\right)\tan\left(\frac{215.27°}{2}\right)\right] = 252.547° = 4.4078$ rad, and the mean angular rate (Eq. 3.12) $n = \frac{2\pi}{T} = \frac{2\pi}{17{,}380\,\text{s}} = 3.615\times10^{-4}\text{s}^{-1}$. This then gives the elapsed time according to Kepler's equation and assuming a prograde orbit of $\Delta t = \frac{1}{n}[(E_2 - e\sin E_2) - (E_1 - e\sin E_1)] = \frac{1}{3.615\times10^{-4}s^{-1}}[(4.4078 - 0.684\sin(4.4078)) - (1.199 - 0.684\sin(1.199))] = 12{,}443.92\,\text{s} = 3.457$ hr. This time is much longer than the measured elapsed time. The eccentricity is iteratively decreased until the correct time is achieved, as tabulated in Table 8.1.

The iteration suggests that a satisfactory solution is $e = 0.324$ and $a = 13{,}237$ km. The periapsis radius is $r_p = a(1-e) = 13{,}237(1-0.324) = 8948$ km.

Table 8.1 Iterative solution to Example Problem 8b (target elapsed time $\Delta t = 5400$ s).

e	θ_1 (deg)	a (km)	Δt (s)
0.68	115.11	14,424	12,276
0.40	93.04	12,701	6322
0.30	43.02	14,605	4633
0.345	81.56	12,999	5670
0.333	77.86	13,118	5521
0.3236	74.36	13,237	5397

8.4 Topocentric-Based Systems

8.4.1 Range, Azimuth, and Elevation (and Their Rates) Known

Ground-based observations, such as radar, are often employed to determine the elements of objects orbiting the Earth. Ground-based observations typically are based on the *SEZ* reference frame, as shown in Fig. 4.10. In principle, sufficient information at a single instance in time can be used to determine an orbit. For a ground-based observation, this could include the following scalar information (see Fig. 4.15):

ρ, the range to the target object,
$\dot{\rho}$, the rate at which the range is changing,
El, the elevation and its rate of change, $\dot{El}$, and
Az, the azimuth and its rate of change, $\dot{Az}$.

Knowing the scalar distance and the elevation and azimuth allows determination of the vector position of the target as:

$$\rho = \rho_S \hat{\mathbf{S}} + \rho_E \hat{\mathbf{E}} + \rho_Z \hat{\mathbf{Z}}, \tag{8.24}$$

with the respective components given by:

$$\begin{aligned} \rho_S &= -\rho\cos(El)\cos(Az) \\ \rho_E &= \rho\cos(El)\sin(Az) \\ \text{and } \rho_Z &= \rho\sin(El). \end{aligned} \tag{8.25}$$

The velocity of the object as seen from the observation site is the time derivative of Eq. 8.24, or

$$\dot{\rho} = \dot{\rho}_S \hat{\mathbf{S}} + \dot{\rho}_E \hat{\mathbf{E}} + \dot{\rho}_Z \hat{\mathbf{Z}} \tag{8.26}$$

with the components from differentiating Eq. 8.25:

$$\begin{aligned} \dot{\rho}_S &= -\dot{\rho}\cos(El)\cos(Az) + \rho\sin(El)\left(\dot{El}\right)\cos(Az) + \rho\cos(El)\sin(Az)\left(\dot{Az}\right), \\ \dot{\rho}_E &= \dot{\rho}\cos(El)\sin(Az) - \rho\sin(El)\left(\dot{El}\right)\sin(Az) + \rho\cos(El)\cos(Az)\left(\dot{Az}\right), \text{and} \\ \dot{\rho}_Z &= \dot{\rho}\sin(El) + \rho\cos(El)\left(\dot{El}\right). \end{aligned} \tag{8.27}$$

The radius and velocity vectors relative to the Earth's center are given respectively by $\boldsymbol{r} = \boldsymbol{\rho} + \mathbf{R}_\oplus$ (Eq. 4.20) and $\boldsymbol{V} = \dot{\rho} + \boldsymbol{\omega}_\oplus \times \boldsymbol{r}$ (Eq. 4.21). To determine the elements of the orbit, the position and velocity vectors need to be converted from *SEZ* coordinates to geocentric coordinates, using the method described in Section 4.3.

It should be mentioned that this method of orbit determination, based on a single point of observation typically contains some uncertainty in the measurements, with a result that can be imprecise. Methods for improving the accuracy of the orbit determination based on multiple measurements are addressed in other texts on the topic.[5,6]

Example Problem 8d

An observing site in Hawaii (latitude 19.8° N, longitude 155.5° W) uses radar to track a passing satellite. Radar measurements at a fixed moment in time give the following information, all in topographic *SEZ* coordinates:

Range $\rho = 5676.5$ km with range rate $\dot{\rho} = 6.3219$ km/s
Elevation $El = 47.83°$ with elevation rate $\dot{El} = 0.06945°/s$
Azimuth $Az = 180.0°$ with azimuth rate $\dot{Az} = -0.07752°/s$

Determine the radius and velocity vectors of the satellite at that point in geocentric (*IJK*) coordinates. At the time of the measurement, the observing site is 20° east of the point of Aries (the $\hat{\boldsymbol{I}}$ axis).

Solution:
The three components of the velocity vector can be determined from Eq. 8.25, as $\rho_S = -\rho\cos(El)\cos(Az) = -(5676.5)\cos(47.83°)\cos(180.0°) = 3810.8$ km, $\rho_E = \rho\cos(El)\sin(Az) = (5676.5)\cos(47.83°)\sin(180.0°) = 0$ km, and $\rho_Z = \rho\sin(El) = (5676.5)\sin(47.83°) = 4207.2$ km. This gives the radius vector, in *SEZ* coordinates (Eq. 24), as $\rho = 3810.8\hat{\boldsymbol{S}} + 4207.2\hat{\boldsymbol{Z}}$ km. Determining the velocity components using Eq. 8.27 first requires converting the measured elevation rates, in degrees, to radians, so $\dot{El} = (0.06945°\mathrm{s}^{-1})\left(\frac{\pi}{180°}\right) = 0.001212\,\mathrm{s}^{-1}$ and $\dot{Az} = (-0.07752°\,\mathrm{s}^{-1})\left(\frac{\pi}{180°}\right) = -0.001353\,\mathrm{s}^{-1}$. The components of the velocity vector are then (noting for simplicity that $\sin(Az) = \sin(180°) = 0$ and $\cos(Az) = \cos(180°) = -1$):

$$\begin{aligned}
\dot{\rho}_S &= -\dot{\rho}\cos(El)\cos(Az) + \rho\sin(El)\left(\dot{El}\right)\cos(Az)\\
&= -\left(6.3219\frac{\mathrm{km}}{\mathrm{s}}\right)\cos(47.83°)(-1) + (5676.5\,\mathrm{km})\sin(47.83°)\left(0.001212\,\mathrm{s}^{-1}\right)(-1)\\
&= -0.8305\,\mathrm{km/s},
\end{aligned}$$

$$\dot{\rho}_E = \rho\cos(El)\cos(Az)\left(\dot{Az}\right) = (5676.5\,\mathrm{km})\cos(47.83°)(-1)\left(-0.001353\,\mathrm{s}^{-1}\right) = 5.1560\,\mathrm{km/s},$$

and

$$\begin{aligned}
\dot{\rho}_Z &= \dot{\rho}\sin(El) + \rho\cos(El)\left(\dot{El}\right) = (6.3219\,\mathrm{km/s})\sin(47.83°)\\
&\quad + (5676.5\,\mathrm{km})\cos(47.83°)\left(0.001212\,\mathrm{s}^{-1}\right) = 9.3042\,\mathrm{km/s}.
\end{aligned}$$

The velocity vector is then from Eq. 8.26 $\dot{\boldsymbol{\rho}} = -0.8305\hat{\boldsymbol{S}} + 5.1560\hat{\boldsymbol{E}} + 9.3042\,\hat{\boldsymbol{Z}}$ km/s.

To convert the radius and velocity vectors into *IJK* coordinates, the transformation matrix given in Eq. 4.25 needs to be applied, with $\phi = 19.8°$ the latitude and $\theta = 20°$ the eastward shift of the observing site from the first point of Aries. This gives:

$$[Q]^{-1} = \begin{bmatrix} \sin\phi\cos\theta & -\sin\theta & \cos\phi\cos\theta \\ \sin\phi\sin\theta & \cos\theta & \cos\phi\sin\theta \\ -\cos\phi & 0 & \sin\phi \end{bmatrix} = \begin{bmatrix} 0.3183 & -0.3420 & 0.8841 \\ 0.1159 & 0.9397 & 0.3218 \\ -0.9409 & 0 & 0.3387 \end{bmatrix}.$$

The radius vector in *IJK* coordinate is then

$$\begin{aligned}
\rho &= \begin{bmatrix} 0.3183 & -0.3420 & 0.8841 \\ 0.1159 & 0.9397 & 0.3218 \\ -0.9409 & 0 & 0.3387 \end{bmatrix}\begin{bmatrix} 3{,}810.8 \\ 0 \\ 4207.2 \end{bmatrix} = \begin{bmatrix} 4932.6 \\ 1795.5 \\ -2160.6 \end{bmatrix}\\
&= 4932.6\hat{\boldsymbol{I}} + 1795.5\hat{\boldsymbol{J}} - 2160.6\hat{\boldsymbol{K}}\ \mathrm{km}.
\end{aligned}$$

Correcting for the distance of the observing site from the Earth's center then gives the geocentric radius vector noting that the Earth's radius vector $\boldsymbol{R}_\oplus = (6378\,\mathrm{km})\left(\cos\phi\cos\theta\hat{\boldsymbol{I}} + \cos\phi\sin\theta\hat{\boldsymbol{J}} + \sin\phi\hat{\boldsymbol{K}}\right) = 5638.7\hat{\boldsymbol{I}} + 2052.4\hat{\boldsymbol{J}} + 2160.2\hat{\boldsymbol{K}}$ so that

$$\begin{aligned}
\boldsymbol{r} = \boldsymbol{\rho} + \boldsymbol{R}_\oplus &= 4932.6\hat{\boldsymbol{I}} + 1795.5\hat{\boldsymbol{J}} - 2160.6\hat{\boldsymbol{K}} + 5638.7\hat{\boldsymbol{I}} + 2052.4\hat{\boldsymbol{J}} + 2160.2\hat{\boldsymbol{K}}\\
&= 10{,}571.3\hat{\boldsymbol{I}} + 3847.9\hat{\boldsymbol{J}}.
\end{aligned}$$

Similarly, the velocity vector rotated into *IJK* coordinates is

$$\dot{\rho} = \begin{bmatrix} 0.3183 & -0.3420 & 0.8841 \\ 0.1159 & 0.9397 & 0.3218 \\ -0.9409 & 0 & 0.3387 \end{bmatrix} \begin{bmatrix} -0.8305 \\ 5.1560 \\ 9.3042 \end{bmatrix} = \begin{bmatrix} 6.1981 \\ 7.7429 \\ 3.9327 \end{bmatrix}$$
$$= 6.1981\hat{\boldsymbol{I}} + 7.7429\hat{\boldsymbol{J}} + 3.9327\hat{\boldsymbol{K}}\ \text{km/s}.$$

In geocentric coordinates, this becomes

$$\boldsymbol{V} = \dot{\rho} + \boldsymbol{\omega}_{\oplus} \times \boldsymbol{r} = 6.1981\hat{\boldsymbol{I}} + 7.7429\hat{\boldsymbol{J}} + 3.9327\hat{\boldsymbol{K}} + \left(0.000729\hat{\boldsymbol{K}}\ \text{s}^{-1}\right) \times \left(10{,}571.3\hat{\boldsymbol{I}} + 3847.9\hat{\boldsymbol{J}}\ \text{km}\right)$$
$$= 3.3924\hat{\boldsymbol{I}} + 15.494\hat{\boldsymbol{J}} + 3.9327\hat{\boldsymbol{K}}\ \text{km/s}.$$

8.4.2 Azimuths and Elevations Known at Multiple Times

The available information for orbit determination might consist only of angle measurements at a series of different times. This could be the case for optical tracking of satellites and is the only practical method for determining the orbits and trajectories of planets of all kinds. A coordinate system relating angles to the observed line-of-sight vector to the object is shown in Fig. 8.3. This coordinate system could be the *SEZ* system used in ground-based observations (see Section 4.2). Alternatively, this could be the geocentric *IJK* system for planetary observations, where the displacement of the observing site from the Earth's center is very small compared to the distance to the remote object. The coordinates shown in Fig. 8.3 are commonly used in cylindrical geometries, where θ would correspond an azimuth (in the *SEZ* frame that would be relative to South), and ϕ the altitude. The directions corresponding to each of several observations are known, but the distance is not, so the vectors $\boldsymbol{L}_i$ are taken to be unit vectors. Using *SEZ* coordinates, the components of each unit vector would be

$$\boldsymbol{L_i} = \begin{bmatrix} L_S \\ L_E \\ L_Z \end{bmatrix} = \begin{bmatrix} \cos\phi_i \cos\theta_i \\ \cos\phi_i \sin\theta_i \\ \sin\phi_i \end{bmatrix}. \tag{8.28}$$

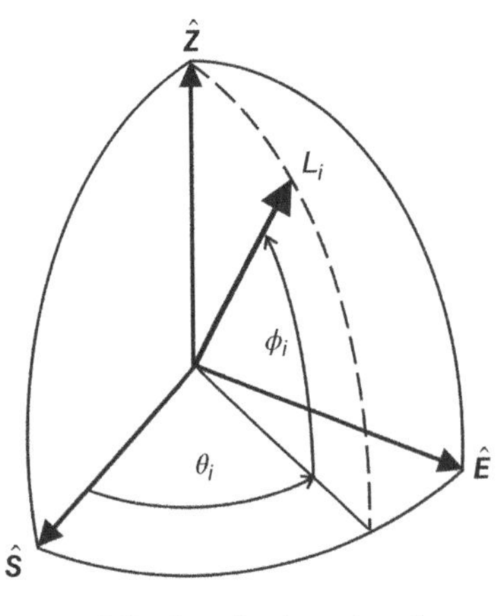

Figure 8.3 Coordinate system for a line-of-sight unit vector $\boldsymbol{L}_i$.

There are several different methods for solving this angles-only problem, each with particular advantages, such as simplicity and suitability for calculations by computer. The presentation here is termed Newton's method and is described in detail by Vallado[5] and BMWS.[6] This method is best suited to planetary observations; it is not best suited to Earth satellite observations. Other methods are presented in the texts by Vallado and Prussing & Conway[7]. The vector radius to the object at a given time in this method is given by (see Eq. 4.20):

$$\boldsymbol{r} = \rho\boldsymbol{L} + \boldsymbol{R}_{\oplus}, \tag{8.29}$$

7 Prussing, J.E. and Conway, B.A. *Orbital Mechanics*, 2nd edition, Oxford University Press, 2013.

where ρ is the unknown scalar distance to the object and $\boldsymbol{R}_{\oplus}$ is the radius vector from the Earth's (or Sun's) center to the observing site. The vector radius can be differentiated with time to give the velocity and acceleration vectors

$$\dot{\boldsymbol{r}} = \dot{\rho}\boldsymbol{L} + \rho\dot{\boldsymbol{L}} + \dot{\boldsymbol{R}}_{\oplus}, \tag{8.30}$$

and

$$\ddot{\boldsymbol{r}} = -\frac{\mu}{r^3}\boldsymbol{r} = 2\dot{\rho}\dot{\boldsymbol{L}} + \ddot{\rho}\boldsymbol{L} + \rho\ddot{\boldsymbol{L}} + \ddot{\boldsymbol{R}}_{\oplus}, \tag{8.31}$$

where the time derivatives of the radius vector $\boldsymbol{R}_{\oplus}$ are due to the Earth's rotation for near-Earth objects. For planetary observations the relevant radius vector is from the Sun to the Earth). From the orbital equation of motion (Eq. 2.11) $\ddot{\boldsymbol{r}} = -\frac{\mu}{r^3}\boldsymbol{r}$. Performing a scalar multiplication of Eq. 8.31 with the vector $\boldsymbol{L} \times \dot{\boldsymbol{L}}$ gives:

$$-\frac{\mu}{r^3}\boldsymbol{r}\cdot\left(\boldsymbol{L}\times\dot{\boldsymbol{L}}\right) = \rho\ddot{\boldsymbol{L}}\cdot\left(\boldsymbol{L}\times\dot{\boldsymbol{L}}\right) + \ddot{\boldsymbol{R}}_{\oplus}\cdot\left(\boldsymbol{L}\times\dot{\boldsymbol{L}}\right). \tag{8.32}$$

The first two terms on the right side of Eq. 8.31dropped out since the vector $\boldsymbol{L} \times \dot{\boldsymbol{L}}$ is normal to both $\boldsymbol{L}$ and $\dot{\boldsymbol{L}}$. Solving Eq. 8.32 for the scalar distance results in:

$$\rho = \frac{-1}{\ddot{\boldsymbol{L}}\cdot\left(\boldsymbol{L}\times\dot{\boldsymbol{L}}\right)}\left[\frac{\mu}{r^3}\boldsymbol{r}\cdot\left(\boldsymbol{L}\times\dot{\boldsymbol{L}}\right) + \ddot{\boldsymbol{R}}_{\oplus}\cdot\left(\boldsymbol{L}\times\dot{\boldsymbol{L}}\right)\right]. \tag{8.33}$$

Noting using Eq. 8.29 that $\boldsymbol{r}\cdot\left(\boldsymbol{L}\times\dot{\boldsymbol{L}}\right) = \left(\rho\boldsymbol{L} + \boldsymbol{R}_{\oplus}\right)\cdot\left(\boldsymbol{L}\times\dot{\boldsymbol{L}}\right) = \boldsymbol{R}_{\oplus}\cdot\left(\boldsymbol{L}\times\dot{\boldsymbol{L}}\right)$ gives:

$$\rho = \frac{-1}{\ddot{\boldsymbol{L}}\cdot\left(\boldsymbol{L}\times\dot{\boldsymbol{L}}\right)}\left[\frac{\mu}{r^3}\boldsymbol{R}_{\oplus}\cdot\left(\boldsymbol{L}\times\dot{\boldsymbol{L}}\right) + \ddot{\boldsymbol{R}}_{\oplus}\cdot\left(\boldsymbol{L}\times\dot{\boldsymbol{L}}\right)\right]. \tag{8.34}$$

Taking the scalar product of both sides of Eq. 8.30 with itself yields a second equation relating r and ρ:

$$r^2 - \left[\rho^2 + R_{\oplus}{}^2 + 2\rho(\boldsymbol{L}\cdot\boldsymbol{R}_{\oplus})\right] = 0. \tag{8.35}$$

Equations 8.34 and 8.35 can then be combined to give a single (if complicated!) equation for the scalar radius, ρ, from which the radius from the system center (the Earth, or the Sun), r, can be determined from Eq. 8.35.

The time-dependent value and time derivatives of the unit vector $\boldsymbol{L}$ needed to solve the problem can be estimated numerically using three unit vectors, $\boldsymbol{L}_1$, $\boldsymbol{L}_2$, and $\boldsymbol{L}_3$, and the corresponding times of each observation, t_1, t_2, and t_3. One way to express these functions is to use Lagrange interpolation as follows:

$$\boldsymbol{L}(t) = \frac{(t-t_2)(t-t_3)}{(t_1-t_2)(t_1-t_3)}\boldsymbol{L}_1 + \frac{(t-t_1)(t-t_3)}{(t_2-t_1)(t_2-t_3)}\boldsymbol{L}_2 + \frac{(t-t_1)(t-t_2)}{(t_3-t_1)(t_3-t_2)}\boldsymbol{L}_3, \tag{8.36}$$

$$\dot{\boldsymbol{L}}(t) = \frac{2t-t_2-t_3}{(t_1-t_2)(t_1-t_3)}\boldsymbol{L}_1 + \frac{2t-t_1-t_3}{(t_2-t_1)(t_2-t_3)}\boldsymbol{L}_2 + \frac{2t-t_1-t_2}{(t_3-t_1)(t_3-t_2)}\boldsymbol{L}_3, \tag{8.37}$$

$$\ddot{\boldsymbol{L}}(t) = \frac{2}{(t_1-t_2)(t_1-t_3)}\boldsymbol{L}_1 + \frac{2}{(t_2-t_1)(t_2-t_3)}\boldsymbol{L}_2 + \frac{2}{(t_3-t_1)(t_3-t_2)}\boldsymbol{L}_3. \tag{8.38}$$

It should be emphasized that the aforementioned expressions are only approximations and work best when the time differences between the observations are small. The approximations can be improved by incorporating additional observations and developing more complex expressions (such as least-squares polynomials or other functions) to determine the needed derivatives.

The radius vector at a given time is simply $\boldsymbol{r} = \rho\boldsymbol{L} + \boldsymbol{R}_{\oplus}$ from Eq. 8.29. The velocity vector at that time can be determined by taking the time derivative of that equation, or:

$$\boldsymbol{V} = \dot{\boldsymbol{r}} = \dot{\rho}\boldsymbol{L} + \rho\dot{\boldsymbol{L}} + \dot{\boldsymbol{R}}_{\oplus}. \tag{8.39}$$

The needed value of $\dot{\rho}$ can be obtained by now taking the scalar product of Eq. 8.31 with the vector $\boldsymbol{L} \times \ddot{\boldsymbol{L}}$ and noting that now the middle two terms on the right side of Eq. 8.31 drop out since the $\boldsymbol{L} \times \ddot{\boldsymbol{L}}$ is normal to both $\boldsymbol{L}$ and $\ddot{\boldsymbol{L}}$. The result is:

$$-\frac{\mu}{r^3}\boldsymbol{r} \cdot \left(\boldsymbol{L} \times \ddot{\boldsymbol{L}}\right) = 2\dot{\rho}\dot{\boldsymbol{L}} \cdot \left(\boldsymbol{L} \times \ddot{\boldsymbol{L}}\right) + \ddot{\boldsymbol{R}}_{\oplus} \cdot \left(\boldsymbol{L} \times \ddot{\boldsymbol{L}}\right) \tag{8.40}$$

or rearranging noting that $\boldsymbol{r} \cdot \left(\boldsymbol{L} \times \ddot{\boldsymbol{L}}\right) = \boldsymbol{R}_{\oplus} \cdot \left(\boldsymbol{L} \times \ddot{\boldsymbol{L}}\right)$:

$$\dot{\rho} = \frac{-1}{2\dot{\boldsymbol{L}} \cdot \left(\boldsymbol{L} \times \ddot{\boldsymbol{L}}\right)}\left[\frac{\mu}{r^3}\boldsymbol{R}_{\oplus} \cdot \left(\boldsymbol{L} \times \ddot{\boldsymbol{L}}\right) + \ddot{\boldsymbol{R}}_{\oplus} \cdot \left(\boldsymbol{L} \times \ddot{\boldsymbol{L}}\right)\right]. \tag{8.41}$$

Inserting this result, along with the value of ρ determined previously from solving Eqs. 8.34 and 8.35, into Eq. 8.39 then gives the velocity vector, $\boldsymbol{V}$ Knowing the radius and velocity vectors at a point in time then allows determining the orbital elements as discussed in Section 4.1.

8.5 Problems

8.1 An Earth orbiting satellite is determined to have the following three radii in geocentric coordinates:

$$\boldsymbol{r}_1 = 36{,}513\hat{\boldsymbol{I}} + 21{,}081\,\hat{\boldsymbol{J}}\text{ km},$$

$$\boldsymbol{r}_2 = -27{,}101\hat{\boldsymbol{I}} + 32{,}398\,\hat{\boldsymbol{J}}\text{ km, and}$$

$$\boldsymbol{r}_3 = -24{,}183\hat{\boldsymbol{I}} - 34{,}537\,\hat{\boldsymbol{J}}\text{ km.}$$

What are the semi-major axis, eccentricity, and inclination of the orbit? What is the orbital period? Hint: It may not be necessary to use Gauss' method (Section 8.2); just carefully consider the given radius vectors.

8.2 An earth orbiting satellite has these three radii in geocentric coordinates:

$$\boldsymbol{r}_1 = 5667\hat{\boldsymbol{I}} + 4986\hat{\boldsymbol{J}} + 3349\hat{\boldsymbol{K}}\text{ km},$$

$$\boldsymbol{r}_2 = 2596\hat{\boldsymbol{I}} + 5767\hat{\boldsymbol{J}} + 4899\hat{\boldsymbol{K}}\text{ km, and}$$

$$\boldsymbol{r}_3 = -15{,}922\hat{\boldsymbol{I}} - 5339\,\hat{\boldsymbol{J}} - 1036\hat{\boldsymbol{K}}\text{ km.}$$

Determine:

(a) the semi-major axis, eccentricity, and perigee and apogee radii, and
(b) all elements of the orbit,[8] including the true anomaly at the location of the first radius. Assuming that the radius vectors are in time-sequence order, is the spacecraft climbing or falling in its orbit at that point?

8 These include the semi-major axis, eccentricity, inclination, longitude of the ascending node, and argument of periapsis.

8.3 An object has these three radius vectors in geocentric coordinates:

$$\boldsymbol{r}_1 = -35{,}354\hat{\boldsymbol{I}} + 35{,}357\hat{\boldsymbol{J}} + 9.07\hat{\boldsymbol{K}}\text{ km},$$

$$\boldsymbol{r}_2 = -5980\hat{\boldsymbol{I}} - 5975\,\hat{\boldsymbol{J}} - 18{,}126\hat{\boldsymbol{K}}\text{ km, and}$$

$$\boldsymbol{r}_3 = 152{,}368\hat{\boldsymbol{I}} - 92{,}607\hat{\boldsymbol{J}} + 90{,}597\hat{\boldsymbol{K}}\text{ km}.$$

Determine:

(a) the semi-major axis, eccentricity, and perigee and apogee radii, and
(b) all elements of the trajectory,[8] including the true anomaly at the location of the first and last radius. What kind of object would this likely be?

8.4 An earth orbiting satellite has the two radius vectors in geocentric coordinates:

$\boldsymbol{r}_1 = -5708\hat{\boldsymbol{I}} + 3296\hat{\boldsymbol{J}} + 3579\hat{\boldsymbol{K}}$ km and $\boldsymbol{r}_2 = -4876\hat{\boldsymbol{I}} - 8443\hat{\boldsymbol{J}} + 0.310\hat{\boldsymbol{K}}$ km.

The time between them is measured to be $\Delta t = 1721$ s $= 0.487$ hr. Determine:

(a) the semi-major axis, eccentricity, and perigee and apogee radii, and
(b) all other elements of the orbit.[8]

Given the inclination of the orbit, what was the likely launch location?

8.5 A spacecraft just launched has these two radius vectors in geocentric coordinates:

$\boldsymbol{r}_1 = 6572\hat{\boldsymbol{I}} - 1.76\hat{\boldsymbol{J}} + 604\hat{\boldsymbol{K}}$ km and $\boldsymbol{r}_2 = 3.87\hat{\boldsymbol{I}} + 14{,}520\hat{\boldsymbol{J}} + 0.089\hat{\boldsymbol{K}}$ km.

The time between these two positions is $\Delta t = 733$ s $= 0.204$ hr. Determine:

(a) the semi-major axis, eccentricity, and perigee radius,
(b) all other elements of the orbit,[8] and
(c) the spacecraft velocity at perigee (km/s). Assuming that the spacecraft was launched on a day near the autumnal equinox, what kind of mission is the spacecraft likely executing? What was the likely launch site?

8.6 Radar position measurement tests of an object are made from a ground station with latitude and longitude 41.85° N, 88.06 °W. Measurements are taken of the Moon over three successive days, each at the time at which the observation station is 45° east of the first point of Aries. This radar system gives high-precision values of the target's radius and direction (but not the rates). The three radius vectors are as follows:

$$\boldsymbol{\rho}_1 = 146{,}619\hat{\boldsymbol{S}} + 256{,}244\hat{\boldsymbol{E}} + 206{,}317\hat{\boldsymbol{Z}},$$

$$\boldsymbol{\rho}_2 = 98{,}897\hat{\boldsymbol{S}} + 313{,}251\hat{\boldsymbol{E}} + 151{,}529\hat{\boldsymbol{Z}},\text{ and}$$

$$\boldsymbol{\rho}_3 = 45{,}168\hat{\boldsymbol{S}} + 351{,}225\hat{\boldsymbol{E}} + 87{,}150\hat{\boldsymbol{Z}}.$$

Determine the semi-major axis, eccentricity, and inclination of the orbit, and verify that those values are consistent with the orbit of the Moon.

8.7 A passing satellite is tracked from a radar station situated at latitude 47.43° N and longitude 121.77° W. The following data are recorded (all in topographic *SEZ* coordinates):

Range $\rho = 802.77$ km with range rate $\dot{\rho} = 6.068$ km/s
Elevation $El = 39.37°$ with elevation rate $\dot{El} = -0.890°/\text{s}$
Azimuth $Az = 171.89°$ with azimuth rate $\dot{Az} = -2.386°/\text{s}$

Determine the radius and velocity vectors of the satellite at that point in geocentric (*IJK*) coordinates. At the time of the measurement, the observing site is 30° west of the point of Aries (the $\hat{\boldsymbol{I}}$ axis). Based on this information, is the satellite (i) a *Starlink* satellite, (ii) the International Space Station, or (iii) the Hubble Space Telescope?

8.8 An observing site situated at latitude 51.11° N and longitude 8.86° E uses radar to track a satellite. The radar measurements give the following information, all in topographic *SEZ* coordinates:

Range $\rho = 13{,}722$ km with range rate $\dot{\rho} = 3.98$ km/s
Elevation $El = 50.93°$ with elevation rate $\dot{El} = -0.075°/\text{s}$
Azimuth $Az = 150.4°$ with azimuth rate $\dot{Az} = -0.0579°/\text{s}$

Determine the radius and velocity vectors of the satellite at that point in geocentric (*IJK*) coordinates. At the time of the measurement, the observing site is exactly aligned with the point of Aries (the $\hat{\boldsymbol{I}}$ axis). Determine the radius and velocity vectors of the satellite at that point in geocentric (*IJK*) coordinates. Calculate the semi-major axis, eccentricity, and inclination of the orbit. Considering the discussion of Earth orbiting satellites in Chapter 7, what kind of satellite is this?

9

Relative Motion and the Restricted 3-Body Problem

Previous chapters have considered the case of a single object on a trajectory about a (much) larger object, such as a planet orbiting the sun, a satellite in Earth orbit, or a spacecraft in various segments of an interplanetary mission. The determination of the elements of the trajectory, the time versus position, ground tracks, and velocity changes needed for maneuvers are all important considerations in mission planning and analysis.

More complex, but equally important, situations arise when multiple bodies are taken into consideration. One example is the relative motion between two different spacecraft orbiting a planet. Understanding this relative motion is vital to enabling one spacecraft to rendezvous with another. A second important situation is when a spacecraft is subject to the gravitational pull of *two* massive bodies simultaneously. This is the case, for example, of a spacecraft on a trajectory from the Earth to the Moon. These "three body"situations are the focus of this chapter. This chapter is outlined as follows:

- Relative motion in orbit
- Restricted 3-body problem

9.1 Relative Motion

An image of a *Dragon 2* spacecraft performing a rendezvous with the International Space Station (ISS) is shown in Fig. 9.1. This is the terminal phase of the rendezvous when the approaching spacecraft can complete the docking maneuver by thrusting directly in the direction of the target. As will be shown later, this direct maneuver is only possible when the two spacecraft are very close (less than approximately 100 m). A more complicated relative motion occurs when the two spacecraft are more widely separated in their respective orbits.

The motion of each spacecraft in its orbit and in the absence of thrusting maneuvers is governed entirely by the central gravitational force of the Earth, as the gravitational forces between the two orbiting bodies are negligible. The trajectory of each spacecraft is thus given by its respective Orbit Equation (as well as the perturbations discussed in Chapter 7) The relative motion between the two spacecraft, however, cannot be accurately determined by simply comparing their positions from the respective orbit equations. One difficulty is that the distances involved in the Orbit Equation, being typically thousands of kilometers, are simply too large to allow for accurate computation of differences in position on the order of a kilometer or less. A second complication is that neither spacecraft

Introduction to Orbital Mechanics: A Concise and Practical Approach, First Edition. James C. Hermanson.

Figure 9.1 Orbital rendezvous in progress. *Source:* SpaceX / NASA / Public Domain. https://medium.com/swlh/explained-how-spaceships-rendezvous-with-the-iss-137f9bcfe124.

is in an inertial frame of reference as relative to the other, given that both spacecraft are in orbit around the Earth.

The geometry of the relative motion between a target spacecraft (here, the ISS) and a chasing vehicle (the Dragon capsule) is shown in Fig. 9.2. The Cartesian xyz coordinates (unit vectors $\hat{\boldsymbol{\imath}}\,\hat{\boldsymbol{\jmath}}\,\hat{\boldsymbol{k}}$) are taken to be fixed on the ISS target, which orbits the Earth with radius r_0 (for simplicity the target orbit is taken to be circular). The position of the chaser spacecraft is the vector $\boldsymbol{r} = \boldsymbol{r_0} + \boldsymbol{\delta r}$, with $\boldsymbol{\delta r}$ being the relative position vector to the chaser, as seen by the target. Determining the relative motion (the change in $\boldsymbol{\delta r}$) requires two considerations: (i) the 2-body, orbital motion of both spacecraft around the Earth; and (ii) the dynamics in the noninertial reference frame of the target spacecraft.

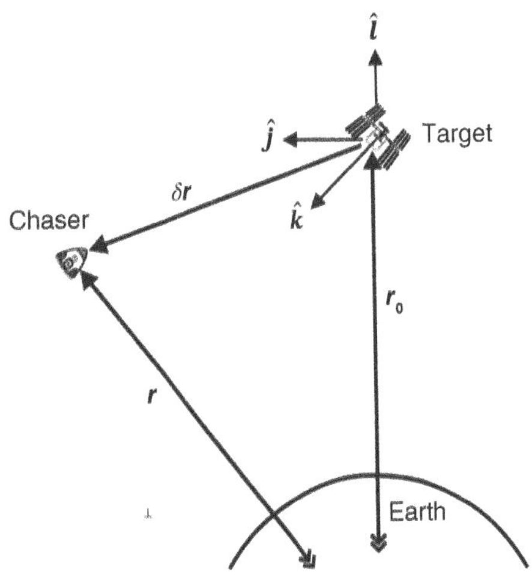

Figure 9.2 Relative motion between two orbiting spacecraft.

For the first consideration, the relative motion can be written using the governing equation for 2-body motion (Eq. 2.11) for each of the two spacecraft to give:

$$\delta\ddot{\boldsymbol{r}} = -\ddot{\boldsymbol{r}}_0 + \ddot{\boldsymbol{r}} = -\ddot{\boldsymbol{r}}_0 - \mu\frac{\boldsymbol{r}}{r^3} = -\ddot{\boldsymbol{r}}_0 - \mu\frac{(\boldsymbol{r}_0 + \delta\boldsymbol{r})}{r^3}. \tag{9.1}$$

The denominator on the right side of Eq. 9.1 can be evaluated by using the following relation:

$$r^2 = \boldsymbol{r}\cdot\boldsymbol{r} = (\boldsymbol{r}_0 + \delta\boldsymbol{r})\cdot(\boldsymbol{r}_0 + \delta\boldsymbol{r}). \tag{9.2}$$

Noting that the distance between the two spacecraft as the rendezvous is approached is much smaller than the distance of either spacecraft from the Earth's center, or $\delta r/r_0 \ll 1$, allows expanding and simplifying Eq. 9.2 to give:

$$r^2 = r_0^2\left(1 + \frac{2\boldsymbol{r}_0\cdot\delta\boldsymbol{r}}{r_0^2}\right) + (\text{smaller terms}).$$

Rearranging then gives the needed denominator term in Eq. 9.1 as $\frac{1}{r^3} = \frac{1}{r_0^3}\left(1 + \frac{2\boldsymbol{r}_0\cdot\delta\boldsymbol{r}}{r_0^2}\right)^{-3/2} = \frac{1}{r_0^3}\left(1 - \frac{3\boldsymbol{r}_0\cdot\delta\boldsymbol{r}}{r_0^2}\right)$, where the binomial expansion is applied since $\frac{2\boldsymbol{r}_0\cdot\delta\boldsymbol{r}}{r_0^2} \ll 1$. Inserting this result into the last term in Eq. 9.1 gives:

$$\delta\ddot{\boldsymbol{r}} = -\ddot{\boldsymbol{r}}_0 - \mu\frac{1}{r_0^3}\left(1 - \frac{3\boldsymbol{r}_0\cdot\delta\boldsymbol{r}}{r_0^2}\right)(\boldsymbol{r}_0 + \delta\boldsymbol{r}) = -\ddot{\boldsymbol{r}}_0 - \mu\left[\frac{1}{r_0^3} - \frac{3\boldsymbol{r}_0\cdot\delta\boldsymbol{r}}{r_0^5}\right](\boldsymbol{r}_0 + \delta\boldsymbol{r}). \tag{9.3}$$

Multiplying out the last term, neglecting again higher-order terms since $(\delta\boldsymbol{r})^2 \ll \delta\boldsymbol{r}\boldsymbol{r}_0$ yields:

$$\delta\ddot{\boldsymbol{r}} = -\ddot{\boldsymbol{r}}_0 - \frac{\mu}{r_0^3}\boldsymbol{r}_0 - \mu\left[\delta\boldsymbol{r} - \frac{3}{r_0^2}(\boldsymbol{r}_0\cdot\delta\boldsymbol{r})\boldsymbol{r}_0\right].$$

The first two terms on the right side cancel since $\ddot{\boldsymbol{r}}_0 = -\frac{\mu}{r_0^3}\boldsymbol{r}_0$ (Eq. 2.11). Therefore,

$$\delta\ddot{\boldsymbol{r}} = -\mu\left[\delta\boldsymbol{r} - \frac{3}{r_0^2}(\boldsymbol{r}_0\cdot\delta\boldsymbol{r})\boldsymbol{r}_0\right]. \tag{9.4}$$

This is one of two key relations needed for the relative motion problem, which describes the effects of Newtonian gravitation by the Earth, with the linearizations as discussed. The derivation leading to Eq. 9.4 was performed in the inertial frame of the Earth. What is, however, needed is the motion of the chaser spacecraft relative to the target object, using the target-fixed, moving coordinate system shown in Fig. 9.2, which rotates about the Earth as the target moves in its orbit. The acceleration of an object in a frame rotating with the angular velocity vector $\boldsymbol{\Omega}$ can be related to the acceleration in the inertial frame using this standard result from dynamics[1]:

$$\ddot{\boldsymbol{r}} = \ddot{\boldsymbol{r}}_0 + \dot{\Omega}\times\delta\boldsymbol{r} + \Omega\times(\Omega\times\delta\boldsymbol{r}) + 2\Omega\times\delta\boldsymbol{V}_{rel} + \boldsymbol{a}_{rel}. \tag{9.5}$$

The terms $\ddot{\boldsymbol{r}}$ and $\ddot{\boldsymbol{r}}_0$ are respectively the accelerations of the chase vehicle and the target spacecraft in the inertial coordinate system. $\delta\boldsymbol{V}_{\text{rel}}$ and $\boldsymbol{a}_{\text{rel}}$ are respectively the velocity and acceleration of the chase vehicle in the target-centered, rotating reference frame. These are the quantities needed in the relative-motion problem. For the case of the target vehicle in a circular orbit, the angular

1 See, for example, Meriam, J.L. and Kraige, L.G. *Engineering Mechanics-Dynamics*, Wiley, 2012.

velocity vector is $\Omega = n\hat{\boldsymbol{k}}$, with $n = \sqrt{\frac{\mu}{r_0^3}}$, the mean motion. Since the angular velocity is constant for a circular orbit, $\dot{\Omega} = 0$ and the second term on the right side of Eq. 9.5 vanishes. The relative velocity and acceleration in the coordinates of the rotating frame are given by:

$$\delta \boldsymbol{V}_{rel} = \delta\dot{x}\hat{\boldsymbol{i}} + \delta\dot{y}\hat{\boldsymbol{j}} + \delta\dot{z}\hat{\boldsymbol{k}} \tag{9.6}$$

and

$$\boldsymbol{a}_{\text{rel}} = \delta\ddot{\boldsymbol{r}} = \delta\ddot{x}\hat{\boldsymbol{i}} + \delta\ddot{y}\hat{\boldsymbol{j}} + \delta\ddot{z}\hat{\boldsymbol{k}}\,. \tag{9.7}$$

Inserting these relations into Eq. 9.5 with some algebra and the use of Vector Identity #7 (Chapter 2, Section 2.4) and collecting terms gives:

$$\delta\ddot{\boldsymbol{r}} = \left(-n^2\delta x - 2n\delta\dot{y} + \delta\ddot{x}\right)\hat{\boldsymbol{i}} + \left(-n^2\delta y + 2n\delta\dot{x} + \delta\ddot{y}\right)\hat{\boldsymbol{j}} + \delta\ddot{z}\hat{\boldsymbol{k}}. \tag{9.8}$$

Noting that the relative distance vector between the two spacecraft is $\delta\boldsymbol{r} = \delta x\hat{\boldsymbol{i}} + \delta y\hat{\boldsymbol{j}} + \delta z\hat{\boldsymbol{k}}$ implies that $\boldsymbol{r}_0 \cdot \delta\boldsymbol{r} = r_0\hat{\boldsymbol{i}} \cdot \left(\delta x\hat{\boldsymbol{i}} + \delta y\hat{\boldsymbol{j}} + \delta z\hat{\boldsymbol{k}}\right) = r_0\delta x$. Inserting this into Eq. (9.4) gives $\delta\ddot{\boldsymbol{r}} = -n^2 \left[\delta x\hat{\boldsymbol{i}} + \delta y\hat{\boldsymbol{j}} + \delta z\hat{\boldsymbol{k}} - \frac{3}{r_0^2} r_0\delta x r_0\hat{\boldsymbol{i}}\right]$, which simplifies to

$$\delta\ddot{\boldsymbol{r}} = 2n^2\delta x\hat{\boldsymbol{i}} - n^2\delta y\hat{\boldsymbol{j}} - n^2\delta z\hat{\boldsymbol{k}}. \tag{9.9}$$

Equation 9.8, which represents the dynamics in the moving (target) reference frame, and Eq. 9.9, which reflects the 2-body orbital motion about the Earth, can then be combined since both are now expressed in the *ijk* coordinate system of the target. Subtracting Eq. 9.9 from Eq. 9.8 to eliminate $\delta\ddot{\boldsymbol{r}}$ then gives:

$$\left(\delta\ddot{x} - 3n^2\delta x - 2n\delta\dot{y}\right)\hat{\boldsymbol{i}} + \left(\delta\ddot{y} + 2n\delta\dot{x}\right)\hat{\boldsymbol{j}} + \left(\delta\ddot{z} + n^2\delta z\right)\hat{\boldsymbol{k}} = 0$$

That this vector equation equals zero implies that each of the three components must also be identically zero, or

$$\boxed{\delta\ddot{x} - 3n^2\delta x - 2n\delta\dot{y} = 0} \tag{9.10a}$$

$$\boxed{\delta\ddot{y} + 2n\delta\dot{x} = 0} \tag{9.10b}$$

$$\boxed{\delta\ddot{z} + n^2\delta z = 0.} \tag{9.10c}$$

These are the Clohessy-Wiltshire equations[2] of relative orbital motion. The solutions for the relative position of the chaser spacecraft have the following forms[3]:

$$\begin{aligned}
\delta x &= \frac{2}{n} C_1 + C_2\cos(nt) + C_3\sin(nt),\\
\delta y &= -3C_1 t - 2C_2\sin(nt) + 2C_3\cos(nt) + C_4,\\
\delta z &= C_5\cos(nt) + C_6\sin(nt).
\end{aligned} \tag{9.11}$$

2 Clohessy, W.H. and Wiltshire, R.S. (1960). "Terminal Guidance System for Satellite Rendezvous". *Journal of the Aerospace Sciences.* **27** (9): 653–658 Alternatively, these have been referred to as Hill's equations, given that G.W. Hill derived essentially the same equations in 1878.

3 Details of the solution can be found, for example, at https://ensatellite.com/hills-equations/

The respective relative velocities by differentiation are:

$$\begin{aligned}\delta\dot{x} &= -C_2 n\cos(nt) + C_3 n\sin(nt),\\ \delta\dot{y} &= -3C_1 - 2C_2 n\cos(nt) - 2C_3 n\sin(nt),\\ \delta\dot{z} &= -C_5 n\sin(nt) + C_6 n\cos(nt).\end{aligned} \tag{9.12}$$

The constants C_1 through C_6 can be determined by applying the initial conditions of relative position and velocity. Solving gives:

$$\begin{aligned}C_1 &= 2n\delta x_0 + \delta\dot{y}_0\\ C_2 &= -3\delta x_0 - \frac{2}{n}\delta\dot{y}_0\\ C_3 &= \frac{1}{n}\delta\dot{x}_0\\ C_4 &= -\frac{2}{n}\delta\dot{x}_0 + \delta y_0\\ C_5 &= \delta z_0\\ C_6 &= \frac{1}{n}\delta\dot{z}_0,\end{aligned}$$

where δx_0, δy_0, δz_0 and $\delta\dot{x}_0$, $\delta\dot{y}_0$, $\delta\dot{z}_0$ are respectively the components of relative position and velocity at time $t = 0$. The complete set of equations for relative motion under the Clohessy-Wiltshire formulation is then:

$$\boxed{\begin{aligned}\delta x &= [4 - 3\cos(nt)]\delta x_0 + \frac{\sin(nt)}{n}\delta\dot{x}_0 + \frac{2}{n}[1 - \cos(nt)]\delta\dot{y}_0,\\ \delta y &= 6[\sin(nt) - nt]\delta x_0 + \delta y_0 + \frac{2}{n}[\cos(nt) - 1]\delta\dot{x}_0 + \frac{1}{n}[4\sin(nt) - 3nt]\delta\dot{y}_0,\\ \delta z &= \cos(nt)\delta z_0 + \frac{1}{n}\sin(nt)\delta\dot{z}_0.\end{aligned}} \tag{9.13}$$

$$\boxed{\begin{aligned}\delta\dot{x} &= 3n\sin(nt)\delta x_0 + \cos(nt)\delta\dot{x}_0 + 2\sin(nt)\delta\dot{y}_0,\\ \delta\dot{y} &= 6n[\cos(nt) - 1]\delta x_0 - 2\sin(nt)\delta\dot{x}_0 + [4\cos(nt) - 3]\delta\dot{y}_0,\\ \delta\dot{z} &= -n\sin(nt)\delta z_0 + \cos(nt)\delta\dot{z}_0.\end{aligned}} \tag{9.14}$$

These two sets of relative motion equations, each with three components, lend themselves to vector/matrix representations, as follows. For the relative position, this would be:

$$\begin{aligned}\delta\mathbf{r}(t) = \begin{bmatrix}\delta x\\ \delta y\\ \delta z\end{bmatrix} &= \begin{bmatrix}4 - 3\cos(nt) & 0 & 0\\ 6[\sin(nt) - nt] & 1 & 0\\ 0 & 0 & \cos(nt)\end{bmatrix}\begin{bmatrix}\delta x_0\\ \delta y_0\\ \delta z_0\end{bmatrix}\\ &+ \begin{bmatrix}\frac{\sin(nt)}{n} & \frac{2}{n}[1 - \cos(nt)] & 0\\ \frac{2}{n}[\cos(nt) - 1] & \frac{1}{n}[4\sin(nt) - 3nt] & 0\\ 0 & 0 & \frac{1}{n}\sin(nt)\end{bmatrix}\begin{bmatrix}\delta\dot{x}_0\\ \delta\dot{y}_0\\ \delta\dot{z}_0\end{bmatrix}.\end{aligned} \tag{9.15}$$

Correspondingly for the relative velocities:

$$\delta\dot{r}(t) = \delta V(t) = \begin{bmatrix} \delta\dot{x} \\ \delta\dot{y} \\ \delta\dot{z} \end{bmatrix} = \begin{bmatrix} 3n\cos(nt) & 0 & 0 \\ 6n[\cos(nt)-1] & 0 & 0 \\ 0 & 0 & -n\sin(nt) \end{bmatrix} \begin{bmatrix} \delta x_0 \\ \delta y_0 \\ \delta z_0 \end{bmatrix} + \begin{bmatrix} \cos(nt) & 2\sin(nt) & 0 \\ -2\sin(nt) & 4\cos(nt)-3 & 0 \\ 0 & 0 & \cos(nt) \end{bmatrix} \begin{bmatrix} \delta\dot{x}_0 \\ \delta\dot{y}_0 \\ \delta\dot{z}_0 \end{bmatrix}. \tag{9.16}$$

These are the complete Clohessy-Wiltshire solutions in three dimensions. For the simpler case where the motion is only in the two dimensions x and y (in the orbital plane of the target spacecraft), only the vector and matrix components above/inside the dashed lines need to be considered.

An example rendezvous trajectory computed using the Clohessy-Wiltshire method is shown in Fig. 9.3. In this case, both spacecraft are initially in circular orbits about the Earth with an altitude of 500 km. The chaser vehicle begins the rendezvous maneuver 4 km behind the target spacecraft and initiates the maneuver by performing a simultaneous velocity change of $\delta\dot{y}_0 = -0.235\,\text{m/s}$ and $\delta\dot{x}_0 = -0.01$ m/s. The initiation maneuver is a braking maneuver, which is needed to raise the velocity of the chaser vehicle, as discussed in Chapter 5, Section 5.5. The resulting motion of the chase vehicle in the fixed coordinate frame of the target spacecraft (Fig. 9.2), is the looping trajectory shown. Eventually, the chase vehicle arrives at the coordinates of the target spacecraft, and a terminal docking maneuver can be performed. If an additional maneuver is not performed at that point, the chase vehicle will continue past the target on a similar trajectory as that of the approach.

Example Problem 9a

In the example trajectory shown in Fig. 9.3, how much time will be required for the chase vehicle to reach the point of rendezvous from the initial maneuver?

Solution:

The angular velocity of the 500 km circular orbit is $n = \sqrt{\frac{\mu}{r^3}} = \sqrt{\frac{398{,}600\frac{\text{km}^3}{\text{s}^2}}{(6378+500)^3\,\text{km}^3}} = 0.001107\,\text{s}^{-1}$.

Combining this with the initial relative positions of $\delta y_0 = -4.00$ km and $\delta x_0 = 0$ km and the initial

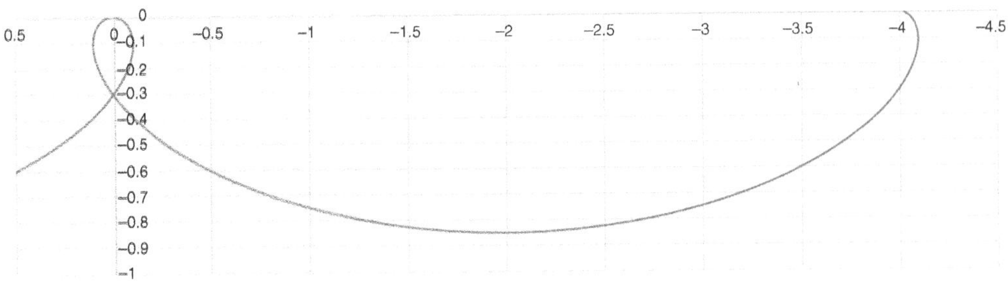

Figure 9.3 A rendezvous trajectory, relative to the target vehicle.

velocity change specified allows determinate of the elapsed time by setting $\delta y_0 = 0$ in Eq. 9.13. This gives the implicit relation:

$$\begin{aligned}\delta y &= 6[\sin(nt) - nt]\delta x_0 + \delta y_0 + \frac{2}{n}[\cos(nt) - 1]\delta\dot{x}_0 + \frac{1}{n}[4\sin(nt) - 3nt]\delta\dot{y}_0 \\ &= -4.00\,\text{km} + \frac{2}{0.001107s^{-1}}\left[\cos\left(0.001107s^{-1}t\right) - 1\right](-0.01)\left(\frac{1}{1000}\right)\frac{\text{km}}{s} \\ &+ \frac{1}{0.001107s^{-1}}\left[4\sin\left(0.001107s^{-1}t\right) - 3\left(0.001107s^{-1}t\right)\right] = 0.\end{aligned}$$

Solving numerically gives $t = 4490\ \text{s} = 1.25\ \text{hr}$.

9.2 Restricted 3-Body Problem and Earth-Moon Trajectories

Another problem of importance in orbital mechanics is the trajectory of a spacecraft under the gravitational influence of two massive bodies. A particular example is a spacecraft on a trans-lunar trajectory from Earth. According to the Sphere of Influence (SOI) concept presented in Chapter 6, Section 6.3, the Earth's SOI amounts to 145 Earth radii. The Moon, however, is at an average distance of approximately 60 Earth radii, which lies well within the SOI of the Earth. So, while the application of the sphere of influence concept is an acceptable approximation for spacecraft on interplanetary trajectories as discussed in Chapter 6, for trans-lunar trajectories, the influence of the Moon's gravitational pull, in addition to that of the Earth, must also be considered.

The geometry of a spacecraft in the Earth-Moon system in shown in Fig. 9.4 using Cartesian coordinates. This problem is the *restricted* 3-body problem, where the motion of the two primary bodies, the Earth and the Moon, are taken to be constant. As an approximation, the Earth-Moon system can modeled as rotating around its center of mass at a constant rate, Ω, with the Earth-Moon distance, r_{12}, taken to be constant. The vector distances to the spacecraft from the Earth, the Moon, and the system center of mass are respectively designated $\boldsymbol{r}_{\oplus}$, $\boldsymbol{r}_{☽}$, and $\boldsymbol{r}$. In a similar fashion to the relative motion analysis in Section 9.1, addressing the restricted 3-body problem requires consideration of both the gravitational forces involved as well as the rotation of the noninertial reference frame, shown in Fig. 9.4 to be fixed on the center of mass of the Earth-Moon system.

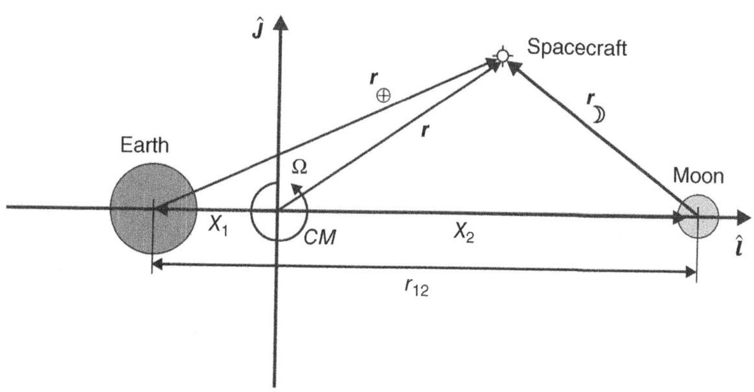

Figure 9.4 A spacecraft in the Earth-Moon system.

The dynamics of the system can be described in the same fashion as discussed in the previous section. Expressing Eq. 9.5 slightly differently as:

$$\boldsymbol{a}_p = \boldsymbol{a}_0 + \dot{\Omega} \times \boldsymbol{r} + \Omega \times (\Omega \times \boldsymbol{r}) + 2\Omega \times \boldsymbol{V} + \boldsymbol{a}, \tag{9.17}$$

where $\boldsymbol{a}_p$ is the acceleration of the spacecraft in the inertial frame of reference. The acceleration of the system, $\boldsymbol{a}_0$, is zero. The constant angular rotation vector can be written as $\Omega = \Omega\,\hat{\boldsymbol{k}}$, where $\hat{\boldsymbol{k}}$ is the unit vector normal to the x-y plane shown in Fig. 9.4 and again, the term $\dot{\Omega} = 0$. The distance, velocity, and acceleration vectors of the spacecraft in Cartesian coordinates in the rotating frame are given by:

$$\boldsymbol{r} = x\hat{\boldsymbol{i}} + y\hat{\boldsymbol{j}} + z\hat{\boldsymbol{k}}, \boldsymbol{V} = \dot{x}\hat{\boldsymbol{i}} + \dot{y}\hat{\boldsymbol{j}} + \dot{z}\hat{\boldsymbol{k}}, \text{and } \boldsymbol{a} = \ddot{x}\hat{\boldsymbol{i}} + \ddot{y}\hat{\boldsymbol{j}} + \ddot{z}\hat{\boldsymbol{k}}. \tag{9.18}$$

Inserting these vectors into Eq. 9.17 gives the acceleration in inertial coordinates as:

$$\boldsymbol{a}_p = \Omega\,\hat{\boldsymbol{k}} \times \left[\Omega\,\hat{\boldsymbol{k}} \times \left(x\hat{\boldsymbol{i}} + y\hat{\boldsymbol{j}} + z\hat{\boldsymbol{k}}\right)\right] + 2\Omega\,\hat{\boldsymbol{k}} \times \left(\dot{x}\hat{\boldsymbol{i}} + \dot{y}\hat{\boldsymbol{j}} + \dot{z}\hat{\boldsymbol{k}}\right) + \left(\ddot{x}\hat{\boldsymbol{i}} + \ddot{y}\hat{\boldsymbol{j}} + \ddot{z}\hat{\boldsymbol{k}}\right). \tag{9.19}$$

This acceleration is due to the combined gravitational pull of the Earth and Moon, or:

$$\boldsymbol{a}_p = -\frac{\mu_\oplus \boldsymbol{r}_\oplus}{r_\oplus^3} - \frac{\mu_☽ \boldsymbol{r}_☽}{r_☽^3}. \tag{9.20}$$

The vectors $\boldsymbol{r}_\oplus$ and $\boldsymbol{r}_☽$ can be expressed in Cartesian coordinates by noting that the origin of the system is located at the center of mass of the Earth-Moon system (Fig. 9.4). The scalar distances along the x-axis are related by $x_2 = x_1 + r_{12}$, where x_1 and x_2 are respectively the distances of the Earth and Moon from the center of mass and r_{12} is the Earth-moon distance (x_1 is negative). In addition, the mass balance about the center of mass dictates that $m_\oplus x_1 + m_☽ x_2 = 0$, where $m_\oplus$ and $m_☽$ are the masses of the Earth and Moon, respectively. This allows writing $x_1 = -\pi_2 r_{12}$ and $x_2 = +\pi_1 r_{12}$, using the reduced masses $\pi_1 = \frac{m_\oplus}{m_\oplus + m_☽}$ and $\pi_2 = \frac{m_☽}{m_\oplus + m_☽}$. The vector distances from the Earth and the Moon to the spacecraft can then be expressed as:

$$\boldsymbol{r}_\oplus = (x - x_1)\hat{\boldsymbol{i}} + y\hat{\boldsymbol{j}} + z\boldsymbol{k} \text{ and } \boldsymbol{r}_☽ = (x - x_2)\hat{\boldsymbol{i}} + y\hat{\boldsymbol{j}} + z\hat{\boldsymbol{k}}. \tag{9.21}$$

Inserting these relations into Eq. 9.20, and equating with Eq. 9.19 (carrying out the vector products), and collecting the terms in each unit vector direction $\hat{\boldsymbol{i}}, \hat{\boldsymbol{j}}$, and $\hat{\boldsymbol{k}}$ gives:

$$\ddot{x} - 2\Omega\dot{y} - \Omega^2 x = -\frac{\mu_\oplus}{r_\oplus{}^3}(x + \pi_2 r_{12}) - \frac{\mu_☽}{r_☽{}^3}(x - \pi_1 r_{12}), \tag{9.22}$$

$$\ddot{y} + 2\Omega\dot{x} - \Omega^2 y = -\frac{\mu_\oplus}{r_\oplus{}^3} y - \frac{\mu_☽}{r_☽{}^3} y, \tag{9.23}$$

and

$$\ddot{z} = -\frac{\mu_\oplus}{r_\oplus{}^3} z - \frac{\mu_☽}{r_☽{}^3} z. \tag{9.24}$$

The scalar distances from the spacecraft to the Earth and Moon are, respectively,

$$r_\oplus = \left[(x + \pi_2 r_{12})^2 + y^2 + z^2\right]^{1/2} \text{and } r_☽ = \left[(x - \pi_1 r_{12})^2 + y^2 + z^2\right]^{1/2}. \tag{9.25}$$

Determining the trajectory of a spacecraft in the plane of the moon's orbit around the Earth requires the simultaneous solution of Eqs. 9.22 and 9.23 (Eq. 9.24 can be neglected in this case).

Unlike the basic 2-body problem, there are no analytical solutions and the two equations must be solved numerically. To-scale solutions for a simple Earth-moon trajectory are shown in Fig. 9.5. It can be seen that the trajectory is strikingly different from the elliptical-transfer trajectories developed in Chapter 6. In addition, the trajectory appears very different depending on whether it is viewed in the inertial versus the rotating reference frame.

A stylized, not-to-scale rendering of the actual trajectory in Fig. 9.5a is shown in Fig. 9.6, where point ① is the point of trans-lunar injection. Atmospheric reentry occurs in the vicinity of point ②.

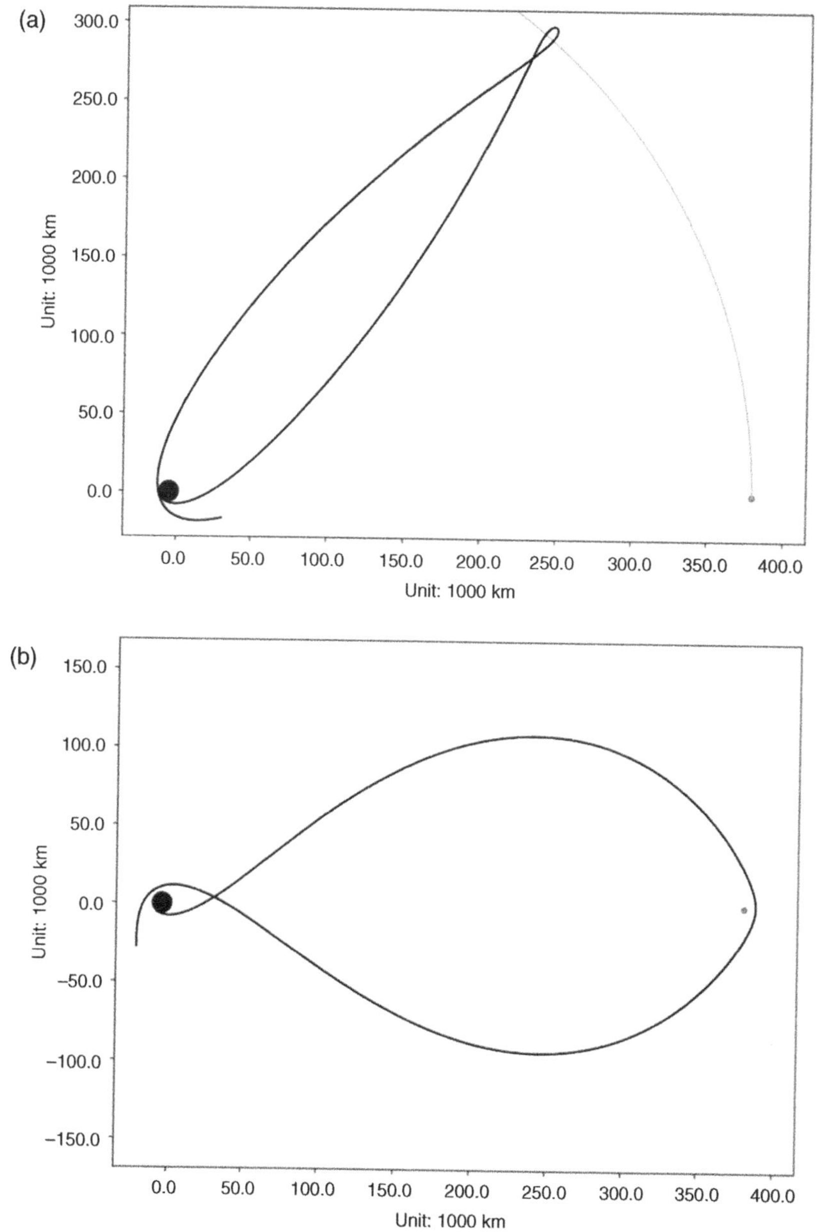

Figure 9.5 Earth-moon trajectory in the restricted 3-body problem. (a) In the inertial reference frame (Moon moving), and (b) in the rotating reference frame (Moon fixed). *Source:* [5]/Nodraak https://blog.nodraak.fr/2019/06/free-return-trajectory-1-flying-to-the-moon-with-apollo/

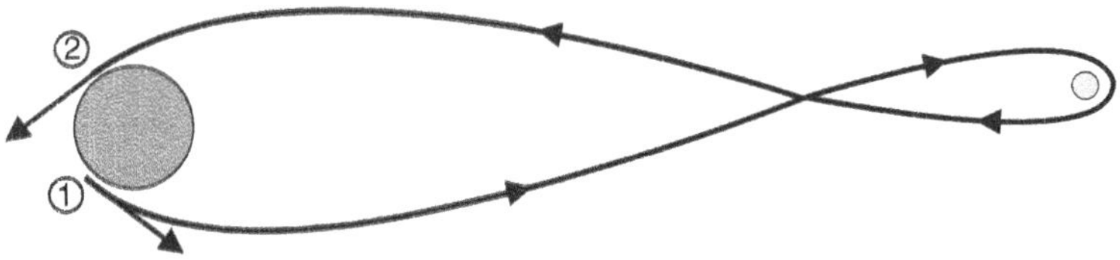

Figure 9.6 Stylized Earth-moon return trajectory.

This is an example of the "free return" trajectory used in Apollo crewed missions to the Moon. It is important to note that mid-course corrections are normally needed to ensure that the incoming spacecraft is within a narrow re-entry corridor which has a height of on the order of 10 km. An entry too high will result in the spacecraft having insufficient braking to reach the Earth's surface; an entry too low results in unacceptably high heating loads. Other, more complicated trajectories between the Earth and the Moon are possible, such as those employed in NASA's *Artemis* Program, which include multiple trajectory-change maneuvers within the Earth-Moon system.[4]

9.3 Restricted 3-Body Problem – Whole-Field Solution

The system of Eqs. 9.22–9.24 can be combined into a single function that describes the relative motion at any point in the Earth-Moon system. First, it can be noted that Eq. 9.20 can be alternatively expressed in ijk coordinates as:

$$\boldsymbol{a}_p = \vec{\nabla}\left(\frac{\mu_\oplus}{r_\oplus} + \frac{\mu_{\rightmoon}}{r_{\rightmoon}}\right) = \left(\frac{\partial}{\partial x}\hat{\boldsymbol{i}} + \frac{\partial}{\partial y}\hat{\boldsymbol{j}} + \frac{\partial}{\partial z}\hat{\boldsymbol{k}}\right)\left(\frac{\mu_\oplus}{r_\oplus} + \frac{\mu_{\rightmoon}}{r_{\rightmoon}}\right) = \frac{\partial}{\partial x}\left(\frac{\mu_\oplus}{r_\oplus} + \frac{\mu_{\rightmoon}}{r_{\rightmoon}}\right)\hat{\boldsymbol{i}} + \frac{\partial}{\partial y}\left(\frac{\mu_\oplus}{r_\oplus} + \frac{\mu_{\rightmoon}}{r_{\rightmoon}}\right)\hat{\boldsymbol{j}} + \frac{\partial}{\partial z}\left(\frac{\mu_\oplus}{r_\oplus} + \frac{\mu_{\rightmoon}}{r_{\rightmoon}}\right)\hat{\boldsymbol{k}}. \tag{9.26}$$

Combining this expression with Eq. 9.19 leads again to three scalar equations:

$$\ddot{x} - 2\Omega\dot{y} - \Omega^2 x = \frac{\partial}{\partial x}\left(\frac{\mu_\oplus}{r_\oplus} + \frac{\mu_{\rightmoon}}{r_{\rightmoon}}\right), \tag{9.27}$$

$$\ddot{y} + 2\Omega\dot{x} - \Omega^2 y = \frac{\partial}{\partial y}\left(\frac{\mu_\oplus}{r_\oplus} + \frac{\mu_{\rightmoon}}{r_{\rightmoon}}\right), \tag{9.28}$$

and

$$\ddot{z} = \frac{\partial}{\partial z}\left(\frac{\mu_\oplus}{r_\oplus} + \frac{\mu_{\rightmoon}}{r_{\rightmoon}}\right). \tag{9.29}$$

These equations can be combined by first multiplying each equation respectively by $2\dot{x}$ (onto Eq. 9.27), $2\dot{y}$ (onto Eq. 9.28), and $2\dot{z}$ (onto Eq. 9.29), then adding all terms on the left side and right side of each equation. The result is:

$$2\dot{x}\ddot{x} + 2\dot{y}\ddot{y} + 2\dot{z}\ddot{z} - 2\Omega^2 x\dot{x} - 2\Omega^2 y\dot{y} = 2\dot{x}\frac{\partial}{\partial x}\left(\frac{\mu_\oplus}{r_\oplus} + \frac{\mu_{\rightmoon}}{r_{\rightmoon}}\right) + 2\dot{y}\frac{\partial}{\partial y}\left(\frac{\mu_\oplus}{r_\oplus} + \frac{\mu_{\rightmoon}}{r_{\rightmoon}}\right) + 2\dot{z}\frac{\partial}{\partial z}\left(\frac{\mu_\oplus}{r_\oplus} + \frac{\mu_{\rightmoon}}{r_{\rightmoon}}\right), \tag{9.30}$$

4 See, for example, this overview of the Artemis I mission: https://appel.nasa.gov/2020/11/23/nasa-publishes-plan-for-lunar-exploration/

noting that the terms $-2\Omega\dot{y}$ and $+2\Omega\dot{x}$ in Eqs. 9.27 and 9.28, respectively, canceled due to the cross multiplication. Equation 9.30 can then be integrated over time by noting that $\int 2\dot{x}\ddot{x}dt = \dot{x}^2 + constant$ and similarly for the terms in y and z. Also on the left side of Eq. 9.30, $\int 2\Omega^2 x\dot{x}dt = \Omega^2 x^2 + constant$ (similarly for the terms in y and z). The right side can be integrated with respect to time by noting that $\dot{x} = \frac{\partial x}{\partial t}$ and $\frac{\partial x}{\partial t}\frac{\partial}{\partial x}\left(\frac{\mu_\oplus}{r_\oplus} + \frac{\mu_☽}{r_☽}\right) = \frac{\partial}{\partial t}\left(\frac{\mu_\oplus}{r_\oplus} + \frac{\mu_☽}{r_☽}\right)$ (similarly for the corresponding terms in y and z). Integrating and combining all the constants of integration then gives:

$$\dot{x}^2 + \dot{y}^2 + \dot{z}^2 - \Omega^2(x^2 + y^2) = 2\left(\frac{\mu_\oplus}{r_\oplus} + \frac{\mu_☽}{r_☽}\right) - 2C. \tag{9.31}$$

The constant C is termed Jacobi's integral:

$$C = \frac{1}{2}(\dot{x}^2 + \dot{y}^2 + \dot{z}^2) - \frac{1}{2}\Omega^2(x^2 + y^2) - \left(\frac{\mu_\oplus}{r_\oplus} + \frac{\mu_☽}{r_☽}\right). \tag{9.32}$$

Equation 9.31 suggests the definition of a useful (potential) function

$$U(x,y,z) = \frac{\mu_\oplus}{r_\oplus} + \frac{\mu_☽}{r_☽} + \frac{1}{2}\Omega^2(x^2 + y^2), \tag{9.33}$$

so that,

$$U(x,y,z) + C = \frac{1}{2}(\dot{x}^2 + \dot{y}^2 + \dot{z}^2). \tag{9.34}$$

For the case of zero spacecraft velocity in the restricted 3-body reference frame, $\dot{x}^2 + \dot{y}^2 + \dot{z}^2 = 0$, so

$$U(x,y,z) \geq C_0, \tag{9.35}$$

where C_0 is a constant (negative) value of the Jacobi integral for the case of zero relative velocity. Different values of C_0 lead to a set of contours of locations of zero relative velocity, as shown in Fig. 9.7. The contours shown do not represent the trajectories of the spacecraft in the restricted 3-body system. Rather, the contours are boundaries across which a spacecraft cannot cross (without executing a maneuver), depending on the value of the Jacobi integral (Eq. 9.32), which depends on the velocity of the spacecraft. In the case of a spacecraft coasting with just enough energy to reach a specific contour with zero velocity, the spacecraft will then "fall back" away from that contour and regain velocity.

Of particular interest are points in the restricted 3-body domain where both the relative velocity and the relative acceleration are zero. These are termed the *Lagrange Points*[5] and represent locations where a spacecraft will in principle remain fixed in space relative to the two other bodies. These points are commonly termed L_1, L_2, L_3, L_4, and L_5, and are shown in Fig. 9.7. Applying the condition of zero relative velocity $(\dot{x} = \dot{y} = \dot{z} = 0)$ and zero relative acceleration $(\ddot{x} = \ddot{y} = \ddot{z} = 0)$to Eqs. 9.22–9.24, gives:

$$-\Omega^2 x = -\frac{\mu_\oplus}{{r_\oplus}^3}(x + \pi_2 r_{12}) - \frac{\mu_☽}{{r_☽}^3}(x - \pi_1 r_{12}), \tag{9.35}$$

5 Joseph-Louis Lagrange (Giuseppe Lodovico Lagrangia, 1736–1813) was an Italian-French mathematician who, among many other notable contributions, solved the 3-body problem and identified the Lagrangian points in 1772.

$$-\Omega^2 y = -\frac{\mu_\oplus}{r_\oplus{}^3}y - \frac{\mu_☽}{r_☽{}^3}y, \tag{9.36}$$

and

$$0 = -\frac{\mu_\oplus}{r_\oplus{}^3}z - \frac{\mu_☽}{r_☽{}^3}z. \tag{9.37}$$

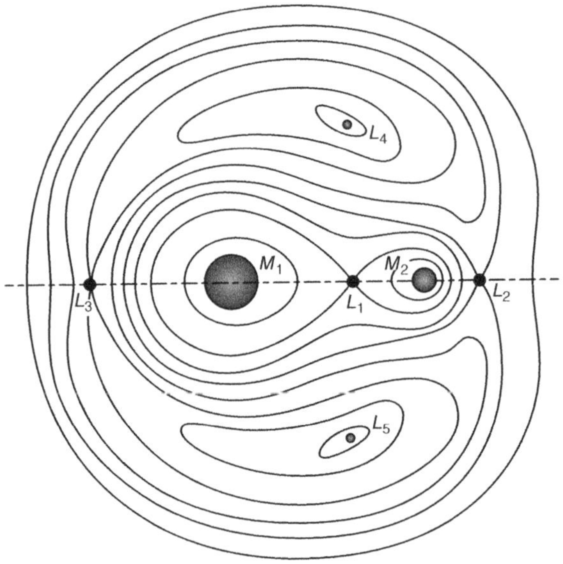

Figure 9.7 Zero-velocity contours in the restricted 3-body problem, showing the locations of the five Lagrange points. *Source:* [6]/HyperPhysics http://hyperphysics.phy-astr.gsu.edu/hbase/Mechanics/lagpt.html

Equation 9.37 shows that the Lagrange points must lie in the plane of the Moon's orbit ($z = 0$). The locations of the three Lagrange points that lie on the x-axis can be determined from Eq. 9.35 only, since $y = 0$ in that case. From Eq. 9.25, the scalar distances from the spacecraft to the Earth and moon in this case are simply $r_\oplus = |x + \pi_2 r_{12}|$ and $r_☽ = |x - \pi_1 r_{12}|$. Some algebra results in this equation for the locations x:

$$x = \frac{1}{\Omega^2}\frac{\mu_\oplus}{|x + \pi_2 r_{12}|^3}(x + \pi_2 r_{12}) + \frac{1}{\Omega^2}\frac{\mu_☽}{|x - \pi_1 r_{12}|^3}(x - \pi_1 r_{12}). \tag{9.38}$$

This relation can be simplified by relating the angular rotation rate to the period of the orbit of the system, or $\Omega = \frac{2\pi}{T} = \sqrt{\frac{\mu}{r_{12}{}^3}}$, where the gravitational constant is here $\mu = G(m_\oplus + m_☽)$. Noting that $\mu_\oplus = Gm_\oplus$ and $\mu_☽ = Gm_☽$ so that $\frac{\mu_\oplus}{\mu} = \pi_1$ and $\frac{\mu_☽}{\mu} = \pi_2$ allows rewriting Eq. 9.38 as follows:

$$x = r_{12}{}^3\frac{\pi_1}{|x + \pi_2 r_{12}|^3}(x + \pi_2 r_{12}) + r_{12}{}^3\frac{\pi_2}{|x - \pi_1 r_{12}|^3}(x - \pi_1 r_{12}). \tag{9.39}$$

This transcendental equation does not have an analytical solution and must be solved numerically. There are generally three positive roots that correspond to the locations of L_1, L_2, and L_3. The locations of the off-axis Lagrange points, L_4 and L_5, can be found by the simultaneous solution of Eqs. 9.35 and 9.36, where x and y are the only independent variables. After some algebra (noting that both $r_\oplus$ and $r_☽$ depend on both x and y) it can be shown that both L_4 and L_5 are located at $x = \left(\frac{1}{2} - \pi_2\right) r_{12}$, but with equal and opposite values of $y_{L_4} = \frac{\sqrt{3}}{2} r_{12}$ and $y_{L_5} = -\frac{\sqrt{3}}{2} r_{12}$. These values, and those implicit in Eq. 9.39, are valid for any restricted 3-body system (such Earth-Moon, Sun-Earth, Sun-Jupiter, etc.).

Although the relative acceleration of a spacecraft at a Lagrange point is theoretically zero, there is a question of stability. Lagrange points L_4 and L_5 are generally stable with objects able to remain near those points. Such objects include the *Trojan asteroids* that are found in the vicinity of those Lagrange points in the orbit of Jupiter (and other planets). The Lagrange points L_1, L_2, and L_3 are however not stable, so a spacecraft will require station keeping to remain in the vicinity of those points. For the Earth-Moon system, the gravitational pull of the Sun (not included in the analysis presented here) also causes the Lagrange points L_4 and L_5 to be unstable.

Example Problem 9b

Determine the locations of the five Lagrange points in the Earth-moon system, both in physical (x, y) and normalized (x/r_{12}, y/r_{12}) coordinates.

Solution:

The x-distance to Lagrange points L_4 and L_5 is given by $x = \left(\frac{1}{2} - \pi_2\right) r_{12}$, where $\pi_2 = \frac{m_☽}{m_\oplus + m_☽} = \frac{1}{\frac{m_\oplus}{m_☽} + 1} = \frac{1}{81.301 + 1} = 0.01215$. The normalized distance is then $\frac{x}{r_{12}} = \left(\frac{1}{2} - \pi_2\right) = \left(\frac{1}{2} - 0.01215\right) = 0.4878$. The physical distance is $x = (0.4878) r_{12} = (0.4878)(384{,}400\,\text{km}) = 187{,}529\,\text{km}$. $\frac{y_{L_4}}{r_{12}} = \frac{\sqrt{3}}{2} = 0.8660$ and $\frac{y_{L_5}}{r_{12}} = -\frac{\sqrt{3}}{2} = -0.8660$ giving the physical distances $y_{L_4} = (0.8660) r_{12} = (0.8660)(384{,}400\ km) = 332{,}890$ km and $y_{L_5} = -332{,}890$ km. Determining the positions of L_1, L_2, and L_3 is more involved.

Normalizing Eq. 9.39 by r_{12} gives: $\frac{x}{r_{12}} = \frac{\pi_1}{\left|\frac{x}{r_{12}} + \pi_2\right|^3}\left(\frac{x}{r_{12}} + \pi_2\right) + \frac{\pi_2}{\left|\frac{x}{r_{12}} - \pi_1\right|^3}\left(\frac{x}{r_{12}} - \pi_1\right)$. The values of the reduced masses are $\pi_1 = \frac{m_\oplus}{m_\oplus + m_☽} = \frac{1}{1 + \frac{m_☽}{m_\oplus}} = \frac{1}{1 + 1/81.301} = 0.9878$ and $\pi_2 = \frac{m_☽}{m_\oplus + m_☽} = \frac{1}{\frac{m_\oplus}{m_☽} + 1} = \frac{1}{81.301 + 1} = 0.01215$. Letting $\xi = x/r_{12}$ and inserting numerical values gives: $\xi - \frac{0.9878}{|\xi + 0.01215|^3}(\xi + 0.01215) - \frac{0.01215}{|\xi - 0.9878|^3}(\xi - 0.9878) = 0$. Numerical solution yields the three real roots $\xi = 0.8369$, 1.156, and -1.005 or $\frac{x_{L_1}}{r_{12}} = 0.8369 r_{12}$, $\frac{x_{L_2}}{r_{12}} = 1.156$, and $\frac{x_{L_3}}{r_{12}} = -1.005$. The physical distances are then $x_{L_1} = (0.8359) r_{12} = (0.8359)(384{,}400\,\text{km}) = 444{,}366\,\text{km}$, $x_{L_2} = (1.156) r_{12} = (1.156)(384{,}400\,\text{km},) = 444{,}366\,\text{km}$, and $x_{L_3} = (-1.005) r_{12} = (-1.005)(384{,}400\,\text{km}) = -386{,}322\,\text{km}$. The Lagrange points L_1, L_2, and L_3 all lie on the Earth-moon axis ($y = 0$). These values are tabulated below in Table 9.1. These values are from the barycenter of the Earth-moon system, which is situated inside the Earth 4700 km from the Earth's center.

Table 9.1 Earth-Moon Lagrange points.

Lagrange point	x (km)	y (km)
L_1	321,320	0
L_2	444,366	0
L_3	−386,322	0
L_4	187,529	332,890
L_5	187,529	−332,890

An interesting peculiarity of the restricted 3-body problem is that comparatively stable orbits are possible about points L_1, L_2, and L_3. These *halo orbits* are roughly (but not necessarily exactly) in the *y-z* plane close to the particular Lagrange point. Such orbits about the Lagrange points require station-keeping. One example is the *James Webb Telescope*, which was placed into a halo orbit behind the Moon around, L_2. Spacecraft have also been parked into halo orbits in the Sun-Earth system, such as the *ISEE-3* probe about point L_1. Two halo orbits are sketched in Fig. 9.8. The nonintuitive motion about an empty point in space is made possible by the complex gravitational field and rotating-system dynamics that characterize the restricted 3-body problem.

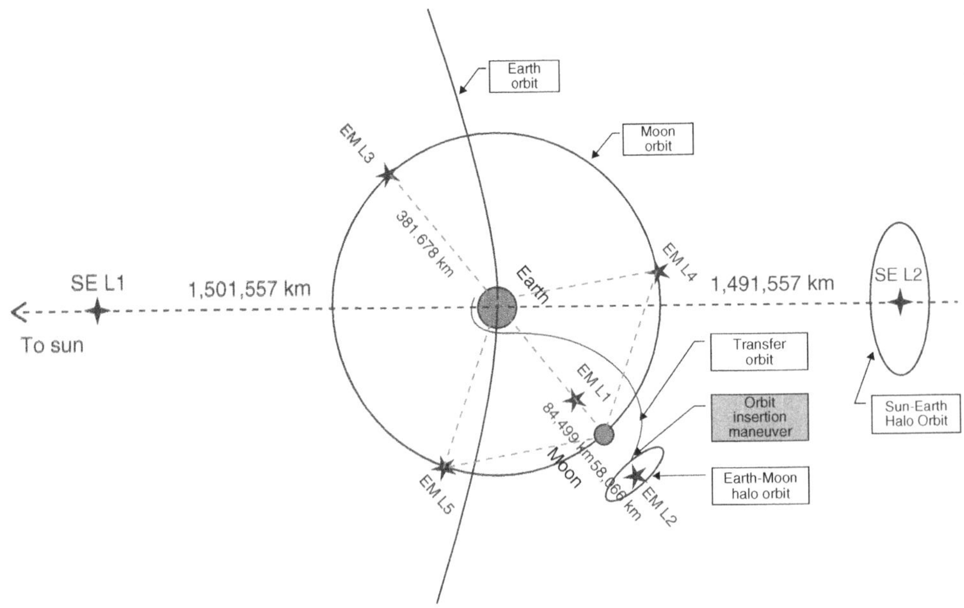

Figure 9.8 Illustration of halo orbits about the Lagrange point L_2 in both the Earth-moon and Sun-Earth systems. *Source:* https://www.researchgate.net/figure/Lagrange-Point-Orbits-and-Halo-orbits_fig2_269693893

9.4 Problems

9.1 A space station is in a circular orbit with an altitude of 500 km. Another spacecraft is directly below the station, in a circular orbit 10 km below the space station. What are, relative to the space station, the (i) vector distance, (ii) vector velocity, and (iii) vector acceleration?

9.2 Plot the velocities (i.e. $\delta\dot{x}_0$ and $\dot{y}_0$) corresponding to Example Problem 9a. What will be the total (scalar) velocity change required from the chase vehicle?

9.3 A spacecraft is within 500 m of a space station in the same circular orbit with an altitude of 400 km. Being apparently "close enough," the spacecraft applies a velocity change of 0.2 m/s in the direction of the space station. Calculate and plot the resulting trajectory relative to the space station. What is the distance between the spacecraft and the space station at closest approach?

9.4 Verify the algebra leading up to Eq. 9.10a, b, and c.

9.5 A spacecraft is on a direct line between the Earth and the Moon. At what point along this axis are the gravitational forces from the Earth and Moon equal? How does this location compare to the Earth's Sphere of Influence, and what does that imply for the restricted 3-body problem?

9.6 Determine the locations L_4 and L_5 in the Sun-Jupiter system. Compare those calculated locations with the known locations[6] of Jupiter's *Trojan asteroids.*

9.7 Show that the distances to the Lagrange points L_4 and L_5, from the centers of the Earth and Moon, form an equilateral triangle with the Earth-moon axis.

9.8 Determine the locations of all five Lagrange points in the Sun-Earth system.

6 One *source:* https://expanse.fandom.com/wiki/Jupiter_Trojans

10

Rocket Fundamentals and Space Propulsion

A very important consideration in any orbital or trajectory maneuver is the amount of velocity change required. As seen in the previous chapters, the magnitude of such velocity changes can be several km/s. How these velocity changes can be accomplished, and the implications and practical limitations for the rocket systems involved, are the subjects of this chapter. It will be seen that the inherent constraints on rocket propulsion can have substantial implications for space missions; in particular, the amount of allowable payload mass. The chapter outline is as follows:

- Principles of rocket thrust
- Rocket performance
- Staging
- Sounding rockets
- Electric space propulsion

10.1 Principles of Rocket Thrust

A rocket operates in a manner consistent with Newton's third law of motion mentioned in Chapter 2 – that for every action, there is an equal and opposite reaction. The "action" in this case is the discharge of momentum at a very high rate from the exit of the rocket nozzle, with the resulting "reaction" being the thrust force generated.

A simplified diagram of a rocket is shown in Fig. 10.1. For simplicity, the rocket is shown fixed on a test stand, so the motion of the rocket does not have to be considered. The momentum changes associated with the system can be assessed by use of a *Conservation Principle*, used very often in engineering analysis to track the development and change of any quantity in a process (such as the momentum, mass flow rate, energy, etc.). The first step in a conservation principle is to define a *Control Volume* (CV) that identifies the system to be analyzed. The CV shown in Fig. 10.1 is taken to enclose the entire rocket assembly. Having defined a suitable CV, the next step is to identify all of the changes in momentum. Taking both the thrust force vector $\boldsymbol{F}$ and the exhaust flow to be in the

Introduction to Orbital Mechanics: A Concise and Practical Approach, First Edition. James C. Hermanson.

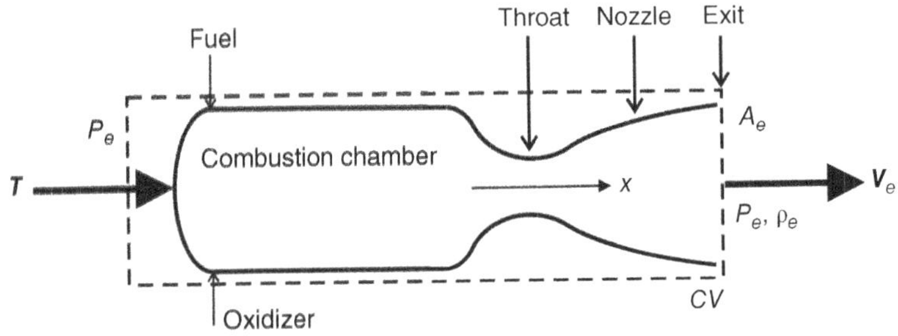

Figure 10.1 Control-volume analysis of a rocket.

x-direction, as shown in Fig. 10.1, only the components of momentum in the x-direction need to be considered here. The changes in x-momentum can be expressed conceptually as follows:

$$\begin{pmatrix} \text{change in} \\ x-\text{momentum} \end{pmatrix}_{CV} = \begin{pmatrix} x-\text{momentum} \\ \text{flow} \end{pmatrix}_{in} - \begin{pmatrix} x-\text{momentum} \\ \text{flow} \end{pmatrix}_{out} + \begin{pmatrix} \text{applied forces in} \\ x-\text{direction} \end{pmatrix} \tag{10.1}$$

The first term, the nonstationary term, reflects the change in momentum of the system due to momentum flowing into, and out of, the CV. If the rocket is in steady-state operation, the nonstationary term is zero. The momentum flow in the x-direction due to the propellants entering the rocket combustion chamber can be made to be exactly zero by taking the propellant inflows to be normal to the x-axis, as shown. In actuality, given that the propellants often enter as liquids, the associated momentum flow is negligible as the flow velocities are very low compared to the velocity of the gases exiting the rocket nozzle.

The outflow of x-momentum from the system is substantial and is the principle action responsible for rocket thrust. Taking all properties at the nozzle exit to be uniform, the momentum outflow is

$$\begin{pmatrix} x-\text{momentum} \\ \text{flow} \end{pmatrix}_{out} = (\rho_e V_e)(V_e A_e) = \dot{m} V_e \tag{10.2}$$

where ρ_e and V_e are, respectively, the mass density and velocity of the exhaust gas stream and A_e is the nozzle exit area. The momentum outflow is the product of the momentum density of the gas (the mass density per unit volume), $\rho_e V_e$ multiplied by the volume flow rate, $V_e A_e$. The mass density multiplied by the volume flow rate is the mass flow rate at the nozzle exit, $\dot{m} = \rho_e V_e A_e$. The nonzero terms in Eq. 10.1 can then be written as follows:

$$\dot{m} V_e = \Sigma F_x \tag{10.3}$$

where ΣF_x represents the sum of *all* forces acting on the CV in the x-direction. In addition to the thrust force, T, this includes the forces due to both the pressure at the nozzle exit, P_e, and that of the surrounding atmosphere, P_a. The CV was conveniently chosen here so that the area at the front end is equal to that at the nozzle exit, A_e though this is not strictly necessary. The forces in the x-direction then sum up to:

$\Sigma F_x = T + P_a A_e - P_e A_e$. Equation 10.3 then becomes

$$\dot{m} V_e = T + P_a A_e - P_e A_e$$

or rewriting:

$$\boxed{T = \dot{m} V_e + (P_e - P_a) A_e.} \tag{10.4}$$

This is the basic equation for the thrust of a rocket, and it illustrates several important aspects of rocket propulsion. One interesting effect to note is that all other variables being equal, the thrust of a rocket will be greater in space, where the ambient pressure P_a is essentially zero, than during launch in the atmosphere. Thus, in space, the rocket thrust equation simplifies to:

$$\boxed{T_{vac} = \dot{m}V_e + P_eA_e.} \tag{10.5}$$

It can be seen from either Eq. 10.4 or 10.5 that achieving a high level of thrust requires a combination of a high mass flow rate and high exit velocity. Very high liquid propellant mass flows typically dictate the use of turbopumps for both the fuel and oxidizer. As one example, the combined mass flow rates of oxygen and hydrogen in a single Space Shuttle Main Engine (SSME) were nominally 515 kg/s. A high nozzle exit velocity (3733.8 m/s = 3.7 km/s in the SSME) can be achieved by a combination of high gas temperatures due to combustion combined with expansion in the nozzle to achieve supersonic exit velocities.[1] Increasing the nozzle exit area, A_e, can lead to an increase in rocket thrust, but this can be offset by the lower exit pressure that would result (increasing the size of the nozzle also increases the weight). An SSME is shown in operation in Fig. 10.2.

Figure 10.2 Space Shuttle Main Engine (Aerojet Rocketdyne RS-25). The RS-25 is currently in use in the Space Launch System. *Source:* NASA / https://commons.wikimedia.org/wiki/File:Shuttle_Main_Engine_Test_Firing.jpg / Public domain / Last accessed on Feb 18, 2026.

Example Problem 10a

The SSME has a thrust in vacuum of 2.28 MN. What would be the nominal thrust at sea level? The nozzle exit diameter is $D_e = 2.4$ m and take the exit pressure to be 17.45 kPa.

Solution:

The difference in thrust can be determined from the general thrust (Eq. 10.4) and that for vacuum conditions (Eq. 10.5). This gives: $T - T_{vac} = \dot{m}V_e + (P_e - P_a)A_e - \dot{m}V_e + P_eA_e = -P_aA_e$, pointing to the expected decrease in thrust at sea level. The nozzle exit area is $A_e = \pi D_e^2/4 = 3.1416\ (2.4\ \text{m})^2/4 = 4.524\ \text{m}^2$. Combining this with the nominal atmospheric pressure of $P_a = 101.3$ kPa gives as the sea-level thrust $T = T_{vac} - P_aA_e = 2.28\ \text{MN} - (101.3\ \text{kN/m}^2)(4.254\ \text{m}^2) = 1.84\ \text{MN}$. This compares well with the published value of 1.86 MN. The comparatively small difference in thrust between sea level and space illistrates that the pressure term in the thrust (Eq. 10.4), $(P_e - P_a)A_e$, is normally much smaller than the momentum flow term, $\dot{m}V_e$.

1 An excellent reference for compressible flow, including supersonic nozzle flow, is Hodge, B.K. and Koenig, K. *Compressible Fluid Dynamics*, Pearson India, 2015.

10.2 Rocket Performance

The overall performance of a rocket is typically characterized by the *specific impulse*, which is defined by relating the thrust to the propellant flow as follows:

$$\boxed{I_{sp} = \frac{T}{\dot{m}g_0},} \tag{10.6}$$

where g_0 is the gravitational acceleration at sea level, 9.81 m/s^2. The incorporation of the gravitational acceleration means that the denominator in Eq. 10.6 is actually weight flow, resulting in the specific impulse having the units of seconds. This historic, but seemingly unusual definition does have the advantage that the resulting value (in seconds) is the same whether SI or Imperial units are employed. Representative ranges of specific impulse for various types of chemical rocket systems are given in Table 10.1.

Monopropellant rockets use a single, reactive propellant, such as hydrazine (H_2NNH_2). The propellant decomposes upon contact with a suitable catalyst such as iridium, releasing heat. Monopropellant thrusters have the advantage of simplicity and are commonly used for attitude control of spacecraft (termed a Reaction Control System or RCS). Solid propellants consist of a fuel (such as aluminum powder) mixed with an oxidizer such as hydroxyl-terminated polybutadiene (HTPB) or ammonium perchlorate (NH_4ClO_4). Solid propellants are storable and are commonly used in missiles for that reason. Many rocket boosters, such as the Space Shuttle Solid Rocket Booster (SRB) and the Delta IV boosters, use solid propellants. Hybrid rockets typically use a solid fuel, such as paraffin wax (HTPB can also be used as a fuel), and a liquid oxidizer (for example, nitrous oxide (N_2O), or liquid oxygen). Hybrid rockets have the advantages of being simpler than bi-propellant rockets and are safer to operate. A hybrid rocket system is employed by the Scaled Composites *SpaceShipTwo* flown by Virgin Galactic. Bi-propellant liquid rockets are commonly used when very high performance is required. Typical fuels include liquid hydrogen and kerosene RP-1, with liquid oxygen as the oxidizer. The Space Shuttle Main Engine and the Space Launch System core stage are example applications of bi-propellant liquids. More easily storable liquid propellants include monomethylhydrazine (MMH, $[CH_3]HN_2H_2$) and dinitrogen tetroxide (N_2O_4), among others. Applications for such storable, non-cryogenic propellants include the Apollo Lunar Model (LM) and Service Module (SM). Both liquid bi-propellant and hybrid rockets can be throttled and restarted.

It can be seen from Table 10.1 that the practical upper limit of specific impulse for chemical rockets is approximately 450 s. Considerably higher specific impulse is possible using electric space propulsion (Section 10.5), but those systems are incapable of providing the high thrust levels needed at launch.

Table 10.1 Approximate specific impulse ranges of chemical rockets.

Rocket type	Specific impulse range (seconds)
Monopropellant	120–325
Solid propellant	230–290
Hybrid	180–240
Bi-propellant	240–450

Source: An essential reference for rocket propulsion is Sutton, G.P.; Biblarz, O.: *Rocket Propulsion Elements*, 8th edition, Wiley & Sons, Hoboken, 2010.

Another parameter commonly employed to characterize rocket performance is the *effective* or *equivalent* velocity, *c*, which takes into account the thrust increment/decrement due to the pressure term to give:

$$c = \dot{m} + \frac{(P_e - P_a)A_e}{\dot{m}} \tag{10.7}$$

which simplifies to

$$c = \dot{m} + \frac{P_e A_e}{\dot{m}} \tag{10.8}$$

for the case of a rocket operating in the vacuum of space. Using the effective exhaust velocity then allows the rocket thrust to be simply expressed as

$$T = \dot{m} g_0 I_{sp} = \dot{m} c. \tag{10.9}$$

In the vacuum of space, the effective velocity is essentially constant for a rocket operating under steady conditions. During a ground launch and ascent, the effective velocity changes somewhat with the decreasing atmospheric pressure but can be taken as a first approximation to be constant during a particular segment of the ascent (rocket staging is discussed in Section 10.3).

Given that the thrust of a rocket depends directly on the mass flow rate of the propellants, a question of key importance for any spacecraft orbital maneuver is how much mass of propellant will be needed for a given change in the velocity of the vehicle. The required propellant mass may be modest for small velocity changes, but for the large changes in velocity needed for many of the orbital maneuvers discussed previously, the propellant mass will be substantial – for ground launch applications, the propellant mass will often be the bulk of the entire mass of the vehicle.

Determining the velocity change brought about by the rocket thrust requires consideration of the momentum of the entire rocket system, taking into account the continuous change in vehicle mass as well as the momentum of the gases. Analysis of the system starts with the control volume shown in Fig. 10.3, which is taken to enclose both the rocket and the exhaust gases. The vehicle has the instantaneous velocity *V* and mass *m*, and mass is exhausted from the rocket nozzle with velocity V_e.

Taking the momentum of the system at some time *t* to be $p_t = mV$ (*p* is not to be confused here with pressure), the momentum at a subsequent time $t + dt$ will be

$$p_{t+dt} = (m - dm)(V + dV) + (V - V_e)dm \tag{10.10}$$

where the first term on the right is the change in momentum of the rocket; the second, the momentum change due to the rocket exhaust. The momentum change over the differential time increment *dt* is then

$$p_{t+dt} - p_t = (m - dm)(V + dV) + (V - V_e)dm - mV = mdV - V_e dm, \tag{10.11}$$

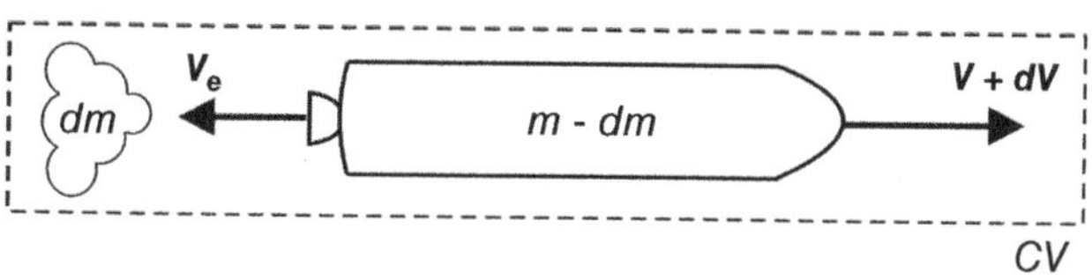

Figure 10.3 Control-volume analysis of a rocket in flight.

neglecting the small $dmdV$ term. The time rate of system momentum change can then be written as follows:

$$\frac{dp}{dt} = \frac{p_{t+dt} - p_t}{dt} = m\frac{dV}{dt} - V_e\frac{dm}{dt} = F_{ext}, \tag{10.12}$$

where F_{ext} represents the sum of the external forces on the rocket. In the absence of aerodynamic drag and gravity losses, the only external forces are due to the pressures acting on the rocket. Noting that $\frac{dm}{dt} = \dot{m}$, Eq. 10.12 can be rewritten as follows:

$$m\frac{dV}{dt} = \dot{m}V_e + (P_e - P_a)A_e = T \tag{10.13}$$

where T is the rocket thrust (Eq. 10.4). This gives that the rocket thrust and vehicle acceleration are simply related by the expression

$$T = m\frac{dV}{dt}, \tag{10.14}$$

which is correct even though the rocket mass changes with time. Noting that the thrust can be expressed as $T = \dot{m}c$ and the change in vehicle mass is $\frac{dm}{dt} = -\dot{m}$ allows rewriting Eq. 10.14 as follows:

$$\frac{dV}{dt} = \frac{T}{m} = -\frac{c}{m}\frac{dm}{dt}. \tag{10.15}$$

Multiplying out the differential time element dt gives this simple relation for the velocity change:

$$dV = -c\frac{dm}{m}. \tag{10.16}$$

Taking the effective exhaust velocity to be constant allows this to be integrated directly to give the velocity change between any two times of interest as

$$V_2 - V_1 = -c[\ln(m_2) - \ln(m_1)] = c\ln\left(\frac{m_1}{m_2}\right). \tag{10.17}$$

Setting $m_0 = m_1$ to be the vehicle mass at the start of the rocket burn and $m_f = m_2$ at rocket burnout then gives this relation for the velocity change:

$$\boxed{\Delta V = I_{sp}g_0\ln\left(\frac{m_0}{m_f}\right)}. \tag{10.18}$$

This is the famous *Tsiolkovsky rocket equation*,[2] which has major implications for any spacecraft relying on rocket propulsion to affect a velocity change. Increasing the specific impulse of the rocket is thus clearly of benefit in achieving a given velocity change. Unfortunately, the dependence on the mass ratio, or ratio of initial to final vehicle mass, is logarithmic, and thus an increasingly large change in mass is needed if the velocity change is to be increased. This explains why large velocity changes can only be achieved with conventional rocket propulsion by very large changes in

2 Published by the Russian mathematician Konstantin Tsiolkovsky in 1903.

spacecraft mass. Therefore, much, if not most, of the spacecraft mass will consist of propellant. The Tsiolkovsky rocket equation (Eq. 10.18) can be rearranged to give:

$$\frac{m_f}{m_0} = \exp\left(\frac{-\Delta V}{c}\right). \tag{10.19}$$

The final mass would be the initial mass minus the mass of propellant consumed, m_p, so $m_f = m_0 - m_p$. Rearranging Eq. 10.19 then gives the fraction of the initial vehicle mass that must consist of propellant as:

$$\frac{m_p}{m_0} = 1 - \exp\left(\frac{-\Delta V}{c}\right). \tag{10.20}$$

Example Problem 10b

A spacecraft in low earth orbit (LEO) is propelled by a rocket with a specific impulse of 450 s. How much of the initial vehicle mass will need to consist of propellant to reach escape velocity? How much more propellant will be needed if the velocity change is 5 or 8 km/s?

Solution:

The orbital velocity in LEO with a radius of 7600 km is $V_c = 7.713$ km/s, with the corresponding escape velocity of $V_{esc} = 10.908$ km/s. The required velocity boost is then $\Delta V = V_{esc} - V_c = 3.195$ km/s. The effective velocity $c = I_{sp}g_0 = (450\ \text{s})(9.81\ \text{m/s}^2) = 4.415$ km/s. The fraction of the initial vehicle mass that must be propellant is given by Eq. 10.20 as $\frac{m_p}{m_0} = 1 - \exp\left(\frac{-\Delta V}{c}\right) = 1 - \exp\left(\frac{-3.195\ \text{km/s}}{4.415\ \text{km/s}}\right) = 0.515$, so more than half of the initial mass must consist of propellant. For the higher velocity boosts needed to achieve a hyperbolic excess velocity, repeating the calculation for $\Delta V = 5$ km/s gives $m_p/m_0 = 0.678$. For $\Delta V = 8$ km/s, the corresponding value is $m_p/m_0 = 0.837$. Given that the vehicle also will have masses associated with its structure, engines, and other systems, that only 16% of the initial mass available for everything else will likely have a substantial impact on the allowable mass on the actual payload carried – the spacecraft.

10.3 Staging

The basic concept behind the use of multiple stages on a vehicle is to discard weight during launch or a trajectory insertion maneuver, increasing overall performance. Staging can be analyzed by applying the Tsiolkovsky rocket equation (Eq. 10.18) to each stage of the vehicle in turn, taking into account all of the masses and changes in mass due to propellant consumption associated with each stage. A launch vehicle with two stages is shown schematically in Fig. 10.4a.

The staged vehicle shown in Fig. 10.4a is a "tandem" or "serial" staging configuration, where each stage is mounted atop the previous stage, with each stage started in sequence. Examples of tandem staging include the *Saturn V* and *Atlas-Centaur* launch vehicles. An alternative arrangement is "parallel" staging (Fig. 10.4b), such as the Space Shuttle, where some or all stages burn simultaneously. Some vehicles, for example, the *Delta III* and the Space Launch System (SLS), consist of a combination of tandem and parallel staging.

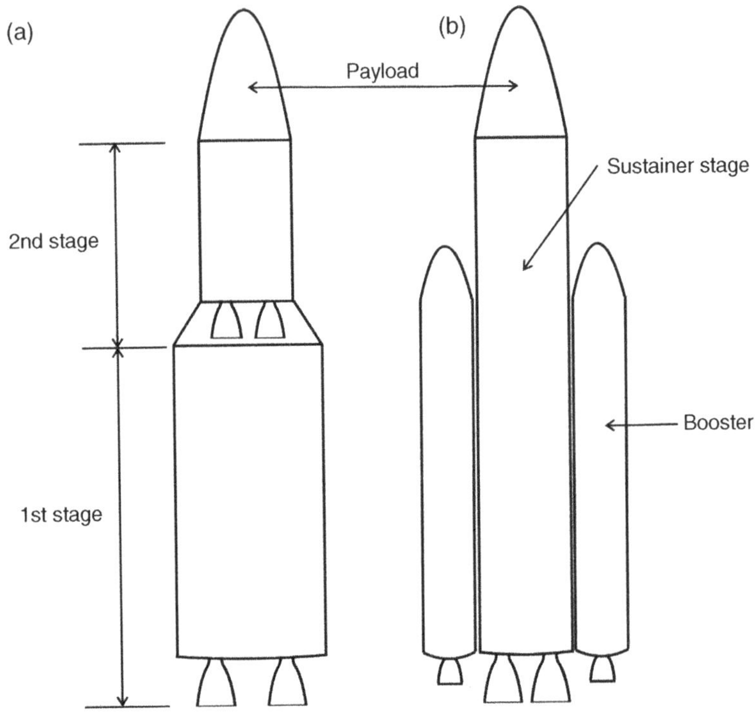

Figure 10.4 Rocket staging. (a) Tandem staging and (b) parallel staging.

The initial mass of each stage, m_0, can be written as

$$m_0 = m_p + m_{dw} + m_{PL}. \tag{10.21}$$

where m_p is the mass of propellant and m_{PL} is the mass of the payload. The deadweight mass, m_{dw}, consists of all other masses associated with the stage, such as structure, engines, propellant tanks, avionics, etc. Assuming that all propellant in each stage is consumed, the final stage mass is then

$$m_f = m_{dw} + m_{PL}. \tag{10.22}$$

The payload of the uppermost, final stage is the spacecraft being launched. In a tandem-staged rocket, each stage carries as its "payload" all of the stages stacked on top of it. Thus, for the arrangement shown in Fig. 10.4a, the second-stage mass consists of the actual payload (normally the spacecraft). The first stage then carries the spacecraft payload plus the mass of the second stage as its payload. The same reasoning applies to any number of serially mounted stages.

There are different systems for accounting for the masses and mass changes in a staged rocket system. This discussion will use the terminology of Mattingly & Boyer,[3] which includes the definitions tabulated in Table 10.2. Deadweight includes the structure, propellant tanks, engines, avionics, etc.

3 Mattingly, J.D. and Boyer, K.M. *Elements of Propulsion – Gas Turbines and Rockets*, 2nd edition, AIAA, 2016.

Table 10.2 Mass accounting terminology.

Parameter	Definition	Meaning
Payload fraction	$\lambda = \frac{m_{PL}}{m_0}$	$\frac{\text{payload}}{\text{initial mass}}$
Deadweight ratio	$\delta = \frac{m_{dw}}{m_0}$	$\frac{\text{deadweight}}{\text{initial mass}}$
Load fraction	$\lambda_L = \frac{m_{PL}}{m_p + m_{dw}}$	$\frac{\text{payload}}{\text{propellant + deadweight}}$
Structural ratio	$\varepsilon = \frac{m_{dw}}{m_p + m_{dw}}$	$\frac{\text{deadweight}}{\text{propellant + deadweight}}$

The mass ratio needed in the Tsiolkovsky rocket equation (Eq. 10.18), m_0/m_f, can then be expressed as follows:

$$MR = \frac{m_0}{m_f} = \frac{m_0}{m_{PL} + m_{dw}} = \frac{1}{\lambda + \delta}, \tag{10.23}$$

obtained by dividing the numerator and denominator by m_0. Alternatively, this can be written as follows:

$$MR = \frac{m_p + m_{dw} + m_{PL}}{m_{PL} + m_{dw}} = \frac{1 + \lambda_L}{\lambda_L + \varepsilon} \tag{10.24}$$

by dividing the numerator and denominator by $(m_{dw} + m_p)$. Inserting Eq. 10.23 into the Tsiolkovsky rocket equation, along with the effective exhaust velocity, c, gives the velocity change for a given stage as:

$$\Delta V = c \ln\left(\frac{1}{\lambda + \delta}\right). \tag{10.25}$$

This indicates (predictably) that minimizing the deadweight is important in maximizing rocket performance. For a fixed deadweight ratio, stage performance can also be increased by reducing the payload fraction (often an undesirable trade-off). The quantity $\lambda + \delta$ can also be expressed in terms of the structural ratio, ε, which after some algebra gives:

$$\lambda + \delta = \varepsilon + (1 - \varepsilon)\lambda. \tag{10.26}$$

The velocity increment for a stage can thus also be written as follows:

$$\Delta V = -c \ln[\varepsilon + (1 - \varepsilon)\lambda]. \tag{10.27}$$

The payload fraction and structural ratio for a given stage j, where $j = 1, 2, 3$, etc. can be defined, respectively, as follows:

$$\lambda_j = \frac{m_{0_{j+1}}}{m_{0_j}}, \varepsilon_j = \frac{m_{dw_j}}{m_{p_j} + m_{dw_j}}, \tag{10.28}$$

where m_{0j+1} is the entire mass of all stages and payload above stage j. The total velocity change for the entire rocket is the sum of the velocity change imparted by each stage:

$$\Delta V_{tot} = \sum_j^N \Delta V_j = \Delta V_1 + \Delta V_2 + \Delta V_3 + \dots, \tag{10.29}$$

where N is the total number of stages, and each velocity increment is calculated from the Tsiolkovsky rocket equation described earlier.

Example Problem 10c

A launch vehicle with a mass of $m_{PL} = 10{,}000$ kg requires a velocity change of $\Delta V = 9000$ m/s. The specific impulse of the rocket is $I_{sp} = 400$ s, and the structural ratio is $\varepsilon = 0.1$.[4] What will be the initial rocket vehicle mass if only one stage is employed? What would be the initial mass for a two-stage system, assuming that each stage has the same specific impulse and structural ratio?

Solution:

The effective exhaust velocity is $c = I_{sp} g_0 = (400\ \text{s})\ (9.81\ \text{m/s}^2) = 3920$ m/s. Equation 10.27 can be rewritten as: $\varepsilon + (1-\varepsilon)\lambda = \exp\left(-\frac{\Delta V}{c}\right)$ or $0.1 + (0.9)\lambda = \exp\left(\frac{-9000\ \text{m/s}}{3920\ \text{m/s}}\right) = 0.1007$. This gives the payload fraction as $\lambda = \frac{m_{PL}}{m_0} = 0.007432$, so the initial mass will be $m_0 = \frac{m_{PL}}{\lambda} = \frac{10{,}000\ \text{kg}}{0.007432} = 13.5 \times 10^6$ kg – hardly a practical design! For the two-stage system, $\Delta V_{tot} = \Delta V_1 + \Delta V_2 = -c\{\ln[\varepsilon_1 + (1-\varepsilon_1)\lambda_1] + \ln[\varepsilon_2 + (1-\varepsilon_2)\lambda_2]\}$. For the optimum staging performance,[5] the payload fractions for the stages will be the same $\lambda_1 = \lambda_2 = \lambda$. The stage structural ratios are also taken to have the same values, $\varepsilon_1 = \varepsilon_2 = \varepsilon$. The total velocity change can then be written as $\Delta V_{tot} = -2c\{\ln[\varepsilon + (1-\varepsilon)\lambda]\}$, so $\varepsilon + (1-\varepsilon)\lambda = 0.1 + 0.9\lambda = \exp\left(-\frac{\Delta V}{2c}\right) = \exp\left(\frac{-9000\ \text{m/s}}{(2)3920\ \text{m/s}}\right) = 0.3173$. Solving for the payload ratio gives $\lambda = \lambda_1 = \lambda_2 = 0.2414$. The second stage initial mass is then $m_{02} = m_{PL}/\lambda = 10{,}000\ \text{kg}/0.2414 = 41{,}421$ kg. The total initial mass (both stages combined) will be $m_{01} = m_{02}/\lambda = 41{,}421\ \text{kg}/0.2414 = 171{,}585$ kg. While still substantial, this initial mass is nearly 80 times smaller than that for the single-stage system, showing the utility and importance of staging rocket vehicles when a mission requires substantial velocity changes.

As an example of the large vehicle masses required for space missions beyond Earth orbit, consider the *Saturn V/Apollo* space vehicle. The nominal masses involved are given in Table 10.3; the payload mass for each stage is that of all upper stages combined.

Table 10.3 *Saturn V/Apollo* masses.

	m_0 (kg)	m_p (kg)	m_{dw} (kg)	m_{PL} (kg)
Stage I	2,879,929	2,077,000	137,000	665,929
Stage II	665,929	456,100	40,100	169,729
Stage III	169,729	107,800	15,200	46,729
Command Module (CM)	5,809	–	–	–
Service Module (SM)	30,329*	18,410	6110	5809
Lunar Module (LM)	16,400	11,480	4920	–

* Considering only the CM as the payload of the SM.

4 As a footnote, 10% is roughly the "structural ratio" of a chicken egg, as optimized in nature.
5 This is derived in detail in Curtiss, H., *Orbital Mechanics for Engineering Students (Aerospace Engineering)*, 4th Edition, Butterworth-Heinemann, 2019.

27/2/2026 6:42:2

Considering as the payload to be the entire spacecraft sent to the Moon (CM + SM + LM) suggests the payload ratio based on the initial launch mass of $\frac{m_{CM+SM+LM}}{m_0} = 0.0162$ with the mass ratio MR = 61.6. Returning to Earth from the Moon, the payload ratio and mass ratio, again based on the initial launch mass, were respectively $\frac{m_{CM+SM}}{m_0} = 0.0105$ and MR = 95.0. Finally, considering the end payload to be the Command Module, the only portion that splashed down on Earth, gives $\frac{m_{CM}}{m_0} = 0.0020$ and MR = 496. This indicates that very large mass ratios will be needed for any lunar/interplanetary mission. One option for improving this might be the use of on-orbit refueling, given that most of required masses will likely be propellants.

10.4 Sounding Rockets

Another insightful example of rocket propulsion is the *sounding rocket*, which is used to launch payloads on suborbital trajectories with maximum altitudes typically between 50 and 1500 km.[6] The name derives from the nautical term "sounding," meaning to take measurements, such as the depth of the ocean. In addition to flying scientific payloads to study the upper atmosphere and solar physics, sounding rockets also allow for conducting experiments in microgravity[7] over a duration of several minutes. An example of a sounding rocket is shown in Fig. 10.5. Sounding rockets are normally launched near vertically from the Earth's surface and use one or more rocket stages. After the burnout of the last stage, the rocket then flies on a nearly parabolic trajectory, reaching its peak altitude before returning to Earth.

Figure 10.5 A Black Brant XII sounding rocket. *Source:* Wallops / NASA / https://commons.wikimedia.org/wiki/File:Black_Brant.jpg / Public domain / Last accessed on Feb 18, 2026.

In any ground launch, it is important to include the effects of both gravity and atmospheric drag. Taking those effects into account gives a more general description of rocket acceleration than Eq. 10.16 as:

$$dV = -c\frac{dm}{m} - \frac{D}{m}dt - g\sin\gamma dt \tag{10.30}$$

The first term on the right side represents the rocket thrust. The quantity D is the aerodynamic drag, and γ is the flight angle from vertical (similar to the flight path angle defined in Chapter 2). Integrating this equation with time gives the velocity change due to the rocket burn as:

$$\Delta V = \int_{t_0}^{t} dV = c\ln\left(\frac{m_0}{m_t}\right) - \int_{t_0}^{t}\frac{D}{m}dt - \int_{t_0}^{t} g\sin\gamma dt, \tag{10.31}$$

6 https://sites.wff.nasa.gov/code810/files/sounding%20rocket%20litho-2013.pdf

7 "Microgravity" refers to gravity levels on the order of 10^{-5}–10^{-6} g/s. This term is preferable to "zero gravity," which cannot be exactly achieved in practice.

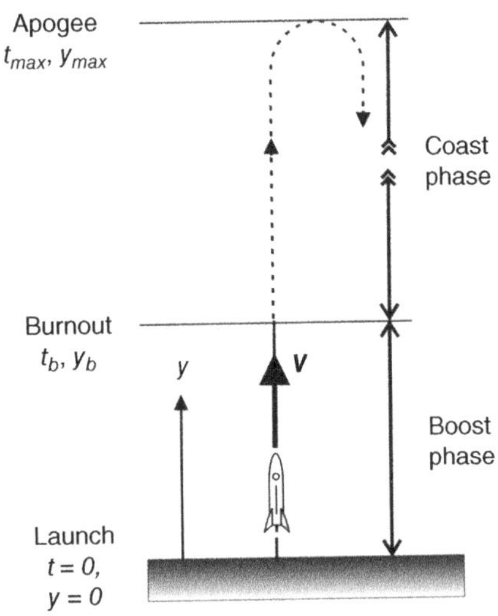

Figure 10.6 Sounding rocket flight profile.

where t is the elapsed time from the time of rocket ignition, t_0. For an in-space maneuver, where there is essentially no aerodynamic drag and for zero flight path angle, this simplifies to the Tsiolkovsky rocket equation (Eq. 10.18) at rocket burnout ($t = t_f$). For the case of a vertical launch ($\gamma = 90°$), the last term in Eq. 10.31 is simply equal to -gt, taking the initial time to be $t_0 = 0$. This term is termed the *gravity penalty.*

Determining the loss in performance due to aerodynamic drag is more complicated. The aerodynamic drag force is usually expressed as follows:

$$D = \frac{1}{2}\rho V^2 C_D A, \tag{10.32}$$

where ρ is the density of the air, and V the velocity. The quantity $\frac{1}{2}\rho V^2$ is the *dynamic pressure* acting on the rocket. C_D is the *drag coefficient*, which combined with a reference area (typically that of the rocket body cross section) is then used to calculate the expected drag force. Determining the drag coefficient depends on many factors involving the complex, high-speed flow around the various surfaces of the rocket vehicle, which change with changes in velocity and altitude.[8]

A simplified flight profile for a sounding rocket is shown in Fig. 10.6, identifying the boost phase and the subsequent unpowered coast to the maximum altitude. Neglecting aerodynamic drag and assuming a vertical launch for simplicity, the velocity of the rocket during the boost phase of a single stage is from Eq. 10.31:

$$V(t) = -c\ln\left(\frac{m(t)}{m_0}\right) - gt \tag{10.33}$$

where the time $t = 0$ corresponds to the moment of launch. Assuming a constant propellant mass flow rate, the rocket mass varies with time as:

$$m(t) = m_0 - (m_0 - m_b)t/t_b, \tag{10.34}$$

where m_b and t_b are, respectively, the mass of the rocket and the time at burnout. This can be rearranged using the mass ratio $MR = m_0/m_b$ to give:

$$V(t) = -c\ln\left[1 - \left(1 - \frac{1}{MR}\right)\frac{t}{t_b}\right] - gt \tag{10.35}$$

At burnout, the velocity is then:

$$V_b = -c\ln\left[\frac{1}{MR}\right] - gt_b. \tag{10.36}$$

8 One of many excellent textbooks on aerodynamics is Anderson, J.D. *Fundamentals of Aerodynamics*, 6th edition, McGraw-Hill, 2016.

Equation 10.35 can be integrated[9] to give the altitude at burnout as follows:

$$y_b = \int_0^{t_b} V(t)dt = c\,t_b \frac{\ln MR}{MR - 1} + ct_b - \frac{1}{2} g t_b^2. \tag{10.37}$$

For the unpowered segment of the ascent, after rocket burnout, the conservation of mechanical energy dictates that:

$$\frac{V^2}{2} + gy = \frac{V_b^2}{2} + gy_b = constant, \tag{10.38}$$

taking the gravitational acceleration to be constant (a reasonable approximation given the modest change in trajectory radius). This directly gives for the altitude at apogee, y_{max}, where the velocity becomes zero, as

$$y_{max} = y_b + \frac{V_b^2}{2g}. \tag{10.39}$$

Example Problem 10d

A sounding rocket with a liftoff mass of 200 kg containing 90 kg of propellant is launched on a vertical trajectory. If the I_{sp} of the rocket is 180 s and the burn time is 15 s, determine the rocket's altitude at apogee. Neglect aerodynamic drag. Will the rocket reach the *Karman Line* (altitude 100 km)?

Solution:

The velocity at burnout is given by Eq. 10.36 as $V_b = -c\ln\left[\frac{1}{MR}\right] - gt_b$, where $c = I_{sp}g_0 = (180\,\text{s})(9.81\,\text{m/s}^2) = 1765.8\,\text{m/s}$ and the mass ratio $MR = \frac{m_0}{m_0 - m_p} = \frac{200\,\text{kg}}{110\,\text{kg}} = 1.818$. This gives $V_b = -(1765.8\,\text{m/s})\ln\left[\frac{1}{1.818}\right] - (9.81\,\text{m/s}^2)(15\,\text{s}) = 980.3\,\text{m/s}$. The burnout altitude from Eq. 10.37 is then $y_b = ct_b\frac{\ln MR}{MR-1} + ct_b - \frac{1}{2}gt_b^2 = (1765.8\text{m/s})(15\text{s})\left[\frac{\ln(1.818)}{1.818-1}\right] + (1765.8\text{m/s}^2)(15\text{s}) - \frac{1}{2}(9.81\text{m/s}^2)(15\text{s})^2 = 44{,}738\text{m}$. The apogee altitude then follows from Eq. 10.39 as $y_{max} = y_b + \frac{V_b^2}{2g} = 44{,}738\text{m} + \frac{(980.3\text{m/s})^2}{2(9.81\text{m/s}^2)} = 93{,}717\text{m} = 93.72\text{km}$. The rocket will fall just short of reaching the Karman Line, suggesting that a rocket with some combination of a lighter payload, more energetic propellant, or more propellant will be needed.

10.5 Electric Spacecraft Propulsion

An alternative method of rocket propulsion using chemically reactive propellants, as discussed in the previous sections, is to use electrical energy to accelerate a suitable propellant. Electric thrusters for space applications can be broadly grouped into three categories (see Table 10.4): electrothermal,

9 Using the integral relation $\int \ln(ax + b)dx = \left(x + \frac{b}{a}\right)\ln(ax + b) - x$ for $a \neq 0$.

Table 10.4 The three primary methods of acceleration utilized in Electric Spacecraft Propulsion.

	Electrothermal	Electrostatic	Electromagnetic
Physics of acceleration	Thermal rocket	Charged particles in an electrostatic field	Lorentz force
Examples currently in use	Resistojets (RJ), arcjet thrusters (AJT)	Ion Thrusters (IT), Hall Current Thrusters(HCT)	Hall Current Thrusters, MPDs, Pulsed Plasma Thrusters (PPTs)
State-of-the-art specific impulse levels	RJ: 300 s (hydrazine) AJT: 600 s (hydrazine)	IT, xenon: 4000–5000 s HCT, xenon: 2000–3000 s	HCT, xenon: 2000–3000 s PPTs 600–1700s

which uses electric heating to raise the temperature of the propellant; electrostatic, which employs an electric field to accelerate charged particles to high exit velocities; and electromagnetic, which uses a magnetic field to accelerate charged particles. Electric propulsion concepts have been discussed since the early years of rocketry, with the first demonstration flights in the 1960s. The advent of larger satellites with available electric power exceeding multiple kilowatts in the late 1980s allowed electric propulsion to gain a foothold on commercial spacecraft since that time period. Several recent sources provide good, recent information on electric spacecraft propulsion, the different thruster technologies, and propulsion systems.[10,11,12]

Electric propulsion (EP) thrusters can yield performance benefits due to considerably higher specific impulses than those that can be achieved by chemical rockets, greatly decreasing the amount of propellant required for a given mission. A drawback of EP systems is that the thrust produced is very low compared to rockets using chemical propellants at the power levels currently available on spacecraft. The low thrust level can lead to very long thruster operating times (weeks, months, or even years) to accomplish a given velocity change.

In addition to the propulsion system components that are common to a chemical propulsion system, i.e. tanks, flow control system, and thrusters, an EP propulsion system must include a power processing unit (PPU). The PPU draws its power from the spacecraft power bus and will provide the voltages, currents, and control circuitry to operate the thruster system.

When comparing EP thrusters to chemical rockets, it is useful to consider the power of the exhaust stream, or the flow rate of kinetic energy. The rocket engine power can be expressed in various forms as follows:

$$P = \frac{1}{2}\dot{m}c_e^2 = \frac{1}{2}Fc_e = \frac{T^2}{2\dot{m}} = \frac{1}{2}Tg_0 I_{sp} \tag{10.40}$$

where c_e is the average exhaust velocity, and the thrust $T = \dot{m}c_e$ for a negligible exit plane pressure, which is typically the case for EP thrusters. As a means to compare different electric propulsion concepts, the efficiency of an EP thruster, i.e. the ratio of the thrust power generated to the electrical power supplied, is commonly defined as:

$$\eta = \frac{\frac{1}{2}\dot{m}c_e^2}{P_{el}} = \frac{Fc_e}{2P_{el}} = \frac{Fg_0 I_{sp}}{2P_{el}} = \frac{T^2}{2\dot{m}\,P_{el}} \tag{10.41}$$

where P_{el} is the input electrical power.

10 Sutton, G.P. and Biblarz, O. *Rocket Propulsion Elements*, 8th edition, Wiley & Sons, Hoboken, 2010.
11 Heister, S., Anderson, W., Pourpoint, T., and Cassady, J. *Rocket Propulsion*, Cambridge University Press, 2019.
12 Goebel, D., Katz, I., and Mikellides, I. *Fundamentals of Electric Propulsion*, 2nd edition, Wiley Science, 2023.

The rocket power given in Eq. 10.40 points to a major drawback of EP systems: the very high power input needed to produce high thrust levels. For example, a SSME producing a thrust of 2×10^6 N with a specific impulse of 420 s generates 4.2 GW of power. This far exceeds the electrical power available on any existing spacecraft – as a point of reference, the International Space Station currently has about 150 kW of electrical power installed through its solar arrays, and a typical large communication satellite has 10–15 kW available. This effectively rules out EP for ground-launch applications.

It can be noted that the available thrust calculated with Eqs. 10.40 and 10.41 decreases with an increase in the specific impulse for a given power level. Thus, missions that consider electric propulsion will not always try to maximize the specific impulse, although Eq. 10.6 appears to suggest otherwise. For a 2 kW arcjet thruster with 600 s of specific impulse and a 35% efficiency, Eq. 10.41 indicates that the thrust cannot be more than 0.24 N – a thrust level that is still of significant interest for orbital maneuvering and station keeping.

Electrothermal EP systems increase the thermal energy of the propellant by the input of electrical power. The maximum theoretical exhaust velocity in an electrothermal thruster is given by:

$$c_e = \sqrt{\frac{2\gamma}{\gamma - 1} \frac{RT_c}{M}}, \tag{10.42}$$

where T_c is the temperature in the chamber upstream of the nozzle, γ the ratio of specific heat capacities for the gas, M the molecular weight of the propellant, and R the universal gas constant.[13] This points to hydrogen as a good propellant for electrothermal thrusters, as the low molecular weight allows for a high exit velocity.

Examples of electrothermal thrusters include resistojets and arcjets. In a resistojet, the propellant gas flows over an electrically heated heat exchanger to raise its temperature without requiring a chemical reaction. The resistojet system is very simple, but the specific impulse achievable is only typically around 300 s for monopropellant hydrazine as a propellant, which still compares favorably to the 200 s of a purely chemical monopropellant system. A resistojet configuration is shown schematically in Fig. 10.7.

A higher degree of propellant heating can be accomplished in an arcjet thruster, which uses a continuously operating electric arc to raise the temperature to the point where the propellant becomes partially ionized. An arcjet is shown schematically in Fig. 10.8. The propellant flows around the cathode, where electrons are emitted into the gas, and through the electric arc formed by the nozzle wall itself, which serves as the anode. When compared to a resistojet, a higher specific

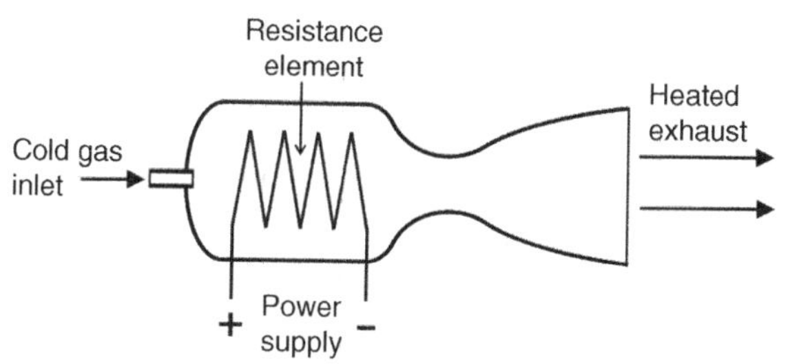

Figure 10.7 Resistojet schematic.

13 Derivations of this relation can be found in Sutton, 8th edition.

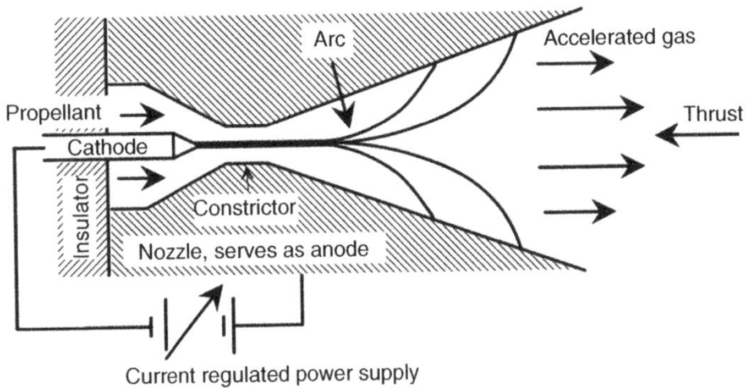

Figure 10.8 Arcjet configuration. *Source:* From Zube, D.M.: Von Laboruntersuchungen zum Entwurf und Bau eines thermischen Lichtbogentriebwerks, PhD thesis, Stuttgart University, Germany 1994.

impulse of around 600 s can be achieved in this fashion for the same propellant, i.e. hydrazine. Arcjets are simple in principle but require special high-temperature refractory materials. The lifetime of an arcjet is limited by erosion during start-up and steady-state operation due to the high thermal loads.

Electrostatic and electromagnetic EP thrusters rely on the interaction between electrically charged particles (ions) and electric and magnetic fields. Electrostatic systems such as Ion Thrusters (ITs) and Hall Current Thrusters (HCTs)[14] use the electric field to directly accelerate the ions. Electromagnetic systems, which include magneto-plasma dynamic (MPD) thrusters, Pulsed Plasma Thrusters (PPTs), and also HCTs, use the Lorentz force[15] generated by the interaction of moving ions with a tailored magnetic field.

The velocity of an ion accelerated in the direction of an electric field with strength E is given by:

$$v = \sqrt{2\frac{qE}{m}}, \tag{10.43}$$

where q and m are, respectively, the electric charge and mass of the ion. Combining this with Eq. 10.41 gives this expression for the thrust of an electrostatic thruster:

$$T = \eta P_{el}\sqrt{\frac{2}{E}\frac{m}{q}}. \tag{10.44}$$

This indicates that for a given desired thrust level, less power will be required by using a propellant with a high molecular mass. More specifically, the desired propellant would have a large mass-to-charge ratio, m/q. Xenon, which has a high molecular weight and is easily ionized, is often used as a propellant in electrostatic EP thrusters for this reason.

14 Depending on the point of view, Hall thrusters are considered either electrostatic or electromagnetic thrusters.
15 The Lorentz force is $\boldsymbol{v} \times \boldsymbol{B}$, where $\boldsymbol{B}$ is the vector magnetic field and $\boldsymbol{v}$ is the velocity vector of a charged particle in that magnetic field.

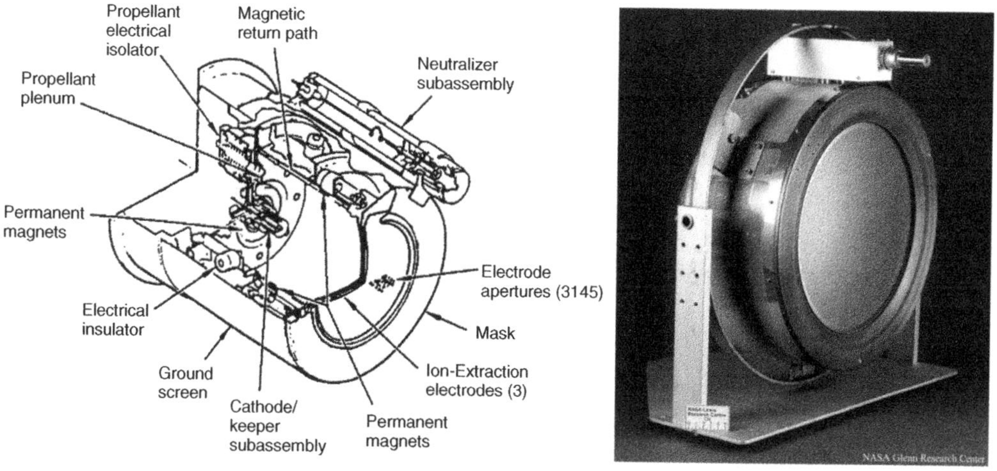

Figure 10.9 An ion thruster. Left, schematic diagram *Source:* Beattie, J.R.; Williams, J.D.; Robson, R.R.: Flight Qualification of an 18 mN Xenon Ion Thruster, IEPC 93-1015, 23rd International Electric Propulsion Conference, Seattle, Sep. 1993; right, actual image of a different thruster. *Source:* NTRS / NASA / https://www.researchgate.net/publication/24285436_Spacecraft_Impacts_with_Advanced_Power_and_Electric_Propulsion / Public domain / Last accessed on Feb 18, 2026. The diameter of the actual thruster is 50 cm.

An ion thruster is shown in Fig. 10.9. The gaseous propellant is ionized by the bombardment of electrons before passing through a pair of grids with a substantial voltage difference (typically kilovolts). An important feature of any ion thruster is the *neutralizer* external to the thruster, which injects electrons into the exhaust stream, preventing the positively charged ions from being attracted back toward the thruster. Ion thrusters can achieve high specific impulses of 5000 s or more. These thrusters need to operate continuously for very long periods, even years, due to their low thrust levels. The very long lifetime requirements lead to erosion of the charged grid surfaces, which substantially degrades the performance or can even lead to thruster failures that need to be addressed with special operating procedures and PPU circuitry for "grid clearance."

An example of an electromagnetic thruster that also uses electrostatic acceleration is the Hall thruster, also known as a Hall Effect thruster (HET), or Hall Current Thruster (HCT). Hall thrusters employ the acceleration of ionized particles in an electric field, similar to ion thrusters. The key difference is that instead of using a fixed electric field formed by the charged grids in the exit plane as in an ion thruster, the Hall thruster uses both electric and magnetic fields. The combination of electric and magnetic fields results in the generation of an azimuthal current within the thruster, which creates an electric field that accelerates the ions to the desired high exit velocities. A schematic diagram and image of a Hall thruster are shown in Fig. 10.10. The neutralizer in this case serves both to neutralize the charge of the exhaust plume as well as to provide electrons to generate the Hall current. An advantage of Hall thrusters is their ability to select operating points with either higher specific impulses but lower thrust or higher thrust with lower specific impulse. The respective operating points are then optimized for specific mission phases. With electric powers of, e.g. 5 kW, Hall thrusters can produce specific impulses of 2200 s for thrust levels approaching 0.30 N. Although these thrusters avoid the problem of the grid erosion inherent to ion thrusters, erosion affects the electrodes, leading to a degradation in performance and eventual life limitation.

A challenge to the design and especially the testing of both ion and Hall current thrusters is that they need to be tested uninterruptedly for many months and years in test facilities that operated at

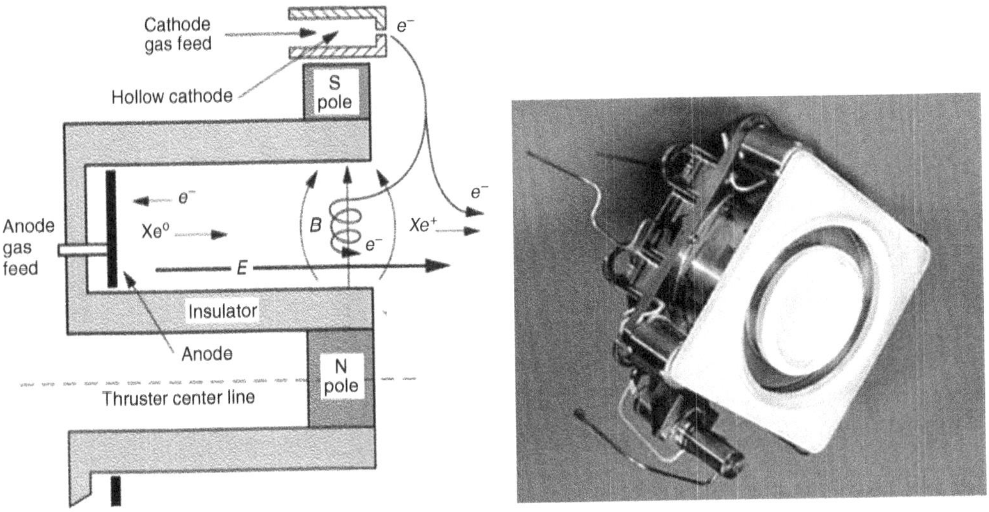

Figure 10.10 A Hall thruster. Left, operational principle; right, actual image of the Aerojet Rocketdyne BPT 4000 Hall thruster. The thruster face is 8 × 8 in. (20 × 20 cm) square. *Source:* De Grys, K; et al.: Development and Testing of a 4500 W Flight Type Hall Thruster and Cathode, Paper IEPC-2001-011, 27th International Electric Propulsion Conference, Passadena, Oct. 2001.

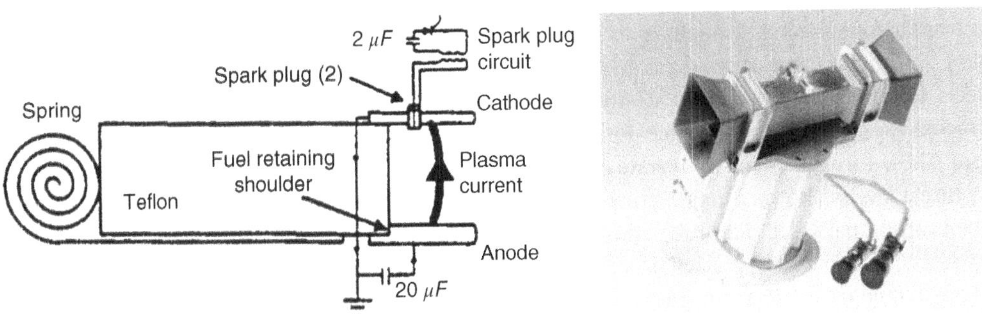

Figure 10.11 Pulsed plasma thruster. The overall length of the actual thruster shown is 12 in. (30 cm). *Source:* Benson, S.W.; Arrington, I.A.; Hoskins, W.A.; Meckel, N.J.: Development of a PPT for the EO-1 Spacecraft AIAA Paper 1999–2276, June 1999.

vacuum chamber pressures below 10^{-5} to 10^{-6} Torr. During ground testing, an ion thruster has been tested at the Jet Propulsion Laboratory uninterruptedly for 30,000 hour or nearly four years.[16]

A final example of an EP system is the PPT, shown in Fig. 10.11. The PPT uses an electric spark to vaporize and ionize a small amount of solid propellant (typically Teflon). The resulting ionized plasma current then creates its own self-induced magnetic field, which leads to the acceleration and ejection of ions, resulting in thrust. Since the plasma consists of both ions and electrons, a neutralizing electron beam is not required. Although the performance of PPTs is low (typical specific impulses around 300 s), they offer the advantages of simplicity and reliability. The PPTs also provide impulse "bits" that are very useful for spacecraft RCS.

16 Sengupta, A.; Brophy, J.; Anderson, J.; Garner, C.; Banks, B; Groh, K.: An Overview of the Results from the 30,000 Hr Life Test of Deep Space 1 Flight Spare Ion Engine, AIAA Paper 2004–3608, 40th AIAA Joint Propulsion Conference, July 2004, Ft. Lauderdale, FL.

Example Problem 10e

Calculate the thrust level (N) of: (i) an arcjet thruster with power $P_{el} = 2.0$ kW, efficiency $\eta = 35\%$, and specific impulse $I_{sp} = 600$ s, and (ii) a Hall thruster with power $P_{el} = 3.0$ kW, efficiency $\eta = 60\%$, and specific impulse $I_{sp} = 2000$ s.

Solution:

Rearranging Eq. 10.41 gives the thrust of an EP thruster as $F = \eta \frac{2P_{el}}{g_0 I_{sp}}$. This gives for the arcjet thruster $F = 0.35 \frac{2(2000\,\mathrm{W})}{(9.81\,\mathrm{m/s^2})(600\,\mathrm{s})} = 0.238\,\mathrm{N}$; for the Hall thruster, $F = 0.60 \frac{2(3000\,\mathrm{W})}{(9.81\,\mathrm{m/s^2})(2000\,\mathrm{s})} = 0.183\,\mathrm{N}$. These low thrust levels are typical of EP systems.

A performance comparison map of both chemical and electric propulsion systems is shown in Fig. 10.12. The overall trade-off is that EP systems can provide high values of specific impulse, but only low levels of thrust. Considering the currently available technologies, applications such as ground launch and impulsive orbital-change maneuvers will continue to rely on chemical propulsion systems, which can provide the large thrust required, but at considerably lower specific impulse than EP thrusters.

Common to most electric propulsion systems is the propellant's need to become electrically conductive, forming a high-temperature plume that can easily exceed 5000 K and more. These high temperatures create significant material challenges for the design of EP thrusters and also require that any attempt to analytically model the fluid mechanics of the plume will need to consider plasma physical behavior and properties.

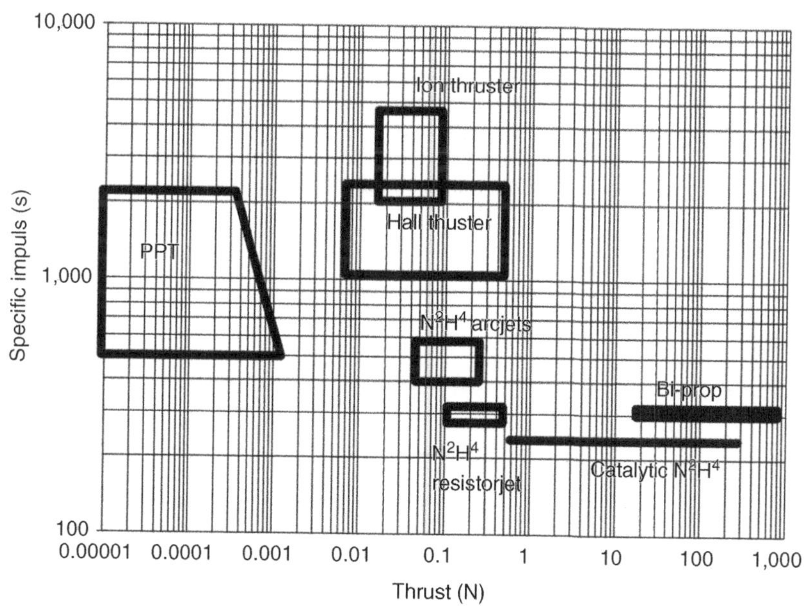

Figure 10.12 Rocket propulsion systems performance comparison. *Source:* Zube, Dieter M.: Introduction to Electric Propulsion, part of the UW AA460 (Introduction to Propulsion) lecture, University of Washington, Seattle, 2004.

10.6 Low-Thrust Maneuvers for Orbit Raising

The continuous, low thrust of a spacecraft using EP will result in a very different trajectory than a spacecraft performing a trajectory change using high-thrust, impulsive maneuvers like the Hohmann transfers discussed in Chapter 5. As one example, consider a satellite to be raised from LEO (with orbital radius r_1) to GEO (r_2). Using a classic Hohmann transfer (see Chapter 5, Eqs. 5.3–5.5), the impulsive velocity changes required are

$$V_1 - V_{c1} = \Delta V_1 = \sqrt{2\mu\left(\frac{1}{r_1} - \frac{1}{r_1 + r_2}\right)} - \sqrt{\frac{\mu}{r_1}} \text{ and } V_{c2} - V_2 = \Delta V_2 = \sqrt{\frac{\mu}{r_2}} - \sqrt{2\mu\left(\frac{1}{r_2} - \frac{1}{r_1 + r_2}\right)},$$

with a transfer time

$$T_t = \pi\sqrt{\frac{a^3}{\mu}} = \pi\sqrt{\frac{\left(\frac{r_1 + r_2}{2}\right)^3}{\mu}}$$

The much lower thrust using EP will result in a gradual raising of the orbital radius from LEO until GEO is reached. The result is the spiral ascending trajectory, shown schematically in Fig. 10.13. The acceleration of the spacecraft using EP will be:

$$a = \frac{F}{m} = \frac{2\eta P_{el}}{g_0 I_{sp} m} \tag{10.45}$$

where m is the spacecraft mass. The total transfer time in this case is given by[17]:

$$T_t = \frac{m_1}{\dot{m}}\left(1 - \exp\left[\frac{\dot{m}}{F}\left(\sqrt{\frac{r_1}{r_2}} - 1\right)\sqrt{\frac{\mu}{r_1}}\right]\right) \tag{10.46}$$

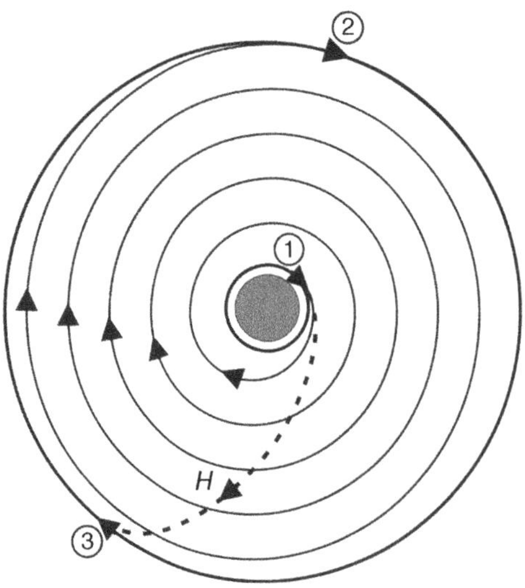

Figure 10.13 A spiral trajectory using EP to raise an orbit. Point ①, departure from LEO; ②, arrival at the target orbit using EP; and ③, arrival at the target orbit on the Hohmann transfer.

17 Messerschmid, E. and Fasoulas, S. *Raumfahrtsysteme*, Springer Verlag, Berlin, 2009.

The transfer time for a spiral, EP-powered ascending trajectory will typically be much longer than the total time from impulsive transfer maneuvers using chemical rockets. The long transfer times associated with EP may however not be a problem for un-crewed missions.

Example Problem 10f

Compare the transfer time from LEO ($r_1 = 6678$ km) to GEO ($r_2 = 42{,}164$ km) for a satellite using a Hohmann transfer versus a spiral, EP-propelled transfer. The EP system is a Hall thruster with an I_{sp} of 1800 s and the mass of the spacecraft in LEO is $m = 1000$ kg. The thrust-to-weight ratio of the satellite using Hall thruster propulsion is 2×10^{-5}. How do the required propellant masses compare for EP versus a chemical thruster with a specific impulse of $I_{sp} = 300$ s?

Solution:

The required time for the Hohmann transfer is (see Chapter 5, Example Problem 5a) $T_{tH} = 5.38$ hour with a total velocity change of $\Delta V = \Delta V_1 + \Delta V_2 = 2.42 + 1.47$ km/s $= 3.98$ km/s. Given the known thrust-to-weight ratio, F/mg_0, the transfer time for the EP-propelled satellite can be determined by rewriting Eq. 10.46 as follows:

$$T_{tEP} = \frac{m}{\dot{m}}\left(1 - \exp\left[\frac{\dot{m}}{F}\left(\sqrt{\frac{r_1}{r_2}} - 1\right)\sqrt{\frac{\mu}{r_1}}\right]\right) = \left(\frac{mg_0}{F}\right) I_{sp}\left(1 - \exp\left[\frac{1}{I_{sp}g_0}\left(\sqrt{\frac{r_1}{r_2}} - 1\right)\sqrt{\frac{\mu}{r_1}}\right]\right) = \left(\frac{1}{2\times 10^{-5}}\right)$$

$$(1800\,\text{s})\left(1 - \exp\left[\frac{1}{(1800\,\text{s})(9.81\,\text{m/s}^2)}\left(\sqrt{\frac{6687\,\text{km}}{42{,}164\,\text{km}}} - 1\right)\sqrt{\frac{398{,}600\,\text{km}^3/\text{s}^2}{6687\,\text{km}}}\right]\right) = 20.81 \times 10^6\,\text{s} = 240.9\,\text{days},$$

clearly much longer than for the Hohmann transfer.

The propellant mass flow rate discharged by the EP thruster is given by $\dot{m} = \left(\frac{m}{I_{sp}}\right)\left(\frac{F}{mg_0}\right) = \left(\frac{1000\,\text{kg}}{1800\,\text{s}}\right)(2 \times 10^{-5}) = 1.11 \times 10^{-5}$ kg/s, which gives a total propellant mass of $(1.11 \times 10^{-5}$ kg/s$)(20.81 \times 10^6$ s$) = 231$ kg.

The required propellant mass for the Hohmann transfer can be determined from Eq. 10.20 as $m_p = m_0\left[1 - \exp\left(\frac{-\Delta V}{I_{sp}g_0}\right)\right] = (1000\,\text{kg})\left[1 - \exp\left(\frac{-3980\,\text{m/s}}{(300\text{s})(9.81\,\text{m/s}^2)}\right)\right] = 741.4$ kg. The much longer transfer time for the EP-propelled satellite will be compensated by a much larger payload than possible for a chemically-propelled satellite.

In virtually all practical applications of spiraling trajectories, the thrusters are not operated continuously, regardless of whether they are interplanetary or in Earth orbit. Reasons ranged from limited power availability to optimized orbital mechanics, thermal considerations of the spacecraft, the thrusters, or its PPU, and intermittent telemetry contact with the ground stations. Thus, these low-thrust orbit transfers are always planned individually for each mission separately.

10.7 Problems

10.1 The Aerojet Rocketdyne RL-10A-4-2 rocket variation has a nozzle exit diameter of $D_e = 1.78$ m and a nominal propellant mass flow rate of $\dot{m} = 22.4$ kg/s. If the nozzle exit velocity and pressure are, respectively, $V_e = 4105.5$ m/s and $P_e = 2.868$ kPa, what would be the nominal thrust and specific impulse in vacuum? What would be the corresponding values at sea-level conditions? Source: https://en.wikipedia.org/wiki/RL10

10.2 A rocket in Earth orbit uses solid propellants with a specific impulse of $I_{sp} = 240$ s. The stage is loaded with $m_p = 10{,}000$ kg of propellant and has a mass of $m_f = 2000$ kg at burnout. What will be the velocity change achieved? Assume that the maneuver is at a zero flight path angle so that there is no gravity loss. To improve on this performance, it is proposed to replace the solid fuel with a bi-propellant system using liquid hydrogen and liquid oxygen, to give an increased specific impulse of $I_{sp} = 450$ s. The same mass of propellant is utilized, but the less-dense liquid fuels require more volume, hence a larger stage. This, combined with the additional pumps, piping, etc., needed for the bi-propellant system results in a higher burnout mass of $m_f = 5500$ kg. What will be the velocity change for the bi-propellant variant, and does it represent an improvement over the baseline solid-propellant rocket?

10.3 A two-stage sounding rocket has a velocity of 2500 m/s at the point of burnout of the second stage, which occurs at an altitude of 30,000 m. Assuming vertical flight, what will be the apogee achieved? Perform the analysis first by taking a constant value of the gravitational acceleration g (reasonable for sounding rockets), which results in a simple dynamics problem. Next, perform the calculations using the exact solution, which takes into account changes in the earth's gravity with altitude. Compare the results from the two approaches and comment.

10.4 A sounding rocket with a liftoff mass of 100 kg containing 50 kg of propellant is launched on a vertical trajectory. If the delivered I_{sp} of the rocket is 150 s, determine the burning time that provides a 20,000 m rocket altitude at apogee. Neglect drag and assume the rocket has constant thrust during the powered portion of the flight.

10.5 A vertically launched sounding rocket has a payload of 150 kg and an empty weight of 1500 kg (excluding the payload). The rocket employs twin liquid engines with a storable propellant that provides an I_{sp} of 280 s. The payload contains sensitive instruments that cannot withstand acceleration levels in excess of 4 g/s.

(a) Assuming negligible drag and that the engines are not throttleable, determine the propellant load, engine burn time, and engine thrust level that will maximize the performance (ΔV) of this vehicle. What is the ΔV achieved under these conditions?

(b) It is suggested that the rocket may have additional performance by shutting down one engine at some appropriate point in the flight. Determine in this case the burning times, propellant consumption, thrust level, and the ΔV achieved during both phases of the flight.

10.6 At time $t = 0$, ignition of a rocket on a launch pad on the Moon occurs. Due to a malfunction, the thrust is equal to only 75% of the initial weight of the rocket. The mass flow rate, $\dot{m}$, is constant.

Determine the time t at which lift-off finally occurs, expressing your answer in terms of I_{sp} and other appropriate physical constants (but *not* $\dot{m}$).

10.7 A launch vehicle is to insert an 8800 kg payload into LEO (altitude 350 km), launching due east from Vandenberg Air Force Base (latitude 34.7° N). An existing first stage of the two-stage launch vehicle uses 800,000 kg of solid propellant with a specific impulse of $I_{sp} = 270$ s. The empty mass of the first stage is 40,000 kg. A new second stage uses liquid methane and liquid oxygen propellants with a specific impulse of 365 s. What will be the required propellant mass for the second stage, if the structural ratio of the second stage is $\varepsilon = 0.12$?

It is suggested that a higher payload mass might be possible by moving the launch site to the Kennedy Space Center (latitude 28.6° N). Determine if this is the case and what would be the new payload mass.

10.8 Consider a two-stage rocket designed to provide an ideal ΔV of 3000 m/s. The following information is known:

Payload mass = 100 kg
1st stage: $\lambda = 0.9$, $I_{sp} = 250$ s, 40% of total ΔV
2nd stage: $\lambda = 0.85$, $I_{sp} = 300$ s, 60% of total ΔV

While the λ values quoted above account for the propulsion systems and structural masses, the interstage structure is treated separately. Assume that the weight of that component is equivalent to 5% of the mass of the Stage 2/payload combination it supports. Determine the:

(a) The Stage 2 propellant and inert masses,
(b) the interstage mass, and
(c) the total vehicle mass.

10.9 The *Delta II* rocket first stage consists of a single, liquid-fueled main engine plus nine strap-on solid rocket motor boosters (SRMs). It is desired to determine the velocity of the Delta II during launch. The total launch mass of the vehicle (first stage, boosters, plus all upper stages and payload) is M_0. The propellant mass in each SRM is M_{pB}. Take the thrust of the main engine to have the constant value F_M. The specific impulse for the SRMs can be taken to have the *constant* value I_{spB}; for the main engine take a different *constant* value, I_{spM}. The burn time for the SRMs is t_b. Hint: It may be helpful to express the performance during the boost phase in terms of *average* quantities (e.g. equivalent exhaust velocity, mass flow rate) for the entire system.

(a) Taking the boost phase to be exactly vertical, write the differential equation for the acceleration (d^2y/dt^2) of the Delta II at a time t after lift-off. Neglect any aerodynamic drag. Express the differential equation in terms of M_0, M_{pB}, F_M, both I_{sp}s and t_b.
(b) The solid rocket boosters burn out at a time t_b after launch. Develop the analytical formula for the vertical velocity of the vehicle at that point, assuming for simplicity that the SRMs remain attached until they are completely burned out and that the burnout is instantaneous. HINT: The use of an "average" equivalent exhaust velocity may be helpful.
(c) Calculate the vehicle velocity at SRM burnout taking $M_0 = 230{,}000$ kg, $M_{pB} = 10{,}000$ kg, $F_M = 1040$ kN, $I_{spB} = 270$ s, $I_{spM} = 300$ s, and $t_b = 63$ s.

10.10 A spacecraft in permanent orbit around the earth is to be used for delivering payloads from low earth orbit (LEO) to geostationary equatorial orbit (GEO). Before each flight from LEO, the spacecraft is refueled with propellant, which it uses up in its round trip to GEO. The outbound leg requires four times as much propellant as the inbound return leg. The ΔV for transfer from LEO to GEO is 4.22 km/s. The specific impulse of the propulsion system is 430 s. If the payload mass is 3500 kg, calculate the empty mass of the vehicle. This is some more mass accounting; in this case for an orbital application (the ΔV required is given to save that step).

10.11 Is a state-of–the art 1 kW electric propulsion system able to provide more than 2 N of thrust on a 750 kg spacecraft?

10.12 To help decide which propulsion system to use, consider the following mission: A geostationary spacecraft requires an annual velocity increment (ΔV) of up to 45 m/s for station keeping. An additional ΔV of 120 m/s is needed to dispose the spacecraft into the graveyard orbit at the end of the mission. Assume a 10-year mission.

The dry spacecraft launch mass is 2000 kg. This includes the orbit-insertion propulsion system mass, but not yet the station-keeping propulsion system mass. Carry out a performance analysis of two candidate propulsion systems below. Which system would you recommend?

(a) Arcjet:
  1) Specific impulse = 600 s,
  2) Thruster mass (four thrusters) = 4 kg,
  3) Power control unit/electronics mass = 12 kg,
  4) Additional tank mass = 15 kg (an existing tank has to be enlarged for the additional station keeping hydrazine),
  5) Power per thruster = 2 kW,
  6) No additional mass is needed to provide the power for the thrusters.

(b) Hall thruster:
  1) Specific impulse = 2500 s,
  2) Thruster mass (four thrusters) = 12 kg,
  3) Power control unit/electronics mass = 4×8 kg,
  4) Additional tank mass = 50 kg (an all-new tank is needed for the xenon propellant),
  5) Power per thruster = 4 kW,
  6) An additional mass of 10 kg is needed for the power system to provide the higher power needed for the thrusters.

Appendix 1

Solar System Properties

Sun[1,2]

Gravitational parameter [km³/s²]	Inclination to Ecliptic [degrees]	Equatorial diameter [km]	Equatorial rotational period* [days]	Velocity (relative to galactic center) [km/s]
1.3271×10^{11}	7.25	1.392×10^{6}	25.1	251

*The rotational period near the solar pole is longer, 34.4 days.

Terrestrial Planets[1,3]

	Gravitational parameter [km^3/s^2]	Semi-major axis [km]	Eccentricity	Inclination of orbit [degrees]	Period [years]	Mean velocity [km/s]
Mercury	22,032	57.9×10^{6}	0.2056	7.00	0.241	47.4
Venus	324,858	108.2×10^{6}	0.0067	3.39	0.615	35.0
Earth	398,600	149.6×10^{6}	0.0167	0	1.000	29.8
Mars	42,828	227.9×10^{6}	0.0935	1.85	1.881	24.1

In Earth Orbit[3,4]

	Gravitational parameter [km^3/s^2]	Semi-major axis [km]	Eccentricity	Inclination of orbit [degrees]	Orbital period	Mean velocity [km/s]
Moon	4902.8	0.384×10^{6}	0.0549	5.15	27.3 days	1.0
ISS	–	6798	0.000748	51.63	93.0 minutes	7.67

1 https://ssd.jpl.nasa.gov/astro_par.html
2 Wikipedia
3 https://physics.info/astronomical/
4 https://heavens-above.com/orbit.aspx?satid=25544

Introduction to Orbital Mechanics: A Concise and Practical Approach, First Edition. James C. Hermanson.

Gas Giant Planets[1,3]

	Gravitational parameter [km^3/s^2]	Semi-major axis [km]	Eccentricity	Inclination of orbit [degrees]	Orbital period [years]	Mean orbital velocity [km/s]
Jupiter	126.71×10^6	778.6×10^6	0.0489	1.30	11.86	13.07
Saturn	37.94×10^6	1.434×10^9	0.0565	2.49	29.42	9.70
Uranus	5.795×10^6	2.872×10^9	0.0457	0.77	83.75	6.83
Neptune	6.837×10^6	4.495×10^9	0.0113	1.77	163.73	5.47

Minor Planets (selected)[5]

	Semi-major axis [km]	Eccentricity	Inclination of orbit [degrees]	Orbital period [years]
Pluto	5.922×10^9	0.2488	17.15	249.09
Sedna	82.21×10^9	0.8613	11.93	12,882
1-Ceres	413.7×10^6	0.0796	10.59	4.599
4-Vesta	353.3×10^6	0.0901	7.14	3.629

Comets (selected)[5]

	Semi-major axis [km]	Eccentricity	Inclination of orbit [degrees]	Orbital period [years]
1P/Halley	2.682×10^9	0.9679	162.19	75.92
2P/Encke	332.1×10^6	0.8477	11.41	3.307
C/1975 V1-A West	-1.828×10^{12}	1.000	43.07	undefined
C/1996 B2 Hyakutake	317.8×10^9	0.9999	124.29	97,943
C/2020 F3 Neowise	53.63×10^9	0.9992	128.94	6,787

Planetary Properties[3,6]

	Equatorial Radius [km]	Polar Radius [km]	Oblateness	J_2 [km^5/s^2]	Equatorial inclination [degrees]	Rotational period [hours]
Mercury	2439.7	2439.7	0.000	60×10^{-6}	0.034	4222.6
Venus	6051.8	6051.8	0.000	4.458×10^{-6}	177.36	2802.0
Earth	6378.1	6356.8	0.003353	1.083×10^{-3}	23.44	24
Mars	3396.2	3376.2	0.00648	1.961×10^{-3}	25.19	24.66
Jupiter	71,492	66,854	0.06487	14.736×10^{-3}	3.13	9.93
Saturn	60,268	54,364	0.09796	16.298×10^{-3}	26.73	10.66
Uranus	25,559	24,973	0.02293	3.343×10^{-3}	97.77	17.24
Neptune	24,764	24,341	0.01708	3.411×10^{-3}	28.32	16.11
Moon	1738.1	1736.0	0.0012	202.7×10^{-6}	6.68	708.7

5 https://ssd.jpl.nasa.gov/tools/sbdb_lookup.html#/
6 Curtis, H. D., Orbital Mechanics for Engineering Students, 4th Ed., Butterworth-Heinemann, 2019

Index

Introduction to Orbital Mechanics: A Concise and Practical Approach, First Edition. James C. Hermanson.

18/2/2026 4:35:0

Printed and bound by CPI Group (UK) Ltd, Croydon, CR0 4YY
17/09/2026
14964313-0001